Consciousness and its Place in Nature

Consciousness and its Place in Nature

Does Physicalism Entail Panpsychism?

Second edition, with new commentaries
and response

Galen Strawson *et al.*
edited by Anthony Freeman and Graham Horswell

imprint-academic.com

Copyright © Imprint Academic, 2006, 2024

No part of any contribution may be reproduced in any form without permission, except for the quotation of brief passages in criticism and discussion.

Published by
Imprint Academic, PO Box 200, Exeter EX5 5YX, UK

ISBN 9781788361187
A CIP catalogue record for this book is available from the British Library and US Library of Congress

Contents

Contributors	vii
Preface to Second Edition, *Graham Horswell*	ix
Editorial Preface, *Anthony Freeman*	1

Target Paper

Realistic Monism: Why Physicalism Entails Panpsychism, *Galen Strawson*	3

Commentaries

Can Panpsychism Bridge the Explanatory Gap? *Peter Carruthers and Elizabeth Schechter,*	32
Being Realistic, *Sam Coleman*	40
Experiences Don't Sum, *Philip Goff*	53
Galen Strawson on Panpsychism, *Frank Jackson*	62
Resisting ?-ism, *William G. Lycan*	65
Property Dualism and the Merits of Solutions to the Mind–Body Problem, *Fiona Macpherson*	72
Hard Questions, *Colin McGinn*	90
Comments on Galen Strawson, *David Papineau*	100
Better to Study Human than World Psychology, *Georges Rey*	110
Experience and the Physical, *David M. Rosenthal*	117
The 'Intrinsic Nature' Argument for Panpsychism, *William Seager*	129
The Seeds of Experience, *Peter Simons*	146
Realistic Panpsychism, *David Skrbina*	151
Ockhamist Comments on Strawson, *J.J.C. Smart*	158
Commentary on Strawson's Target Article, *H.P. Stapp*	163
Comments on Galen Strawson, *Daniel Stoljar*	170
Commentary on Galen Strawson, *Catherine Wilson*	177

Reply

Panpsychism? Reply to Commentators with a Celebration of Descartes, *Galen Strawson*	184

2024 Postscripts

A Panqualityist Manifesto, *Sam Coleman*	285
Can Experiences Sum?, *Philip Goff*	299
Postscript, *Georges Rey*	304
From Panpsychism to Neutral Monism . . . and Back Again(?), *William Seager*	312
On the Present and Future of Panpsychism, *David Skrbina*	328
Underestimating the World, *Daniel Stoljar*	336
Blockers and Laughter: Panpsychism, Archepsychism, Pantachepsychism, *Galen Strawson*	343
Index	365

Contributors

Galen Strawson <gstrawson@mac.com> is Professor of Philosophy at the University of Texas at Austin.

Anthony Freeman was the managing editor of the *Journal of Consciousness Studies* from 1995 to 2010.

Peter Carruthers <pcarruth@umd.edu> is Distinguished University Professor of Philosophy at the University of Maryland. His most recent book is *Human Motives: Hedonism, Altruism, and the Science of Affect.*

Sam Coleman <s.coleman@herts.ac.uk> studied at Oxford and Cambridge Universities and received his PhD for a thesis on the 'knowledge argument' against physicalism at Birkbeck College, University of London. He is Reader in Philosophy at the University of Hertfordshire.

Philip Goff <philip.a.goff@durham.ac.uk> is a philosophy professor at Durham University. His most recent book is *Why? The Purpose of the Universe.*

Graham Horswell <graham.jcs@gmail.com> is the managing editor of the *Journal of Consciousness Studies*.

Frank Jackson <fcjack321@gmail.com> is Professor Emeritus at the Australian National University. He is the author of publications in the philosophy of mind, philosophy of language, and ethics.

William G. Lycan <william.lycan@uconn.edu> is Professor Emeritus of Philosophy at the University of North Carolina. He is author of many books including *Consciousness and Experience* (MIT Press, 1996) and *Philosophy of Language* (Routledge, 2000).

Fiona Macpherson <fiona.macpherson@glasgow.ac.uk> is Professor of Philosophy and Director of the Centre for the Study of Perceptual Experience at the University of Glasgow.

David Papineau <david.papineau@kcl.ac.uk> is Professor of Philosophy of Science at King's College London. He has a BSc in mathematics from the University of Natal and a BA and PhD in philosophy from Cambridge University. His books include *Knowing the Score* (2017) and *The Metaphysics of Sensory Experience* (2021).

Georges Rey <georey2@gmail.com> is a Professor in the Dept of Philosophy, University of Maryland.

Elizabeth Schechter <lizschechter@gmail.com> is an Associate Professor in Philosophy at the University of Maryland. She is the author of *Self-Consciousness and 'Split' Brains: The Minds' I*.

Peter Simons FBA <psimons@tcd.ie> is Professor Emeritus of Philosophy at Trinity College Dublin. After studying mathematics and philosophy at Manchester he taught at Bolton and at Salzburg. He is the author of two books and hundreds of articles on philosophical topics and Austrian and Polish philosophy.

David Skrbina <skrbina@umich.edu> was a Lecturer in Philosophy at the University of Michigan at Dearborn from 2003–2018, and was then at the University of Helsinki from 2020–2022. He received his PhD, from the University of Bath, in 2001. His book, *Panpsychism in the West*, was published by MIT Press in 2005.

J.J.C. Smart was Emeritus Professor of Adelaide and Australian National Universities, and an honorary fellow of Monash University and Corpus Christi College, Oxford. His main interests were in metaphysics, philosophy of mind, philosophy of science, and also ethics.

William Seager <bill.seager@utoronto.ca> is Professor Emeritus of Philosophy at the University of Toronto at Scarborough. Despite having written a number of books and many articles on the problem of consciousness, he remains quite perplexed.

Henry Stapp <hpstapp@lbl.gov> is a theoretical physicist at the University of California's Lawrence Berkeley National Laboratory. He has written extensively about, the ontological implications of quantum theory, and the mind–brain connection. He is the author of *Mind, Matter, and Quantum Mechanics* (1993, expanded 2004).

Daniel Stoljar <daniel.stoljar@anu.edu.au> is Professor in Philosophy and Director of the Centre for Consciousness at the Australian National University. He is the author of a number of books and articles in philosophy of mind, metaphysics, and metaethics.

Catherine Wilson is Emerita Professor of Philosophy at the University of York. She is the author of *Moral Animals* (2004) and articles and papers in early modern philosophy and history and philosophy of science.

Graham Horswell

Preface to the Second Edition

The original edition of this collection, published in 2006, was put together by my predecessor as managing editor of the *Journal of Consciousness Studies*, Reverend Anthony Freeman. At that time, and as Anthony's editorial introduction to the first edition makes clear, it was still considered relatively crazy to take panpsychism seriously. Nonetheless, a few important and dissenting voices were doing just that, thinking seriously and deeply about the idea that mentality may be a fundamental feature of the universe. This idea has had a long and illustrious history, arguably for centuries, but for the last 100 years or so until the very recent past, philosophy has been dominated by physicalist metaphysics, while other ideas have had to take a back seat.

These days panpsychism, in one form or another, has worked its way into mainstream thinking in philosophy of mind and in the science of consciousness more generally.[1] This is in no small part due to Galen Strawson and other like-minded scholars, who have been on the front line of the battlefield on this topic since the 1990s. It would not be too much of a stretch to argue that Strawson's paper on 'realistic monism', contained in this volume, is now considered something of a classic, and perhaps a catalyst for panpsychism's resurgence.

This long overdue new edition of the book aims to give the original commentators, where they felt they had something more to add, an opportunity to update their thinking on the topic. Six took up the challenge, as did Strawson himself, and the result is a set of additional postscripts that enhance the original collection

---

[1] It was also the topic of a 2024 edition of BBC Radio 4's celebrated *In Our Time* series: https://www.bbc.co.uk/programmes/m001vl96

and push the discussion onwards, reflecting on the first edition's position within the current debate.

Some of the commentators could be said to have changed their minds on the topic (Philip Goff perhaps being the most striking example), others not so much (e.g. Georges Rey, who remains trenchantly opposed to the view), but all have added further clear and distinctive ideas to the ongoing conversation.

Strawson's own postscript not only continues to push forward his own view of panpsychism (as he says himself, quoting the political consultant Frank Luntz):

> You say it again, and you say it again and you say it again, and you say it again, and you say it again, and then again and again and again and again, and about the time that you're absolutely sick of saying it is about the time that your target audience has heard it for the first time

he also adds further to his original thesis, bringing in more recent work on the topic and clarifying some misapprehensions about his original target article and response. He wants the take-home message to be, after thirty-odd years of banging the drum, that

> given the empirical evidence, panpsychism in some form is the most parsimonious, conservative, down-to-earth, hard-nosed, plausible, realistic, even plodding view there is about the fundamental nature of reality.

Overall, we hope that the additional content in this new edition serves to bring this important volume up to date with the state of the art, and we thank each of the contributors for their insightful thoughts on where panpsychism in general, and Galen Strawson's realistic monism in particular, stand in contemporary debates in the philosophy of mind. In another fifteen or so years we may have to expand the volume further, and who knows, if the field continues on a similar trajectory, we may all be wondering why it took so long for panpsychism to emerge from the shadows of academia.

Anthony Freeman
Editorial Introduction

Philosophers of mind find themselves drawn in many contradictory directions, almost as though there were unseen forces at work to lure them into paths they would rather not travel. Certain destinations retain their power to attract and repel in almost equal measure, and no matter how often their absurdity is alleged by the majority, a few brave souls are always to be found ignoring the no-entry signs and flirting with these 'dangerous doctrines'. Solipsism is one such tempter; another is epiphenomenalism; but nowhere is the ambiguity of a *mysterium tremendum et fascinans* more apparent than in the interpretation of the mental–material relation known as *panpsychism*.

Panpsychism, at its simplest, is the belief that everything having a physical aspect also has a mental or conscious aspect. It is regarded by many as either plain crazy, or else a direct route back to animism and superstition. The apparent claim that a hunk of rock has a conscious thinking mind is so easy to ridicule: why should anyone take such an idea seriously? Yet David Skrbina (2003; 2005) has convincingly demonstrated that panpsychism has been an underlying theme in Western philosophy over many centuries.

Galen Strawson has always held that realism about consciousness requires one to accept that consciousness 'is among the fundamental properties that must be adverted to in a completed or optimal physics', if physicalism or materialism is true (Strawson, 1994, pp. 61–2), and that a 'panpsychist version of materialism could handle the idea that experiential properties might be fundamental physical properties' (ibid.), although it certainly need not imply that things like rocks are conscious subjects of experience. He used to find this conclusion 'very alarming' (ibid.), as also did Chalmers, who spoke of the 'threat' of panpsychism (Chalmers, 1997, p. 29, quoted on p. 186 below). Now, however, Strawson embraces the position with enthusiasm. Whereas Chalmers concedes the possibility of panpsychism reluctantly, as a price worth paying for his non-reductive and ultimately dualistic ontology, Strawson presents it as a necessary and not unwelcome consequence of his thorough-going physicalism, or realistic monism, as he now calls it.

Strawson's target paper setting out this position, and the impressive array of commentaries upon it, would alone suffice to make this a notable discussion of the hard problem of consciousness. But there is more: we are treated to Strawson's hundred-page reply to his commentators (pp. 184–280), with its celebration of an unlikely hero in today's philosophical climate: 'the magnificent, contumacious Descartes' (p. 199). This is 'the real Descartes, not the "Descartes" of present-day non-historical philosophy,' and among the first things to note about him, 'given that we (the generality of philosophers) refer to him so much, and so freely, and so inaccurately' is that he is '*not a substance dualist* in any conventional understanding of this term' (pp. 201–2, italics original).

Making us look afresh at this great figure in the philosophy of mind is part of a larger campaign to reassess the whole question of consciousness, in the course of which Strawson lists 40 or so theses which together constitute 'the basic framework' within which the problem has to be tackled (pp. 221–34).

When I first heard a version of the target paper at the Toward a Science of Consciousness 2005 conference in Copenhagen, and conceived the idea of publishing it in the *Journal of Consciousness Studies* — and even when I sent out the target paper to commentators — I little imagined that it would result in so prodigious an undertaking. I am grateful to Galen Strawson and all his collaborators for the great amount of effort that has gone into this collection, which I am sure will prove to be the launch pad of an even wider debate.

References

Chalmers, D.J. (1997), 'Moving forward on the Problem of Consciousness', *Journal of Consciousness Studies* **4** (1), pp. 3–46.
Skrbina, D. (2003), 'Panpsychism as an underlying theme in Western philosophy', *Journal of Consciousness Studies*, **10** (3), pp. 4–46.
Skrbina, D. (2005), *Panpsychism in the West* (Cambridge, MA: MIT Press).
Strawson, G. (1994), *Mental Reality* (Cambridge, MA: MIT Press).

Galen Strawson

Realistic Monism
Why Physicalism Entails Panpsychism[1]

1. Physicalism

I take physicalism to be the view that every real, concrete phenomenon in the universe is ... physical. It is a view about the actual universe, and I am going to assume that it is true. For the purposes of this paper I will equate 'concrete' with 'spatio-temporally (or at least temporally) located', and I will use 'phenomenon' as a completely general word for any sort of existent. Plainly all mental goings on are concrete phenomena.[2]

What does physicalism involve? What is it, really, to be a physicalist? What is it to be a *realistic* physicalist, or, more simply, a *real* physicalist? Well, one thing is absolutely clear. You're certainly not a realistic physicalist, you're not a real physicalist, if you deny the existence of the phenomenon whose existence is more certain than the existence of anything else: experience, 'consciousness', conscious experience, 'phenomenology', experiential 'what-it's-likeness', feeling, sensation, explicit conscious thought as we have it and know it at almost every waking moment. Many words are used to denote this necessarily occurrent (essentially non-dispositional) phenomenon, and

---

[1] This paper recasts and expands parts of 'Agnostic materialism' (Strawson, 1994, pp. 43–105, especially pp. 59–62, 72, 75–7) and 'Real materialism' (Strawson, 2003a) and inherits their debt to Nagel (1974). I have replaced the word 'materialism' by 'physicalism' and speak of 'physical stuff' instead of 'matter' because 'matter' is now specially associated with mass although energy is just as much in question, as indeed is anything else that can be said to be physical, e.g. spacetime — or whatever underlies the appearance of spacetime.

[2] More strictly, 'concrete' means 'not abstract' in the standard philosophical sense of 'abstract', given which some philosophers hold that abstract objects — e.g. numbers or concepts — exist and are real objects in every sense in which concrete objects are. I take 'spatio-temporal' to be the adjective formed from 'spacetime', not from the conjunction of space and time.

in this paper I will use the terms 'experience', 'experiential phenomena' and 'experientiality' to refer to it.

Full recognition of the reality of experience, then, is the obligatory starting point for any remotely realistic version of physicalism. This is because it is the obligatory starting point for any remotely realistic (indeed any non-self-defeating) theory of what there is. It is the obligatory starting point for any theory that can legitimately claim to be 'naturalistic' because experience is itself the fundamental given natural fact; it is a very old point that there is nothing more certain than the existence of experience.

It follows that real physicalism can have nothing to do with *physicSalism*, the view — the faith — that the nature or essence of all concrete reality can in principle be fully captured in the terms of *physics*. Real physicalism cannot have anything to do with physicSalism unless it is supposed — obviously falsely — that the terms of physics can fully capture the nature or essence of experience.[3] It is unfortunate that 'physicalism' is today standardly used to mean physicSalism because it obliges me to speak of 'real physicalism' when really I only mean 'physicalism' — realistic physicalism.

Real physicalism, then, must accept that experiential phenomena are physical phenomena. But how can experiential phenomena be physical phenomena? Many take this claim to be profoundly problematic (this is the 'mind–body problem'). This is usually because they think they know a lot about the nature of the physical. They take the idea that the experiential is physical to be profoundly problematic *given what we know about the nature of the physical*. But they have already made a large and fatal mistake. This is because we have no good reason to think that we know anything about the physical that gives us any reason to find any problem in the idea that experiential phenomena are physical phenomena. If we reflect for a moment on the nature of our knowledge of the physical, and of the experiential, we realize, with Eddington, that 'no problem of irreconcilability arises'.[4]

---

[3] For a standard argument that this is impossible in principle, see e.g. Strawson (1994), pp. 62–5.

[4] Eddington (1928, p. 260); the thought was not new. In the background stood Arnauld (1641), Locke (1681), Hume (1739), Priestley (1777), and many others — see Strawson (2003a, §12). Kant makes the point very clearly, on his own special terms. See e.g. Kant (1781/7), A358–60, A380 and B427–8, where he remarks that the 'heterogeneity' of mind and body is merely 'assumed' and not known.

A very large mistake. It is perhaps Descartes's, or perhaps rather 'Descartes's', greatest mistake,[5] and it is funny that in the past fifty years it has been the most fervent revilers of the great Descartes, the true father of modern materialism, who have made the mistake with most intensity. Some of them — Dennett is a prime example — are so in thrall to the fundamental intuition of dualism, the intuition that the experiential and the physical are utterly and irreconcilably different, that they are prepared to deny the existence of experience, more or less (c)overtly, because they are committed to physicalism, i.e. physicSalism.[6]

'They are prepared to deny the existence of experience.' At this we should stop and wonder. I think we should feel very sober, and a little afraid, at the power of human credulity, the capacity of human minds to be gripped by theory, by faith. For this particular denial is the strangest thing that has ever happened in the whole history of human thought, not just the whole history of philosophy. It falls, unfortunately, to philosophy, not religion, to reveal the deepest woo-woo of

---

[5] I think that, in his hidden philosophical heart, he did not make it (he is certainly not a 'substance dualist' as this expression is currently understood; see Clarke, 2003). Arnauld saw the problem clearly, and Hume (1739, p. 159 (1.3.14.8)) diagnosed the mistake definitively in two lines, with specific reference to the Cartesians, but the twentieth century — philosophical division — wasn't listening.

[6] Dennett conceals this move by *looking-glassing* the word 'consciousness' (his term for experience) and then insisting that he does believe that consciousness exists (to looking-glass a term is to use a term in such a way that whatever one means by it, it excludes what the term means — see Strawson, 2005). As far as I can understand them, Dretske, Tye, Lycan and Rey are among those who do the same. It seems that they still dream of giving a reductive analysis of the experiential in non-experiential terms. This, however, amounts to denying the existence of experience, because the nature of (real) experience can no more be specified in wholly non-experiential terms than the nature of the (real) non-experiential can be specified in wholly experiential terms. In the normal case, of course, reductive identification of X with Y is not denial of the existence of X. The reductive claim is 'X exists, but it is really just this (Y)'. In the case of experience, however, to say that it exists but is really just something whose nature can be fully specified in wholly non-experiential, functional terms is to deny its existence. 'But what is this supposed thing you say we're denying?' say the deniers. It's the thing to which the right reply to the question 'What is it?' is, as ever, the (Louis) Armstrong-Block reply: 'If you gotta ask, you ain't never gonna get to know' (Block, 1978). It's the thing whose deniers say that there is no non-question-begging account of it, to which the experiential realist's correct reply is: 'It's question-begging for you to say that there must be an account of it that's non-question-begging in your terms'. Such an exchange shows that we have reached the end of argument, a point further illustrated by the fact that reductive idealists can make exactly the same 'You have no non-question-begging account' objection to reductive physicalists that reductive physicalists make to realists about experience: 'By taking it for granted that the physical is something that can (only) be specified in non-mental terms, you (reductive physicalists) simply beg the question against reductive idealists.' It's striking that the realist notion of the physical that present-day physicalists appeal to was thought to be either without warrant or unintelligible by many of the leading philosophers of the twentieth century. Many were reductive idealists about the physical, and Quine famously compared belief in physical objects to belief in the gods of Homer (Quine, 1951, p. 44).

the human mind. I find this grievous, but, next to this denial, every known religious belief is only a little less sensible than the belief that grass is green.[7]

Realistic physicalists, then, grant that experiential phenomena are real concrete phenomena — for nothing in life is more certain — and that experiential phenomena are therefore physical phenomena. It can sound odd at first to use 'physical' to characterize mental phenomena like experiential phenomena,[8] and many philosophers who call themselves materialists or physicalists continue to use the terms of ordinary everyday language, that treat the mental and the physical as opposed categories. It is, however, precisely physicalists (real physicalists) who cannot talk in this way, for it is, on their own view, exactly like talking about cows and animals as if they were opposed categories. Why? Because every concrete phenomenon is physical, according to them. So all mental (experiential) phenomena are physical phenomena, according to them; just as all cows are animals. So when physicalists — real ones — talk as if the mental (experiential) and the physical were entirely different all they can really mean to be doing is to distinguish, within the realm of the physical, which is the only realm there is, according to them, between mental (experiential)

---

[7] Dennett has suggested that 'there is no such thing [as] . . . phenomenology' and that any appearance of phenomenology is, somehow, wholly the product of some cognitive faculty, the 'judgment module' or 'semantic intent module' that does not itself involve any phenomenology. '*There seems to be phenomenology*,' he concedes, 'but it does *not* follow from this undeniable, universally attested fact that *there really is* phenomenology' (Dennett, 1991b, pp. 365–6). It is unclear what Dennett means by 'phenomenology', but whatever he means this move fails immediately if it is taken as an objection to the present claim that we can be certain both that there is experience and that we can't be radically in error about its nature. It fails for the simple reason that for there to seem to be rich phenomenology or experience just is for there to be such phenomenology or experience. To say that its apparently sensory aspects (say) are in some sense illusory because they are not the product of sensory mechanisms in the way we suppose, but are somehow generated by merely cognitive processes, is just to put forward a surprising hypothesis about part of the *mechanism* of this rich seeming that we call experience or consciousness. It is in no way to put in question its existence or reality. Whatever the process by which the seeming arises, the end result of the process is, as even Dennett agrees, at least this: that it *seems* as if one is having phenomenally rich experience of Beethoven's eighth quartet or an Indian wedding; and if there is this seeming, then, once again, there just is phenomenology or experience (adapted from Strawson, 1994, pp. 51–2).

In denying that experience can be physical, Dennett and his kind find themselves at one with many religious believers. This seems at first ironic, but the two camps are deeply united by the fact that both have unshakable faith in something that lacks any warrant in experience. That said, the religious believers are in infinitely better shape, epistemologically, than the Dennettians.

[8] For purposes of argument I make the standard assumption that while all experiential phenomena are mental phenomena, the converse is not true.

features of the physical, and non-mental (non-experiential) features of the physical.

As a real physicalist, then, I hold that the mental/experiential is physical, and I am happy to say, along with many other physicalists, that experience is 'really just neurons firing', at least in the case of biological organisms like ourselves. But when I say these words I mean something completely different from what many physicalists have apparently meant by them. I certainly don't mean that all characteristics of what is going on, in the case of experience, can be described by physics and neurophysiology or any non-revolutionary extensions of them. That idea is crazy. It amounts to radical 'eliminativism' with respect to experience, and it is not a form of real physicalism at all.[9] My claim is different. It is that experiential phenomena 'just are' physical, so that there is a lot more to neurons than physics and neurophysiology record (or can record). No one who disagrees with this is a real physicalist, in my terms.

In a paper called 'Real materialism' I considered some objections to the claim that the position I have just outlined can really be called a physicalist position. I did my best to answer them and ended concessively, allowing that one might better call the position 'experiential-and-non-experiential monism' rather than 'real physicalism'. It is, in any case, the position of someone who (a) fully acknowledges the evident fact that there is experiential being in reality, (b) takes it that there is also non-experiential being in reality, and (c) is attached to the 'monist' idea that there is, in some fundamental sense, only one kind of stuff in the universe.

The objectors then picked on the word 'monist', and I considered a further concession. You can call my position 'experiential-and-non-experiential ?-ism', if you like, and opt out of the monism-dualism-pluralism oppositions of classical metaphysics. Perhaps you can simply call it '?-ism'.[10] But then you will have to allow that the existence of experiential being at least is certain, and is not put in question by the '?' — so that it would be better to call it 'experiential ?-ism'. If you then want to insist, in line with all standard conceptions of the physical, that non-experiential being also exists, then you will also need to signal the fact that the non-experiential is not put in question by the '?'. In which case you may as well go back to calling the position 'experiential-and-non-experiential ?-ism'.

---

[9] This follows from the fact that current physics contains no predicates for experiential phenomena, and that no non-revolutionary extension of it (no currently conceivable extension of it — see note 3) could do so.

[10] A suggestion made by Sebastian Gardner, nearly twenty years ago.

I persist in thinking that 'physicalism', 'real physicalism', is a good name for my position in the current context of debate, but it's already time to admit that in my understanding real physicalism doesn't even rule out panpsychism — which I take to be the view that the existence of every real concrete thing involves experiential being even if it also involves non-experiential being. If this seems a little colourful then it's time to read Locke on substance again.[11]

Surely I've pushed myself over the edge? How can I say that 'physicalism' is an acceptable name for my position? Because I take 'physical' to be a natural-kind term whose reference I can sufficiently indicate by drawing attention to tables and chairs and — as a realistic physicalist — experiential phenomena.[12] The physical is whatever general kind of thing we are considering when we consider things like tables and chairs and experiential phenomena. It includes everything that concretely exists in the universe. If everything that concretely exists is intrinsically experience-involving, well, that is what the physical turns out to be; it is what energy (another name for physical stuff) turns out to be. This view does not stand out as particularly strange against the background of present-day science, and is in no way incompatible with it.

I don't *define* the physical as concrete reality, as concrete-reality-whatever-it-is; obviously I can't rule out the possibility that there could be other non-physical (and indeed non-spatiotemporal) forms of concrete reality. I simply fix the reference of the term 'physical' by pointing at certain items and invoking the notion of a general kind of stuff. It is true that there is a sense in which this makes my use of the term vacuous, for, relative to our universe, 'physical stuff' is now equivalent to 'real and concrete stuff', and cannot be anything to do with the term 'physical' that is used to mark out a position in what is usually taken to be a substantive debate about the ultimate nature of concrete reality (physicalism vs immaterialism vs dualism vs pluralism vs . . .). But that is fine by me. If it's back to Carnap, so be it.[13]

Have I gone too far? It seems to me that to go this far is exactly the right thing to do at this point in the debate. It's worth it if it helps us to get back to a proper (realistic) openmindedness. But anyone who prefers

---

[11] Locke (1689), 2.23 and 4.3.6.

[12] It's striking that analytic philosophers and psychologists have talked so much about natural-kind terms but have failed to see that 'physical' is a paradigmatic example of such a term in every sense in which 'gold' is.

[13] See Carnap (1950).

to call my position 'realistic monism' instead of 'real physicalism' should feel free to do so.[14]

2. 'It seems rather silly . . .'

This may all seem a little giddy, so I will now rein things in a little by making three conventional substantive *assumptions* about the physical strictly for purposes of argument, using the term 'ultimate' to denote a fundamental physical entity, an ultimate constituent of reality, a particle, field, string, brane, simple, whatever:

(1) there is a plurality of ultimates (whether or not there is a plurality of types of ultimates)[15]
(2) everything physical (that there is or could be) is constituted out of ultimates of the sort we actually have in our universe
(3) the universe is spatio-temporal in its fundamental nature.[16]

I do not, however, think that I need these assumptions in order to show that something akin to panpsychism is not merely one possible form of realistic physicalism, real physicalism, but the only possible form. (It should always be possible to replace references to ultimates by a field-theoretic picture of things.) Eddington is one of those who saw this clearly, and I am now going to join forces with him and ask you to be as tolerant of his terminological loosenesses and oddities as I hope you will be of my appeals to intuition.[17]

One thing we know about physical stuff, given that (real) physicalism is true, is that when you put it together in the way in which it is put together in brains like ours, it regularly constitutes — is, literally is — experience like ours. Another thing we know about it, let us grant, is everything (true) that physics tells us. But what is this second kind of knowledge like? Well, there is a fundamental sense in which it is

---

[14] It is less certain that there is non-experiential stuff than that there is experiential stuff, and in most ears 'real physicalism' signals commitment to the existence of non-experiential stuff in a way that 'realistic monism' does not.

[15] A powerful rival view is that there is at bottom just one thing or substance, e.g. spacetime, or whatever underlies all spacetime appearances. But (1) does not beg any important questions. If anything, it makes things more difficult for me.

[16] This is in doubt in present-day physics and cosmology, for 'rumors of spacetime's impending departure from deep physical law are not born of zany theorizing. Instead, this idea is strongly suggested by a number of well-reasoned considerations' (Greene, 2004, p. 472; see also pp. 473–91). Note that if temporality goes, i.e. not just spacetime^{TM} but temporality in any form, then experience also goes, given that experience requires time. One of the fine consequences of this is that there has never been any suffering. But no theory of reality can be right that has the consequence that there has never been any suffering.

[17] I came upon *The Nature of the Physical World* in a holiday house in Scotland in 1999.

'abstract', 'purely formal', merely a matter of 'structure', in Russell's words.[18] This is a well established but often overlooked point.[19] 'Physics is mathematical', Russell says, 'not because we know so much about the physical world' — and here he means the non-mental, non-experiential world, in my terms, because he is using 'mental' and 'physical' conventionally as opposed terms —

> but because we know so little: it is only its mathematical properties that we can discover. For the rest, our knowledge is negative . . . The physical world is only known as regards certain abstract features of its space-time structure — features which, because of their abstractness, do not suffice to show whether the physical world is, or is not, different in intrinsic character from the world of mind.[20]

Eddington puts it as follows. 'Our knowledge of the nature of the objects treated in physics consists solely of readings of pointers (on instrument dials) and other indicators.' This being so, he asks, 'what knowledge have we of the nature of atoms that renders it at all incongruous that they should constitute a thinking object?' Absolutely none, he rightly replies: 'science has nothing to say as to the intrinsic nature of the atom'. The atom, so far as physics tells us anything about it,

> is, like everything else in physics, a schedule of pointer readings [on instrument dials]. The schedule is, we agree, attached to some unknown background. Why not then attach it to something of a spiritual [i.e. mental] nature of which a prominent characteristic is *thought* [=experience, consciousness]. It seems rather **silly** to prefer to attach it to something of a so-called 'concrete' nature inconsistent with thought, and then to wonder where the thought comes from. We have dismissed all preconception as to the background of our pointer readings, and for the most part can discover nothing as to its nature. But in one case — namely, for the pointer readings of my own brain — I have an insight which is not limited to the evidence of the pointer readings. That insight shows that they are attached to a background of consciousness

in which case

> I may expect that the background of other pointer readings in physics is *of a nature continuous with that revealed to me in this way*,

---

[18] Russell (1927a), pp. 392, 382; (1956), p. 153; (1927b), p. 125.

[19] It takes time to assimilate it fully. It cannot be simply read off the page.

[20] Russell (1948), p. 240; see also p. 247. Russell's overall view is that 'we know nothing about the intrinsic quality of physical events except when these are mental events that we directly experience' (Russell, 1956, p. 153), and that 'as regards the world in general, both physical and mental, everything that we know of its intrinsic character is derived from the mental side' (1927a, p. 402). See Lockwood (1981; 1989), Strawson (2003a).

even while

> I do not suppose that it always has the more specialized attributes of consciousness.

What is certain is that

> in regard to my one piece of insight into the background no problem of irreconcilability arises; I have no other knowledge of the background with which to reconcile it ... *There is nothing to prevent the assemblage of atoms constituting a brain from being of itself a thinking (conscious, experiencing) object in virtue of that nature which physics leaves undetermined and undeterminable.* If we must embed our schedule of indicator readings in some kind of background, at least let us accept the only hint we have received as to the significance of the background — namely, that it has a nature capable of manifesting itself as mental activity.[21]

This all seems intensely sensible and Occamical. Eddington's notion of silliness is extremely powerful. Why then — on what conceivable grounds — do so many physicalists simply assume that the physical, in itself, is an essentially and wholly non-experiential phenomenon?

I write this and think 'Do they really?', and this rapid inner question is not rhetorical or aggressive, meaning 'They must be pretty stupid if they really think, and think they know, that physical stuff is, in itself, and through and through, an essentially non-experiential phenomenon'. It is, rather, part of a feeling that I must be wrong. I must be doing what philosophers are famous for doing — setting up strawman opponents who do not really exist while erasing awareness of my real audience, who will protest that of course they aren't so foolish as to claim to know that physical stuff is, in itself, in its root nature, a wholly non-experiential phenomenon.

My next thought, however, is that I am not wrong. It looks as if many — perhaps most — of those who call themselves physicalists or materialists really are committed to the thesis that

> [NE] physical stuff is, in itself, in its fundamental nature, something wholly and utterly non-experiential.

---

[21] Eddington (1928), pp. 258–60; my emphasis on 'silly'. It is remarkable that this line of thought (so well understood by Russell, Whitehead, Eddington, Broad, Feigl and many others, and equally, in a number of slightly different guises, by Spinoza, Locke, Hume, Kant, Priestley and many others) disappeared almost completely from the philosophical mainstream in the wake of Smart's 1959 paper 'Sensations and brain processes', although it was well represented by Chomsky (see e.g. Chomsky, 1968, 1995). At this point analytical philosophy acquired hyperdualist intuitions even as it proclaimed its monism. With a few honourable exceptions it out-Descartesed Descartes (or 'Descartes') in its certainty that we know enough about the physical to know that the experiential cannot be physical.

I think they take it, for a start, that ultimates are in themselves wholly and essentially non-experiential phenomena. And they are hardly going out on a limb in endorsing NE, for it seems to be accepted by the vast majority of human beings. I do not, however, see how physicalists can leave this commitment unquestioned, if they are remotely realistic in their physicalism, i.e. if they really do subscribe to the defining thesis of real physicalism that

> [RP] experience is a real concrete phenomenon and every real concrete phenomenon is physical.

For if they are real physicalists they cannot deny that when you put physical stuff together in the way in which it is put together in brains like ours, it constitutes — is — experience like ours; all by itself. All by itself: there is on their own physicalist view nothing else, nothing non-physical, involved.

The puzzle, for me, is that I'm sure that some at least of those who call themselves physicalists are realistic physicalists — real realists about experiential phenomena. Yet they do, I think, subscribe to NE — even when they are prepared to admit, with Eddington, that physical stuff has, in itself, 'a nature capable of manifesting itself as mental activity', i.e. as experience or consciousness.

3. Emergence

Is this a possible position? Can one hold RP and NE together? I don't think so, but one defence goes like this:

> Experiential phenomena are *emergent* phenomena. Consciousness properties, experience properties, are emergent properties of wholly and utterly non-conscious, non-experiential phenomena. Physical stuff *in itself*, in its basic nature, is indeed a wholly non-conscious, non-experiential phenomenon. Nevertheless when parts of it combine in certain ways, experiential phenomena 'emerge'. Ultimates in themselves are wholly non-conscious, non-experiential phenomena. Nevertheless, when they combine in certain ways, experiential phenomena 'emerge'.

Does this conception of emergence make sense? I think that it is very, very hard to understand what it is supposed to involve. I think that it is incoherent, in fact, and that this general way of talking of emergence has acquired an air of plausibility (or at least possibility) for some simply because it has been appealed to many times in the face of a seeming mystery.[22] In order to discuss it I am going to take it that any position that combines RP with NE must invoke some notion of

---

[22] Compare the way in which the word 'immaterial' comes to seem to have some positive descriptive meaning although it quite explicitly has none. For a recent helpful taxonomy

emergence, whether or not it chooses to use the word. I will start on familiar ground.

Liquidity is often proposed as a translucent example of an emergent phenomenon, and the facts seem straightforward. Liquidity is not a characteristic of individual H_2O molecules. Nor is it a characteristic of the ultimates of which H_2O molecules are composed. Yet when you put many H_2O molecules together they constitute a liquid (at certain temperatures, at least), they constitute something liquid. So liquidity is a truly emergent property of certain groups of H_2O molecules. It is not there at the bottom of things, and then it is there.

When heat is applied evenly to the bottom of a tray filled with a thin sheet of viscous oil, it transforms the smooth surface of the oil into an array of hexagonal cells of moving fluid called Bénard convection cells.[23] This is another popular example of an emergent phenomenon. There are many chemical and physical systems in which patterns of this sort arise simply from the routine workings of basic physical laws, and such patterns are called 'emergent'.

This is all delightful and true. But can we hope to understand the alleged emergence of experiential phenomena from non-experiential phenomena by reference to such models? I don't think so. The emergent character of liquidity relative to its non-liquid constituents does indeed seem shiningly easy to grasp. We can easily make intuitive sense of the idea that certain sorts of molecules are so constituted that they don't bind together in a tight lattice but slide past or off each other (in accordance with van de Waals molecular interaction laws) in a way that gives rise to — is — the phenomenon of liquidity. So too, with Bénard convection cells we can easily make sense of the idea that physical laws relating to surface tension, viscosity, and other forces governing the motion of molecules give rise to hexagonal patterns on the surface of a fluid like oil when it is heated. In both these cases we move in a small set of conceptually homogeneous shape-size-mass-charge-number-position-motion-involving physics notions with no sense of puzzlement. Using the notion of reduction in a familiar loose way, we can say that the phenomena of liquidity reduce without remainder to shape-size-mass-charge-etc. phenomena — I'll call these 'P' phenomena for short, and assume for now that they are, in themselves, utterly non-experiential phenomena. We can see that the phenomenon of liquidity arises naturally out of, is *wholly dependent on*,

---

of types of emergence, see van Gulick (2001); see also Broad (1925) and McLaughlin (1992).

[23] Velarde and Normand (1980).

phenomena that do not in themselves involve liquidity at all. We can with only a little work suppress our initial tendency to confuse liquidity as it appears to sensory experience (how, we may think, could *this* arise from individual non-liquid molecules?) with the physical phenomenon of liquidity considered just as such, and see clearly that it is just and wholly a matter of P phenomena.

This notion of total dependence looks useful. It seems plain that there must be a fundamental sense in which any emergent phenomenon, say Y, is wholly dependent on that which it emerges from, say X. It seems, in fact, that this must be true by definition of 'emergent'; for if there is not this total dependence then it will not be true after all, not true without qualification, to say that Y is emergent from X. For in this case at least some part or aspect of Y will have to hail from somewhere else and will therefore not be emergent from X. Plainly this is not how it is with liquidity.[24]

It is the dependence requirement that causes the problem when it comes to relating the supposedly emergent phenomena of experience to the supposedly wholly non-experiential phenomena from which they supposedly emerge. For it now seems that if experiential phenomena — colour-experiences, for example — really are somehow (wholly) dependent on non-experiential phenomena, as they must be if they are to be truly emergent from them, that is, emergent from them and from them alone, then there must (to quote myself in a former century) be

> a correct way of describing things . . . given which one can relate [the experiential phenomenon of] color-experience, considered just as such, to the non-experiential phenomena on which it is supposed to depend, in such a way that the dependence is as intelligible as the dependence of the liquidity of water on the interaction properties of individual molecules. The alternative, after all, is that there should be total dependence that is not intelligible or explicable in any possible physics, dependence that must be unintelligible and inexplicable even to God, as it were.[25]

---

[24] Here, then, I reject the commonly embraced but little examined and seemingly wholly mystical notion of emergence that van Gulick (2001) calls 'Radical Kind Emergence' and defines as follows: 'the whole has features that are both (a) different in kind from those had by the parts, and (b) of a kind whose nature is not necessitated by the features of its parts, their mode of combination and the law-like regularities governing the features of its parts.' (Liquidity, in van Gulick's scheme, is a case of 'Modest Kind Emergence': 'the whole has features that are different in kind from those of its parts (or alternatively that could be had by its parts). For example, a piece of cloth might be purple in hue even though none of the molecules that make up its surface could be said to be purple.')

[25] Strawson (1994), p. 69.

I wouldn't put it this way now. The notions of explicability and intelligibility are in origin epistemological, and are potentially misleading because the present claim is not epistemological. It is not, for example, touched by the reply that there is a sense in which all *causal* dependence relations, at least, are ultimately unintelligible to us, even those that seem most intuitively understandable. For although there is a sense in which this is true, in as much as all our explanations of concrete phenomena come to an end in things that are simply given, contingent, not further explicable, it has no bearing here. 'Intelligible to God' isn't really an epistemological notion at all, it's just a way of expressing the idea that there must be something about the nature of the emerged-from (and nothing else) in virtue of which the emerger emerges as it does and is what it is.

You can get liquidity from non-liquid molecules as easily as you can get a cricket team from eleven things that are not cricket teams. In God's physics, it would have to be just as plain how you get experiential phenomena from wholly non-experiential phenomena. But this is what boggles the human mind. We have, once again, no difficulty with the idea that liquid phenomena (which are wholly P phenomena) are emergent properties of wholly non-liquid phenomena (which are wholly P phenomena). But when we return to the case of experience, and look for an analogy of the right size or momentousness, as it were, it seems that we can't make do with things like liquidity, where we move wholly within a completely conceptually homogeneous (non-heterogeneous) set of notions. We need an analogy on a wholly different scale if we are to get any imaginative grip on the supposed move from the non-experiential to the experiential.

What might be an analogy of the right size? Suppose someone — I will call him pseudo-Boscovich, at the risk of offending historians of science — proposes that all ultimates, all real, concrete ultimates, are, in truth, wholly unextended entities: that this is the truth about their being; that there is *no* sense in which they themselves are extended; that they are real concrete entities, but are none the less true-mathematical-point entities. And suppose pseudo-Boscovich goes on to say that when collections of these entities stand in certain (real, concrete, natural) relations, they give rise to or constitute truly, genuinely extended concrete entities; real, concrete extension being in this sense an *emergent property* of phenomena that are, although by hypothesis real and concrete, wholly unextended.

Well, I think this suggestion should be rejected as absurd. But the suggestion that when non-experiential phenomena stand in certain (real, natural, concrete non-experiential) relations they *ipso facto* instantiate or constitute experiential phenomena, experience being an emergent property of wholly and utterly non-experiential phenomena, seems exactly on a par. That's why I offer unextended-to-extended emergence as an analogy, a destructive analogy that proposes something impossible and thereby challenges the possibility of the thing it is offered as an analogy for. You can (to use the letter favoured by the German idealists when either stating or rejecting the law of non-contradiction) get A from non-A for some substitutions for A, such as liquidity, but not all.

> — My poor friend. The idea that collections of concrete entities that are truly, genuinely unextended can give rise to or constitute concrete entities that are truly, genuinely extended is actually scientific orthodoxy, on one widely received view of what ultimates are. It's an excellent candidate for being an analogy of the right size.

But this won't do. It won't do when one is being metaphysically straight, not *metaphysically* instrumentalist, or positivist, or operationalist, or phenomenalist, or radical-empiricist, or verificationist, or neo-verificationist or otherwise anti-realist or Protagorean (alas for the twentieth century, in which all these epistemological notions somehow got metaphysicalized). If one is being metaphysically straight, the intuition that nothing (concrete, spatio-temporal) can exist at a mathematical point, because there just isn't any room, is rock solid.[26] It may be added that anything that has, or is well understood as, a field, or that has any sort of attractive or repulsive being or energy, or any area of influence or influencability, *ipso facto* has extension — extension is part of its being — and that although there are plenty of ultimates that have no charge in what physicists call 'the standard model', there are I believe none that are not associated with a field.[27] So if the idea of unextended-to-extended emergence is offered as an analogy for non-experiential-to-experiential emergence, I don't think it can help.

---

[26] Do not be cowed by physicists or philosophers of physics. If we had been metaphysically sensible we might have got to something like string theory (M-theory or brane theory) decades sooner. It seems intuitively obvious, by the grace of mathematics, that to introduce real, concrete entities that are infinitely small and therefore metaphysically impossible into one's theory will lead to infinite largenesses popping up in protest elsewhere in one's equations. And so it came to pass. And quantum mechanics could give no account of gravity.

[27] As I understand it, every particle in the standard model feels a force, even the photon (i.e. photon–photon forces, mediated by — virtual — pair creation/annihilation processes for the sources of the photon). This sort of point no longer seems required, however, given that all the ultimates of M-theory have extension.

REALISTIC MONISM   17

I'll take this a little further. Suppose someone proposes that there are real, concrete, intrinsically, irreducibly and wholly *non-spatial* phenomena ('wholly non-S phenomena'), and that when they stand in certain wholly non-spatial relations they give rise to or constitute real, concrete, intrinsically and irreducibly spatial phenomena ('S phenomena'), these being emergent features of wholly non-S phenomena. Those who claim to find no difficulty in the idea that genuinely unextended concrete entities can give rise to or constitute genuinely extended concrete entities may like to consider this case separately, because they presumably take it that their putative mathematical-point entities are at least spatial entities, at least in the sense of being spatially located. My hope is that even if they think they can make sense of the emergence of the extended from the unextended, they won't think this about the more radical case of the emergence of the spatial from the non-spatial.

But what do I know about this? Almost nothing. With this kind of speculation 'we are got into fairy land', as Hume says, or rather I am, and any impossibility claim on my part, or indeed anyone else's, may seem rash.[28] And some may now propose that the 'Big Bang' is precisely a case in which S phenomena are indeed emergent features of wholly non-S phenomena.

Don't believe it, I say, falling back on the *argumentum a visceris*. S phenomena, i.e. real, concrete, intrinsically and irreducibly spatial phenomena (bear in mind that we are seeking an analogy for experiential phenomena that we know to be real, concrete, intrinsically and irreducibly experiential) *can't* be emergent properties of wholly non-S phenomena. This is a case where you can't get A from non-A. The spatial/non-spatial case may look like an analogy of the right size for the experiential/non-experiential case, but all it turns up, I suggest, is impossibility. If there is any sense in which S phenomena can be said to emerge from wholly non-S phenomena, then they must fall back into the category of mere appearance, and they are then (by definition, see above) not S phenomena at all. Experiential phenomena, however, cannot do this. They cannot be mere appearance, if only because all appearance depends on their existence.[29] If it were to turn out that real S phenomena can after all emerge from wholly non-S phenomena, all that would follow would be that the spatial case did not after all constitute an analogy of the right size. The experiential/non-experiential divide,

---

[28] Hume (1748), p. 72. It is quite plain, in any case, that people can think (or think they think) anything.

[29] See note 7. One current view of the 'Big Bang' is that it occurred everywhere in an already existing infinite space.

assuming that it exists at all, is the most fundamental divide in nature (the only way it can fail to exist is for there to be nothing non-experiential in nature).[30]

The claim, at least, is plain, and I'll repeat it. If it really is true that Y is emergent from X then it must be the case that Y is in some sense wholly dependent on X and X alone, so that all features of Y trace intelligibly back to x (where 'intelligible' is a metaphysical rather than an epistemic notion). *Emergence can't be brute.* It is built into the heart of the notion of emergence that emergence cannot be brute in the sense of there being absolutely no reason in the nature of things why the emerging thing is as it is (so that it is unintelligible even to God). For any feature Y of anything that is correctly considered to be emergent from X, there must be something about X and X alone in virtue of which Y emerges, and which is sufficient for Y.

I'm prepared to allow for argument that an ultimate's possession of its fundamental properties could be brute in the sense of there being no reason for it in the nature of things, so long as it is agreed that *emergence* cannot be brute. One problem is that brute emergence is by definition a miracle every time it occurs, for it is true by hypothesis that in brute emergence there is absolutely nothing about X, the emerged-from, in virtue of which Y, the emerger, emerges from it. This means that it is also a contradiction in terms, given the standard assumption that the emergence of Y from X entails the 'supervenience' of Y on X,[31] because it then turns out to be a strictly lawlike miracle. But a miracle is by definition a violation of a law of nature![32] If someone says he chooses to use the word 'emergence' in such a way that the notion of brute emergence is not incoherent, I will know that he is a member of the Humpty Dumpty army and be very careful with him.

How did the notion of brute emergence ever gain currency? By one of the most lethal processes of theory formation, or term formation, that there is. The notion of brute emergence marks a position that seemingly has to exist if one accepts both RP (or, more simply, the reality of experience) and NE. And since many are irredeemably committed to both RP and NE, the notion of brute emergence comes to feel substantial to them by a kind of reflected, holographical energy. It has

---

[30] The viscera are not unsophisticated organs. They can refuse the getting of A from non-A for some substitutions for A even while they have no difficulty with the strangest quantum strangenesses (see e.g. Strawson, 2003a, p. 65).

[31] The supervenience thesis states that if Y is supervenient on X then whenever you have an X-type phenomenon you must also have a Y-type phenomenon.

[32] This is Hume's definition of a miracle (I'm assuming that there is no *deus ex machina*). It is often said that this definition requires an absolute, non-statistical notion of a law of nature, but this is not so (see Mackie, 1982, ch. 4).

to be there, given these unquestioned premises, so it is felt to be real. The whole process is underwritten by the wild radical-empiricism-inspired metaphysical irresponsibilities of the twentieth century that still linger on (to put it mildly) today and have led many, via a gross misunderstanding of Hume, to think that there is nothing intrinsic to a cause in virtue of which it has the effect it does.[33]

I'll say it again. For Y truly to emerge from X is for Y to arise from or out of X or be given in or with Y given how x *is*. Y must arise out of or be given in X in some essentially non-arbitrary and indeed wholly non-arbitrary way. X has to have something — indeed everything — to do with it. That's what emerging is (that's how liquidity arises out of non-liquid phenomena). It is essentially an in-virtue-of relation. It cannot be brute. Otherwise it will be intelligible to suppose that existence can emerge from (come out of, develop out of) non-existence, or even that concrete phenomena can emerge from wholly abstract phenomena. Brutality rules out nothing.[34] If emergence can be brute, then it is fully intelligible to suppose that non-physical soul-stuff can arise out of physical stuff — in which case we can't rule out the possibility of Cartesian egos *even if we are physicalists*. I'm not even sure we can rule out the possibility of a negative number emerging from the addition of certain positive numbers. We will certainly have to view with equanimity all violations of existing laws of (non-experiential) physics, dross turning adventitiously into gold, particles decaying into other particles whose joint charge differs from that of the original particle.

Returning to the case of experience, Occam cuts in again, with truly devastating effect. Given the undeniable reality of experience, he says, why on earth (our current location) commit oneself to NE? Why insist that physical stuff in itself, in its basic nature, is essentially non-experiential, thereby taking on

---

[33] Here I make the common assumption that it is legitimate to segment the world into causes and effects. Hume's wholly correct, strictly epistemological claim — that so far as we consider things *a priori* 'any thing may produce any thing' — came to be read as the metaphysical claim that anything may produce anything. For a discussion of this error see e.g. Craig (1987), ch. 2; Strawson (2000). It is worth noting that the epistemological restriction is usually explicitly stated in Hume's *Treatise*, in spite of his youthful liking for dramatic abbreviation: 'I have inferr'd from these principles, that *to consider the matter a priori*, any thing may produce any thing, and that we shall never discover a reason, why any object may or may not be the cause of any other, however great, or however little the resemblance may be betwixt them' (*Treatise*, p. 247); '*for ought we can determine by the mere ideas*, any thing may be the cause or effect of any thing' (pp. 249–50; my emphasis). Brute emergence does indeed license the non-Humean, ontological version of 'any thing may produce any thing'.

[34] Even if a universe could just come into existence when nothing existed, it certainly couldn't emerge from non-existence in the relevant sense of 'emerge'. *Ex nihilo nihil fit*, whatever anyone says (Nobel Prize winners included).

(a) a commitment to something — wholly and essentially non-experiential stuff — for which there is *absolutely no evidence whatever*

along with

(b) the wholly unnecessary (and incoherent) burden of brute emergence

otherwise known as magic? That, in Eddington's terms, is silly.

> — What about the emergence of life? A hundred years ago it seemed obvious to many so-called 'vitalists' that *life* could not emerge from utterly lifeless matter (from P phenomena), just as it seems obvious to you now that *experience* could not emerge from utterly non-experiential matter (from P phenomena). Today, however, no one seriously doubts that life emerged from matter that involved no life at all. The problem of life, that seemed insuperable, simply dissolved. Why should it not be the same with consciousness, a hundred years from now?

This very tired objection is always made in discussions of this sort, and the first thing to note is that one cannot draw a parallel between the perceived problem of life and the perceived problem of experience in this way, arguing that the second problem will dissolve just as the first did, unless one considers life completely apart from experience. So let us call life considered completely apart from experience 'life*'. My reply is then brief. Life* reduces, experience doesn't. Take away experience from life and it (life*) reduces smoothly to P phenomena. Our theory of the basic mechanisms of life reduces to physics via chemistry. Suppose we have a machine that can duplicate any object by a process of rapid atom-by-atom assembly, and we duplicate a child. We can explain its life* functions in exquisite detail in the terms of current sciences of physics, chemistry and biology. We cannot explain its experience at all in these terms.

One of the odd things about the supposed problem of life* is that although it was popular at the end of the nineteenth century it would not have been thought very impressive in the seventeenth and eighteenth centuries. The problem of *experience* seemed as acute then as it does today, but many found little difficulty in the idea that animals including human beings were — except insofar as they had experience — simply physical machines.[35] It should be added that many were quite unmoved by the problem of life* even when it was at the height of its popularity, but found the problem of experience as acute

---

[35] A considerable number also took it that experience, too, was just a physical phenomenon, although we could not understand how. Joseph Priestley made the point that we know nothing about the physical that gives us reason to think that the experiential is not physical with its full force in 1777; Locke had already made it, somewhat circumspectly, in the 1690s.

as their seventeenth- and eighteenth-century predecessors and twentieth- and twenty-first century successors.[36]

4. 'Proto-experiential'

Some may insist again that they find nothing intolerable in the idea that S phenomena can be emergent properties of something wholly non-S, and they may add that they feel the same about the experiential emerging from the wholly non-experiential.

What should one do? Encourage them, first, to see — to allow — that if S phenomena can be emergent properties of wholly non-S phenomena then the stuff emerged-from, the non-spatial whatever-it-is, must at the very least be somehow *intrinsically suited* to constituting spatial phenomena, on their view; it must be 'proto-spatial' in that sense.

> — Quite so. And exactly the same may be true of experiential phenomena. Experiential phenomena can indeed emerge from wholly and utterly non-experiential phenomena. This is possible because these non-experiential phenomena are intrinsically suited to constituting experiential phenomena in certain circumstances, and are 'proto-experiential' in that sense, although ultimately non-experiential in themselves.

This doesn't escape the problem, it merely changes the terms. 'Proto-experiential' now means 'intrinsically suited to constituting certain sorts of experiential phenomena in certain circumstances', and clearly — necessarily — for X to be intrinsically suited to or for constituting Y in certain circumstances is for there to be something about X's nature *in virtue of which* X is so suited. If there is no such in-virtue-of-ness, no such intrinsic suitability, then any supposed emergence is left brute, in which case it is not emergence at all, it is magic, and everything is permitted, including, presumably, the emergence of the (ontological) concrete from the (ontological) abstract. If on the other hand there is such intrinsic suitability, as there must be if there is to be emergence, how can this be possessed by wholly, utterly, through-and-through non-experiential phenomena? (This is the unargued intuition again. Bear in mind that the intuition that the non-experiential could not emerge from the wholly experiential is exactly parallel and unargued.) If you take the word 'proto-experiential' to mean 'not actually experi-

---

[36] See e.g. James (1890), and references there.

ential, but just what is needed for experience',[37] then the gap is unbridged. If you take it to mean 'already intrinsically (occurrently) experiential, although very different, qualitatively, from the experience whose realizing ground we are supposing it be', you have conceded the fundamental point.[38]

> — You're waving your arms around. H_2O molecules are, precisely, 'proto-liquid', and are at the same time, in themselves, wholly and utterly non-liquid.

To offer the liquidity analogy is to see its inadequacy. Liquidity is a P phenomenon that reduces without remainder to other P phenomena. Analysed in terms of P properties, liquid bodies of water and H_2O molecules have exactly the same sorts of properties, and they are made of exactly the same stuff (ultimates). This is not the case when it comes to experiential phenomena and non-experiential phenomena, for it is built into our starting point, set by NE, that they do not have the same sorts of properties at all in this sense. The analogy is not of the right size or kind. What we need, to put it now in terms of P properties, is, precisely, an analogy that could give us some idea of how non-P properties could emerge from P properties — and of how things with only P properties could be proto-non-P phenomena.[39]

---

[37] Compare Chalmers' (1996) use of 'protophenomenal'. Chalmers is a realist about experience but he gives central place to an idea that rules out real physicalism: the idea that there could be creatures that have no experiential properties although they are 'perfect physical duplicates' of experiencing human beings. These creatures, *Australian zombies*, have done a lot of damage in recent discussion, blotting out classical philosophical zombies, who are outwardly and behaviourally indistinguishable from human beings but with unknown and possibly non-biological insides. Chalmers holds that Australian zombies are a real possibility. This, however, is not something that can be shown, if only because there is a great deal we do not know about the nature of the physical, and it is fabulously implausible to suppose that an atom-for-atom duplicate of an experiencing human being could be produced and not have experience (note that one obviously could not produce an atom-for-atom duplicate of one of us while varying the laws of nature).

[38] It is not clear what the import of the phrase 'in certain circumstances' is, but the circumstances must presumably themselves be wholly non-spatial and non-experiential, and they cannot in any case make any contribution to the spatiality or the experientiality if it is to emerge wholly and only from the wholly non-spatial and non-experiential phenomena that are being taken to be distinct from the circumstances in which they find themselves.

[39] Objections to (a) standard physicalism and (b) the rejection of radical emergence are sometimes based on the fact that conventional phenomena — phenomena essentially involving conventions — may plausibly be said to arise from wholly and utterly non-conventional phenomena. There is, however, no difficulty in the idea that all concretely existing conventional phenomena are wholly physical phenomena, and the emergence of conventional phenomena from non-conventional phenomena is easily explicable in general terms by real physicalism, which acknowledges, of course, the existence of experiential phenomena.

It may be said that the analogy can still help indirectly by pointing to a version of what is sometimes called 'neutral monism'. The central idea of neutral monism is that there is a fundamental, correct way of conceiving things — let us say that it involves conceiving of them in terms of 'Z' properties — given which all concrete phenomena, experiential and non-experiential, are on a par in all being equally Z phenomena. They are on a par in just the same way as the way in which, according to NE physicalism, all concrete phenomena are on a par in being P phenomena. The claim is then that if one duly conceives all concrete phenomena as Z phenomena, thereby acknowledging their fundamental uniformity, (i) the emergence of experiential phenomena from non-experiential phenomena is as unsurprising as (ii) the emergence of liquid phenomena from non-liquid phenomena is when one conceives things in terms of P phenomena. For both non-experiential P phenomena and experiential phenomena are Z phenomena, so really all we find is the emergence of Z phenomena from Z phenomena.

This proposal, however, merely confirms the current position. For what we do, when we give a satisfactory account of how liquidity emerges from non-liquidity, is show that there aren't really any new properties involved at all. Carrying this over to the experiential case, we get the claim that what happens, when experientiality emerges from non-experientiality, is that there aren't really any new properties involved at all. This, however, means that there were experiential properties all along; which is, precisely, the present claim. One cannot oppose it by appealing to 'neutral monism' in any version that holds that really only the Z properties are ultimately real, if this involves the view that experiential and non-experiential properties are at bottom only appearances or seemings. Such a view is incoherent, because experience — appearance, if you like — cannot itself be only appearance, i.e. not really real, because there must be experience for there to be appearance (see note 7).

Some may reject 'intrinsically suited to *constituting* Y' as a gloss on 'proto-X'. In place of 'constituting' they may want to substitute 'giving rise to' or 'producing' and this may for a moment seem to open up some great new leeway for the idea of radical emergence. The idea will be that X remains *in itself* wholly and utterly non-experiential, but *gives rise to* something wholly ontologically distinct from itself, i.e. Y. But real physicalists can't make this substitution. For everything real and concrete is physical, on their view, and experiential phenomena are real and concrete, on their view, and none of them will I think want to throw away the conservation principles and say that brand new physical stuff (mass/energy) is produced or given rise to when experiences are emergent from the non-experiential, i.e. all the time, as we and other animals live our lives. That is magic again, and I am assured that nothing like this happens with liquidity and Bénard convection cells.

Quite independently of these examples, and the laws of physics, the relevant metaphysical notion of emergence is I think *essentially* conservative in the sense of the conservation principles.

5. Micropsychism

I have been trying to see what can be done for those who want to combine NE and RP and (therefore) hold that the experiential may emerge from the wholly and utterly non-experiential. I looked for other examples of emergence, in case they could help us to understand the possibility, at least, of such a thing, but examples like liquidity seemed wholly inadequate, not the right size. I then looked for cases of emergence that promised to be of the right size, but they seemed to describe impossibilities and so backfire, suggesting that there really could not be any such thing as radical non-experiential-to-experiential emergence.

That is what I believe: experiential phenomena cannot be emergent from wholly non-experiential phenomena. The intuition that drives people to dualism (and eliminativism, and all other crazy attempts at wholesale mental-to-non-mental reduction) is correct in holding that you can't get experiential phenomena from P phenomena, i.e. shape-size-mass-charge-etc. phenomena, or, more carefully now — for we can no longer assume that P phenomena as defined really are wholly non-experiential phenomena — from *non-experiential* features of shape-size-mass-charge-etc. phenomena. So if experience like ours (or mouse experience, or sea snail experience) emerges from something that is not experience like ours (or mouse experience, or sea snail experience), then that something must already be experiential in some sense or other. It must already be somehow experiential in its essential and fundamental nature, however primitively or strangely or (to us) incomprehensibly; whether or not it is also non-experiential in its essential nature, as conventional physicalism supposes.

Assuming, then, that there is a plurality of physical ultimates, some of them at least must be intrinsically experiential, intrinsically experience-involving. Otherwise we're back at brutality, magic passage across the experiential/non-experiential divide, something that, *ex hypothesi*, not even God can understand, something for which there is no reason at all as a matter of ultimate metaphysical fact, something that is, therefore, objectively a matter of pure chance every time it occurs, although it is at the same time perfectly lawlike.[40]

I conclude that real physicalists must give up NE.[41] Real physicalists must accept that at least some ultimates are intrinsically experience-involving.[42] They must at least embrace *micropsychism*. Given that everything concrete is physical, and that everything physical is constituted out of physical ultimates, and that experience is part of concrete reality, it seems the only reasonable position, more than just an 'inference to the best explanation'. Which is not to say that it is easy to accept in the current intellectual climate.

*Micro*psychism is not yet *pan*psychism, for as things stand realistic physicalists can conjecture that only some types of ultimates are intrinsically experiential.[43] But they must allow that panpsychism may be true, and the big step has already been taken with micropsychism, the admission that at least some ultimates must be experiential. 'And were the inmost essence of things laid open to us'[44] I think that the idea that some but not all physical ultimates are experiential would look like the idea that some but not all physical ultimates are spatio-temporal (on the assumption that spacetime is indeed a fundamental feature of reality). I would bet a lot against there being such radical heterogeneity at the very bottom of things. In fact (to disagree with my earlier self) it is hard to see why this view would not count as a form of dualism.[45] So I'm going to assume, for the rest of this article at least, that the best form of micropsychism is panpsychism.

So now I can say that physicalism, i.e. real physicalism, entails panexperientialism or panpsychism. All physical stuff is energy, in one form or another, and all energy, I trow, is an experience-involving phenomenon. This sounded crazy to me for a long time, but I am quite used to it now that I know that there is no alternative short of 'substance

---

[40] Note again that this is not a version of the merely epistemological point that all concrete connection (e.g. causal connection) is ultimately unintelligible to us (ultimately 'epistemologically brute' for us).

[41] Part of being realistic, evidently, is that one does not treat experience as objectively miraculous every time it occurs.

[42] The most ingenious attempt to get round this that I know of is Broad's — see Broad (1925), ch. 14; and McLaughlin (1992) — but it does not, in the end, work.

[43] They may for example propose (after assuming that the notion of charge has application to ultimates) that only those with charge are intrinsically experiential.

[44] Echoing Philo, who speaks for Hume in his *Dialogues*: 'And were the inmost essence of things laid open to us, we should then discover a scene, of which, at present, we can have no idea. Instead of admiring the order of natural beings, we should clearly see, that it was absolutely impossible for them, in the smallest article, ever to admit of any other disposition' (Hume, 1779, pp. 174–5).

[45] Strawson (1994), p. 77.

dualism', a view for which (as Arnauld saw) there has never been any good argument. Real physicalism, realistic physicalism, entails panpsychism, given that radical emergence is impossible, and whatever problems are raised by this are problems a real physicalist must face.

They seem very large, these problems (so long as we hold on to the view that there is indeed non-experiential reality). To begin with, 'experience is impossible without an experiencer', a subject of experience.[46] So we have, with Leibniz, and right at the start, a rather large number of subjects of experience on our hands — if, that is, there are as many ultimates as we ordinarily suppose. I believe that this is not, in fact, a serious problem, however many ultimates there are,[47] but we will also need to apply our minds to the question whether the class of subjects of experience contains only ultimates, on the one hand, and things like ourselves and other whole animals, on the other hand, or whether there are other subjects in between, such as living cells. Panpsychism certainly does not require one to hold the view that things like stones and tables are subjects of experience — I don't believe this for a moment, and it receives no support from the current line of thought — but we will need to address William James's well known objection to the idea that many subjects of experience can somehow constitute a single 'larger' subject of experience.[48] In general, we will have to wonder how macroexperientiality arises from microexperientiality, where by microexperientiality I mean the experientiality of particles relative to which all evolved experientiality is macroexperientiality.[49]

---

[46] Frege (1918), p. 27. No sensible Buddhist rejects such a claim, properly understood.

[47] For reasons I lay out in Strawson (2003b).

[48] James (1890), Vol. 1, ch. 6. The following passage precedes his statement of the objection: 'We need to try every possible mode of conceiving the dawn of consciousness so that it may not appear equivalent to the irruption into the universe of a new nature, non-existent until then. Merely to call the consciousness 'nascent' will not serve our turn. It is true that the word signifies not yet quite born, and so seems to form a sort of bridge between existence and nonentity. But that is a verbal quibble. The fact is that discontinuity comes in if a new nature comes in at all. The quantity of the latter is quite immaterial. The girl in "Midshipman Easy" could not excuse the illegitimacy of her child by saying, "it was a very small one". And Consciousness, however small, is an illegitimate birth in any philosophy that starts without it, and yet professes to explain all facts by continuous evolution. If evolution is to work smoothly, consciousness in some shape must have been present at the very origin of things. Accordingly we find that the more clear-sighted evolutionary philosophers are beginning to posit it there. Each atom of the nebula, they suppose, must have had an aboriginal atom of consciousness linked with it; and, just as the material atoms have formed bodies and brains by massing themselves together, so the mental atoms, by an analogous process of aggregation, have fused into those larger consciousnesses which we know in ourselves and suppose to exist in our fellow-animals' (1890, Vol. 1, pp. 148–9).

[49] As Nick White reminded me, we certainly don't have to suppose that microexperientiality is somehow weak or thin or blurry (this is perhaps how some people imagine the most

We also have to wonder how the solution to the 'problem of mental causation' is going to drop out of all this. We know, though, that different arrangements of a few types of fundamental ultimates give rise to entities (everything in the universe) whose *non*-experiential properties seem remarkably different from the non-experiential properties of those fundamental ultimates, and we have no good reason not to expect the same to hold true on the experiential side. It may be added that there is no more difficulty in the idea that the experiential quality of microexperientiality is unimaginable by us than there is in the idea that there may exist sensory modalities (qualitatively) unimaginable by us.

It is at this point, when we consider the difference between macroexperiential and microexperiential phenomena, that the notion of emergence begins to recover some respectability in its application to the case of experience. For it seems that we can now embrace the analogy with liquidity after all, whose pedagogic value previously seemed to lie precisely in its inadequacy. For we can take it that human or sea snail experientiality emerges from experientiality that is not of the human or sea snail type, just as the shape-size-mass-charge-etc. phenomenon of liquidity emerges from shape-size-mass-charge-etc. phenomena that do not involve liquidity. Human experience or sea snail experience (if any) is an emergent property of structures of ultimates whose individual experientiality no more resembles human or sea snail experientiality than an electron resembles a molecule, a neuron, a brain, or a human being. Once upon a time there was relatively unorganized matter, with both experiential and non-experiential fundamental features. It organized into increasingly complex forms, both experiential and non-experiential, by many processes including evolution by natural selection. And just as there was spectacular enlargement and fine-tuning of non-experiential forms (the bodies of living things), so too there was spectacular enlargement and fine-tuning of experiential forms.[50]

This is not to advance our detailed understanding in any way. Nor is it to say that we can ever hope to achieve, in the experiential case, the

---

primitive Leibnizian monads). It can be as vivid as an experience of bright red or an electric shock (both of which are 'confused' and 'indistinct' in Leibniz's terms). Compare Rosenberg (2005), ch. 5.

[50] The heart of experience, perhaps, is electromagnetism in some or all its forms; but electromagnetism in all its forms is no doubt just one expression of some single force whose being is intrinsically experiential, whatever else it is or is not. I do not, however, foresee any kind of scientific research programme.

sort of feeling of understanding that we achieve in the liquid case.[51] The present proposal is made at a very high level of generality (which is not a virtue); it merely recommends a general framework of thought in which there need be no more sense of a radically unintelligible transition in the case of experientiality than there is in the case of liquidity. It has nothing to offer to scientific test.

One can I think do further work on this general framework, by working on one's general metaphysics. The object/process/property/state/event cluster of distinctions is unexceptionable in everyday life but it is hopelessly superficial from the point of view of science and metaphysics, and one needs to acquire a vivid sense that this is so. One needs a vivid sense of the respect in which (given the spatio-temporal framework) every object is a process; one needs to abandon the idea that there is any sharp or categorial distinction between an object and its propertiedness.[52] One needs to grasp fully the point that 'property dualism', applied to intrinsic, non-relational properties, is strictly incoherent (or just a way of saying that there are two very different kinds of properties) insofar as it purports to be genuinely distinct from substance dualism, because there is nothing more to a thing's being than its intrinsic, non-relational propertiedness.

We are as inescapably committed to the discursive, subject-predicate form of experience as we are to the spatio-temporal form of experience, but the principal and unmistakable lesson of the endlessness of the debate about the relation between objects and their propertiedness is that discursive thought is not adequate to the nature of reality: we can see that it doesn't get things right although we can't help persisting with it. There is in the nature of the case a limited amount that we can do with such insights, for they are, precisely, insights into how our understanding falls short of reality, but their general lesson — that the nature of reality is in fundamental respects beyond discursive grasp — needs always to be borne in mind.

I have argued that there are limits on how different X and Y can be (can be intelligibly supposed to be) if it is true that Y emerges from X. You can get A from non-A for some substitutions for A but not all. The extended, I have proposed, can't emerge from the intrinsically wholly non-extended (except on pain of being a mere appearance and so not really real). The spatial can't emerge from the intrinsically wholly non-spatial (except on the same pain). The experiential can't emerge

---

[51] Feelings of understanding are just that; they are essentially subjective things with no metaphysical consequences.

[52] See e.g. Strawson (2003b), pp. 299–302, following Nagarjuna, Nietzsche, James, Ramsey and many others.

from the intrinsically wholly non-experiential, and it doesn't have the option of being a mere appearance. You can make chalk from cheese, or water from wine, because if you go down to the subatomic level they are both the same stuff, but you can't make experience from something wholly non-experiential. You might as well suppose — to say it once again — that the (ontologically) concrete can emerge from the (ontologically) abstract.[53] I admit I have nothing more to say if you question this 'can't', but I have some extremely powerful indirect support from Occam's razor and Eddington's notion of silliness.

I finish up, indeed, in the same position as Eddington. 'To put the conclusion crudely', he says, 'the stuff of the world is mind-stuff' — something whose nature is 'not altogether foreign to the feelings in our consciousness'. 'Having granted this', he continues,

> the mental activity of the part of the world constituting ourselves *occasions no surprise*; it is known to us by direct self-knowledge, and we do not explain it away as something other than we know it to be — or, rather, it knows itself to be. It is the physical aspects (i.e. non-mental aspects) of the world that we have to explain.[54]

Something along these general panpsychist — or at least micropsychist — lines seems to me to be the most parsimonious, plausible and indeed 'hard-nosed' position that any physicalist who is remotely realistic about the nature of reality can take up in the present state of our knowledge.[55]

References

Arnauld, A. (1641/1985), 'Fourth set of objections', in *The Philosophical Writings of Descartes*, Vol. 2, trans. J. Cottingham *et al.* (Cambridge: Cambridge University Press).

Block, N. (1978), 'Troubles with functionalism', in *Minnesota Studies in the Philosophy of Science*, **9**.

---

[53] 'The comparison is false because the experiential and the non-experiential are two categories within the concrete.' Well, but the concrete and the abstract are two categories within the real.

[54] Eddington (1928), pp. 276–7. 'Mind-stuff' is William James's term: 'The theory of "mind-stuff" is the theory that our mental states . . . are composite in structure, made up of smaller (mental) states conjoined. This hypothesis has outward advantages which make it almost irresistibly attractive to the intellect, and yet it is inwardly quite unintelligible' (James (1890), Vol. 1, p. 145).

[55] I am grateful to participants in the 2002 University of London one-day conference on consciousness and, since then, to audiences at the University of Reading, Copenhagen University, University of California at Irvine, Trinity College Dublin and Columbia University, including in particular Nick White, Alva Noë, Bill Lyons and David Albert. I am especially grateful to members of the 2002 Konstanz Workshop on 'Real materialism' for their constructive scepticism.

Broad, C.D. (1925), *The Mind and Its Place in Nature* (London: Routledge and Kegan Paul).
Carnap. R. (1950), 'Empiricism, semantics and ontology', *Revue Internationale de Philosophie*, **4**, pp. 20–40.
Chalmers, D. (1996), *The Conscious Mind* (New York: Oxford University Press).
Chomsky, N. (1968), *Language and Mind* (New York: Harcourt, Brace & World).
Chomsky, N. (1995), 'Language and nature', *Mind*, **104**, pp. 1–60.
Clarke, D. (2003), *Descartes's Theory of Mind* (Cambridge: Cambridge University Press).
Craig, E.J. (1987), *The Mind of God and the Works of Man* (Cambridge: Cambridge University Press).
Dennett, D.C. (1991), Consciousness Expalined (Boston, MA: Little, Brown & Co.).
Eddington, A. (1928), *The Nature of The Physical World* (New York: Macmillan).
Frege, G. (1918/1967), 'The thought: a logical inquiry', in *Philosophical Logic*, ed. P.F. Strawson (Oxford: Oxford University Press).
Greene, B. (2004), *The Fabric of the Cosmos* (New York: Knopf).
Hume, D. (1748/1975), *Enquiries Concerning Human Understanding*, ed. L.A. Selby-Bigge (Oxford: Oxford University Press).
Hume, D. (1739–40/2000) *A Treatise of Human Nature*, edited by D. F. Norton and M. Norton (Oxford: Clarendon Press).
Hume, D. (1779/1947), *Dialogues Concerning Natural Religion*, second edition, ed. N. Kemp Smith (Edinburgh: Nelson).
James, W. (1890/1950), *The Principles of Psychology*, Vol. 1 (New York: Dover).
Locke, J. (1689/1975), *An Essay concerning Human Understanding*, edited by P. Nidditch (Oxford: Clarendon Press).
Lockwood, M. (1981), 'What *Was* Russell's Neutral Monism?', *Midwest Studies in Philosophy*, **VI**, pp. 143–58.
Lockwood, M. (1989), *Mind, Brain, and the Quantum* (Oxford: Blackwell).
Mackie, J.L. (1982), *The Miracle of Theism* (Oxford: Oxford University Press).
McLaughlin, B. (1992), 'The rise and fall of British emergentism', in *Emergence or Reduction? Essays on the Prospects of Nonreductive Physicalism*, ed. A. Beckermann, A.H. Flohr and J. Kim (Berlin: Walter de Gruyter).
Nagel, T. (1979), 'Panpsychism', in *Mortal Questions* (Cambridge: Cambridge University Press).
Priestley, J. (1777/1818), *Disquisitions Relating to Matter and Spirit*, in *The Theological and Miscellaneous Works of Joseph Priestley*, volume 3, edited by J.T. Rutt (London).
Rosenberg, G. (2005), *A Place for Consciousness* (New York: Oxford University Press).
Quine, W.V. (1951/1961), 'Two dogmas of empiricism', in W.V. Quine, *From a Logical Point of View*, second edition (New York: Harper and Row).
Russell, B. (1927a/1992a), *The Analysis of Matter* (London: Routledge).
Russell, B. (1927b/1992b), *An Outline of Philosophy* (London: Routledge).
Russell, B. (1948/1992c), *Human Knowledge: Its Scope And Limits* (London: Routledge).
Russell, B. (1956/1995), 'Mind and Matter', in *Portraits from Memory* (Nottingham: Spokesman).
Strawson, G. (1994), *Mental Reality* (Cambridge, MA: MIT Press).
Strawson, G. (2000/2002), 'David Hume: Objects and Power' in *Reading Hume on Human Understanding*, edited by P. Millican (Oxford: Oxford University Press), pp 231–57.

Strawson, G. (2003a), 'Real materialism', in *Chomsky and his Critics*, ed. L. Antony and N. Hornstein (Oxford: Blackwell), pp. 49–88.

Strawson, G. (2003b), 'What is the relation between an experience, the subject of the experience, and the content of the experience?', *Philosophical Issues*, **13**, pp. 279–315.

Strawson, G. (2005), 'Intentionality and experience: terminological preliminaries', in *Phenomenology and Philosophy of Mind*, ed. David Smith and Amie Thomasson (Oxford/New York: Oxford University Press), pp. 41–66.

van Gulick, R. (2001), 'Reduction, emergence and other recent options on the mind-body problem: a philosophical overview', *Journal of Consciousness Studies*, **8** (9–10), pp. 1–34.

Velarde, M.G. and Normand, C. (1980), 'Convection', *Scientific American*, **243**, pp. 92–108.

Peter Carruthers
and Elizabeth Schechter[1]

Can Panpsychism Bridge the Explanatory Gap?

Strawson (2006) claims that so long as we take the physical ultimates of the world to be non-experiential in nature, we will never be capable of explaining or understanding how human conscious experience can emerge from physical processes. He therefore urges physicalists to embrace a panpsychist metaphysics. This is because, he says, we can better understand how macro-experientiality (the conscious experience of creatures like us) might arise from micro-experientiality (the conscious experientiality of the physical 'ultimates' of the universe) than we could ever understand how experientiality arises from the non-experiential. Strawson's view is that the experiential features of the ultimates can make it intelligible that there should be a scientific explanation or reduction of human consciousness. For he says that once we have accepted panpsychism, 'the notion of emergence begins to recover some respectability in its application to the case of experience' (p. 27),[2] and that he is proposing a 'general framework of thought in which there need be no more sense of a radically unintelligible transition in the case of experientiality than there is in the case of liquidity' (p. 28).

The panpsychist metaphysics that Strawson proposes is intended, therefore, to help us with the mind/body problem: the problem of how human conscious experience relates to the physical matter of the brain. He says that many physicalists take the claim that the experiential is also physical as nevertheless:

> profoundly problematic *given what we know about the nature of the physical.* But they have already made a large and fatal mistake. This is

[1] The ordering of the authors' names is alphabetical.
[2] All page references are to Strawson (2006).

because we have no good reason to think that we know anything about the physical that gives us any reason to find any problem in the idea that experiential phenomena are physical phenomena (p. 4, emphasis in original).

In fact, Strawson blames physicalists for creating an insoluble problem (insoluble within the physicalist's current metaphysical framework) where there really is none, via their dualistic tendency to set up a dichotomy between the 'mental' (i.e. conscious, experiential) and the 'physical'. This is a false dichotomy, Strawson says: it is all physical, and we have no reason to think that it isn't all experiential as well. (Physics itself is silent on this matter, he believes.) If we only accepted that the physical was in and of itself essentially experiential, then there would be no mind/body problem *per se*, and scientific explanations of human consciousness (of macro-experientiality) would become intelligible.

One major manifestation of the mind/body problem is the explanatory gap that, it is claimed, will always exist between any description of the physical and/or functional properties of the human body and brain, on the one hand, and consciousness described in phenomenal or experiential terms, on the other. Those who take the explanatory gap to be a serious problem do so precisely because the gap doesn't seem to exist between other higher-level properties (non-consciousness involving properties) and the phenomena that *they* reduce to. Strawson uses the example of liquidity to illustrate the intelligibility of most purported cases of emergence, and to provide a contrast between such cases and the case of consciousness. Intelligible scientific explanations and reductions of such phenomena provide us with a feeling of necessity: given the various chemical properties of H_2O molecules, and given that water, a liquid, is composed of H_2O molecules, and given that liquids are forms of matter in which the molecules that compose them slide off each other instead of gripping each other tightly in a lattice, we see that the property of liquidity *must* emerge from the properties of H_2O molecules, even though those molecules aren't themselves liquids. We see that water's property of liquidity is entailed by the properties of the (non-liquid) H_2O molecules that compose it. But no non-experiential properties seem to compel or entail the qualitative aspects of any particular conscious experience. Whatever the non-experiential features of my body and brain are, it will always appear to me that the experiential features of my consciousness could still have been different, or even absent altogether.

Anti-physicalists like Chalmers (1996) often take this explanatory gap as demonstrating that there is a *metaphysical* gap between the experiential and the (non-experiential) physical. Such philosophers — who say that the mind/body problem is insoluble precisely because experiential properties really *aren't* composed of non-experiential ones — are often called 'property dualists'. Strawson is eager to distance himself from this kind of anti-physicalist, however, and is at pains to point out that he believes that all properties, including experiential properties, are physical. But he thinks that if the ultimate entities in the universe are, while physical, at the same time experiential, then the explanatory gap can be closed, in principle.

One of us (Carruthers, 2000, 2005) has previously suggested, however — along with Loar (1990), Tye (2000) and others — that the existence of the explanatory gap is best explained not by giving up on standard forms of physicalism, but by focusing on the distinction between phenomenal *concepts* and other kinds of concept. Most concepts are, in a broad sense, functional, defined by their relationships with other concepts and/or embedded in bodies of belief about the world that modulate their use. In contrast, recognitional concepts, like 'red', can be applied directly on the basis of perceptual or quasi-perceptual acquaintance with their instances. And phenomenal concepts like 'feels itchy' or 'seems red' are *purely* recognitional concepts. In contrast to non-purely-recognitional concepts like 'red', concepts like 'seems red' aren't tied to any other concepts or beliefs about the world. While we wouldn't apply the concept 'red' to a wall that looked red, for example, if we knew that we were standing in a room illuminated by a red light, our purely recognitional concept 'seems red' would still apply. Regardless of what one knew about the lighting conditions in the room — and in fact regardless of what one knew about *anything* in the world or our own bodies — the concept 'seems red' would still apply solely on the basis of our phenomenal experience of the wall.

When we think about the qualitative nature of our experience we use phenomenal concepts, and these concepts are always isolable from whatever physical concepts might be employed in any proposed physicalist explanation of consciousness (whether neurological, functional or representational). Our phenomenal descriptions of experience — laden, as they are, with phenomenal concepts — can therefore never be entailed by descriptions of our experience that don't use those very same concepts;[3] our phenomenal concepts are free to drift, as it were, unanchored. Accordingly it will always be possible to feel

---

[3] There exist some *a priori* relations amongst phenomenal concepts that permit a weak sort of explanation-by-constitution; but these are extremely limited. For example, if I undergo a feels-red experience then I know *a priori* that I am undergoing a feels-visual experience.

that phenomenal consciousness *itself* isn't anchored by, and that it doesn't depend upon, any other features of this world. But that is just because we have ways of thinking about our experience that don't depend on the application of any non-recognitional concepts. The explanatory gap is here revealed as 'a cognitive illusion' (Tye, 2000), resulting from the conceptual isolation of the concepts that we employ when thinking about our own experience.

Strawson makes no mention of this sort of indirect strategy for dissolving the explanatory gap. (The strategy is indirect because it doesn't seek to explain why such-and-such a physical system should feel like *this*; rather, it seeks to explain why no such explanation is forthcoming, in terms that will make the continued demand for an explanation evaporate.) And he is committed to claiming that the mind/body problem can't be solved within the framework provided by standard forms of physicalism. While disagreeing, we propose to set our reservations aside in the discussion that follows. Our purpose here is rather to evaluate whether Strawson's panpsychism leaves us any better off in the search for a direct, full-frontal solution to the mind/body problem. We will argue that it does not.

Before we can embark on that discussion, however, we need to get a sense of what it would mean to accept that the physical ultimates of this world are also experiential in character. If one takes seriously Block's (1995) distinction between access consciousness (consisting of functionally defined states that can enter into a subject's reasoning, have an impact on belief, and/or be reportable in speech, etc.) and phenomenal consciousness (states that have a feel), then there are two conceptually distinct ways for a phenomenon to be experiential. An access-conscious state would be experiential in the sense of being somebody's (or something's) experience, by virtue of playing a special functional role for some conscious being. A phenomenally conscious state, in contrast, would be experiential in the sense of possessing a feel. A feel that isn't felt by any conscious subject then becomes a conceptual possibility, at least. Indeed some philosophers believe that our experiences possess intrinsic, irreducible, non-representational, non-relationally-defined properties, called *qualia*. Many philosophers argue that qualia are metaphysically impossible, or, at least, that they don't exist in this universe. But those who believe in the existence of qualia assert that the universe contains not only

---

Accordingly, I can explain how it is that I am undergoing a feels-visual experience by saying that I am undergoing a feels-red one. It should be plain, however, that very little about my experience can be explained in this way, i.e. even by using other phenomenal concepts.

people who feel pain and felt pains, but also pain-feely properties, pain qualia, themselves.

Strawson's proposed panpsychism is one according to which the ultimates (quarks or strings or whatever) are tiny subjects of experience, for he says that 'experience is impossible without an experiencer', and that panpsychism means that we have 'a rather large number of subjects of experience on our hands' (p. 26). However, he might instead have proposed a metaphysics according to which the physical ultimates of this world aren't conscious subjects, but rather have feel-properties attached to them. On this version of panpsychism, the ultimates are experiential entities in the sense that they possess irreducible properties of experience, or qualia, but are not themselves subjects of experience.

We believe that the version of panpsychism Strawson advocates, according to which the ultimates are themselves subjects of experience, is the more extreme and more problematic version, and lays itself open to a greater number of objections. We will proceed, then, by evaluating the weaker and more plausible version, according to which the ultimates of the world possess qualia. Any objections that we present to this weaker version will apply also to the stronger version (to which additional objections apply as well).[4]

Can macro-experientiality be reductively explained in terms of micro-experientiality, then? It is hard to see how it can be. How could trillions of particles, whatever their experiential nature, constitute what feels like (and what Strawson [1997] has argued we have phenomenological warrant to believe *is*) a single subject of experience? One problem here is epistemological: I can't know anything about the experientiality of the ultimates that compose me. Certainly I can't know the experiential properties of the individual atoms and molecules that constitute my brain on the basis of introspection. And it is hard to see how one could, even in principle, get any evidence as to the nature of those properties. In which case we can see in advance that we could never be in a position to mount a reductive explanation of our experience. We shall never be able to start from the known experiential properties of the ultimates that compose us, deducing that our experience should have the character that is does, in the way that we *can* start from the known properties of H_2O molecules and deduce

---

[4] As Strawson himself admits, we would 'need to address William James's well known objection to the idea that many subjects of experience can somehow constitute a single "larger" subject of experience'. Although Strawson protests that there is 'no more difficulty in the idea that the experiential quality of micro-experientiality is unimaginable by us than there is in the idea that there may be sensory modalities (qualitatively) unimaginable by us' (p. 27), at least we know that bats, for example, *have* sensory modalities.

the liquidity of water. If we can't know the experiential properties of the ultimates, then we can't ever provide a reductive explanation of phenomenal consciousness in terms of such properties, either.

Strawson will probably reply that he is concerned with a metaphysical rather than an epistemic sense of 'explain'. At issue is not whether we can know, or have reason to believe in, some particular reductive explanation of the properties of our experience. It is rather whether there *exists* such an explanation, in the world. Strawson might say that we can see in advance that there can't (on metaphysical grounds) be any explanation of phenomenal consciousness in terms of standard physical properties like mass, electric charge and so forth. For these properties and the properties of our experience are too heterogeneous to admit of explanation of the latter by the former. Whereas if panpsychism is adopted, and the physical ultimates are at the same time experiential, then at least we shall be explaining like with like. Even if we don't, and can't, know enough about the experiential properties of the ultimates to construct a detailed explanation, we can at least see that one isn't ruled out in principle by the metaphysics of the situation.

Strawson might also reply (either instead, or in addition) that the experiential properties of the ultimates might in principle be known on the basis of an inference to the best explanation, where the target of explanation is the character of our own conscious experience. If by postulating that the atoms in my brain possess such-and-such qualia, and by adding that those atoms with those properties are interacting thus-and-so, we can explain why my experience should feel to me the way it does, then this would provide us with good reason to believe that the atoms in my brain do indeed possess those properties. This sort of indirect inference to properties that can't be accessed directly is just an application of standard abductive inference in science.

While we have doubts about each of these two lines of response to our challenge, we shan't press them here. Rather, we will show that even in a best-case scenario — in which the phenomenal properties of the ultimates are known in complete detail — panpsychism still wouldn't help us with the mind/body problem. For the explanatory gap would remain in place, untouched and wide open as ever: panpsychism does nothing to close it.

To begin to see this, just ask yourself whether, if everyone in the world were a committed panpsychist, children would cease wondering whether their best friends shared their phenomenal colour experiences when visually confronted with the same scenes. Or ask yourself if Tye and Chalmers would co-author papers on the impossibility of zombies. Why would they? If zombies are conceivable now, then they will remain conceivable no matter what they, and we, are made of. It will remain conceivable that there should be a zombie made out of *exactly the same stuff as I am* (whatever that stuff is like), but without undergoing an experience like *this* one (like the one I am currently having). In fact it will always be possible for the panpsychist to think thoughts like the following:

> Couldn't my brain be in the exact (non-experientially described) state that it is in now, as I look at this tree, and couldn't the particles that compose me possess the very same (*experientially* described) properties that they have, and yet couldn't *I* be in a *different* experiential state? Even if every ultimate particle in my body possessed a green-feely experiential property, couldn't *I* be undergoing an experience with the phenomenal properties of an experience of red? And likewise, couldn't there be someone who was composed of particles exactly like mine arranged in exactly the same way, each of which possessed the very same qualia as do the particles that compose me, yet who lacked phenomenal consciousness altogether (just as, according to Strawson's proposed metaphysics, a stone is composed of the exact same experiential ultimates yet lacks experientiality or consciousness of its own)?

The point is a simple one, and was made long ago by Block (1978) in his population-of-China counter-example to functionalism. If one imagines millions of components interacting together in ways that mirror the interactions of my own components, it is still possible to conceive that the resulting complex being lacks phenomenal consciousness altogether, even if each of those components possesses phenomenal properties. In which case no amount of knowledge of the feely nature of the ultimate physical particles could ever explain phenomenal consciousness in the way that knowledge of the component particles of water and the manner in which they interact can explain liquidity.

There are familiar avenues of reply to the population-of-China counter-example, of course. For example, Strawson might retort that the problem is merely one of the limits of our imagination. Since we cannot really conceive, in detail, of millions and millions of experiential entities interacting in as-yet-to-be-specified and highly complex ways, we cannot really tell whether some such story mightn't constitute a successful reductive explanation of our own experientiality. But this sort of reply is equally available to defenders of standard (non-experiential) forms of physicalism, of course (Dennett, 1991). So if it works, it would only serve to undermine Strawson's own argument for panpsychism.

But isn't there at least *some* explanatory gain to be had from panpsychism? Doesn't the fact that the phenomenal properties attributed to the ultimate particles of the universe are of the same metaphysical *type* as the target to be explained (our own phenomenal consciousness) at least indicate that the latter might be explicable *in principle*? In fact not. For we have not the faintest idea how the phenomenal properties possessed by one entity or set of entities might contribute to a reductive explanation of the phenomenal properties possessed by another, as the thought experiments described above indicate.

Panpsychism was urged on us as the one move that might enable us to see how the explanatory gap could one day be bridged. But it turns out to be a blind alley. Even if the ultimates of the universe are experiential in nature, the explanatory gap remains untouched. It is better, then, to remain an old-fashioned (non-panpsychic) physicalist, and to accommodate or circumvent the explanatory gap by other means. Indeed, it is worth noting that the sort of indirect approach to explaining the explanatory gap that we mentioned earlier also has the resources to explain why a panpsychist metaphysic leaves the explanatory gap intact. The claim is that phenomenal concepts (the concepts by means of which we think about phenomenal properties) are conceptually isolated ones, lacking *a priori* connections with other concepts, and lacking any embedding in a wider theory. In which case no story — no matter how detailed — about the component parts of a creature and their modes of interaction will entail an application of such a concept, even, as it turns out, if the story utilizes phenomenal concepts themselves to ascribe feely properties to the components.

References

Block, N. (1978), 'Troubles with functionalism', in *Minnesota Studies in the Philosophy of Science*, ed. C. Savage, **9**, pp. 261–325.

Block, N. (1995), 'A confusion about the function of consciousness', *Behavioral and Brain Sciences*, **18**, pp. 227–47.

Carruthers, P. (2000), *Phenomenal Consciousness* (Cambridge: Cambridge University Press).

Carruthers, P. (2005), *Consciousness: Essays from a Higher-Order Perspective* (Oxford: Oxford University Press).

Chalmers, D. (1996), *The Conscious Mind* (New York: Oxford University Press).

Dennett, D. (1991), *Consciousness Explained* (Allen Lane).

Loar, B. (1990), 'Phenomenal states', in *Philosophical Perspectives: Action Theory and Philosophy of Mind*, ed. J. Tomberlin (Ridgeview).

Strawson, G. (1997), 'The self', *Journal of Consciousness Studies*, **4** (5–6), pp. 405–28.

Strawson, G. (2006), 'Realistic monism: Why physicalism entails panpsychism', *Journal of Consciousness Studies*, **13** (10–11), pp. 3–31.

Tye, M. (2000), *Consciousness, Color, and Content* (MIT Press).

Sam Coleman

Being Realistic
Why Physicalism May Entail Panexperientialism[1]

1. Introduction

In this paper I first examine two important assumptions underlying the argument that physicalism entails panpsychism. These need unearthing because opponents in the literature distinguish themselves from Strawson in the main by rejecting one or the other. Once they have been stated, and something has been said about the positions that reject them, the onus of argument becomes clear: the assumptions require careful defence. I believe they are true, in fact, but their defence is a large project that cannot begin here. So, in the final section I comment on what follows if they are granted. I agree with Strawson that — broadly — 'panpsychism' is the direction in which philosophy of mind should be heading; nevertheless, there are certain difficulties in the detail of his position. In light of these I argue for changes to the doctrine, bringing it into line with the slightly — but significantly — different *panexperientialism*.

2. Smallism

The first assumption is a thesis I call *smallism* — the view that all facts are determined by the facts about the smallest things, those existing at the lowest 'level' of ontology. Strawson is not to be blamed for assuming its truth. It is pervasive amongst philosophers of mind, especially physicalists, who tend to express it by saying that facts about the *micro*physical determine facts about the chemical, the biological and so on. And for many the debate about whether physicalism can

---

[1] Thanks are due to Galen Strawson, Frederik Willemarck and Jennifer Hornsby for helpful comments.

encompass the mental is precisely over whether the mental facts could be fixed in this bottom-up way by the items described by physics. Smallism goes hand in hand with a 'levelled' view of existence, and is the conceptual bread and butter over which *supervenience* has commonly been spread. It is important to note that it is in the first instance a *metaphysical* thesis only, a thesis about upward necessitation of all things by the small things.[2] *No* claim need yet be made about whether this necessitation is, for example, a priori. Such — epistemological — issues crop up later.

That Strawson assumes smallism is clear when he discusses the doctrine he labels 'emergence' — the view that higher-level properties can *emerge from* lower-level bases from which such properties are entirely absent. His critique of emergence springs from the principle that where X emerges from Y 'there must be something about X and X alone in virtue of which Y emerges, and which is sufficient for Y' (Strawson, 2006, p. 18). The example Strawson gives to illustrate acceptable emergence, *liquidity*, is a locus classicus of smallist thinking: although no individual molecule of a liquid body can itself be said to be liquid, the liquidity of the body is determined without remainder by the molecular set-up. Roughly, liquidity occurs because molecules slide over one another. The case of the experiential seems to sit in contrast with that of liquidity, and to violate the aforementioned principle: Strawson claims that there can be nothing about *non-experiential ultimates* (e.g. the microphysical entities as conceived in contemporary physics) in virtue of which experience could emerge, when such ultimates are arranged to make a brain. So, the view that experience emerges in this way is mistaken.

Whether Strawson's criticism of the thesis that non-experiential ultimates generate experience is well made is something we will come back to. The thing to note here is that it is certainly beside *a* point, if not beside *the* point. When he formulates emergence, it is to capture a position that combines two ideas: first, that physical stuff is fundamentally, and wholly, non-experiential, and second, that experience is physical. But a position combining these claims need have no truck with smallism, nor with emergence as defined. Consider the *macro non-reductive physicalist*: she rejects the picture of reality as split into levels, with 'higher' levels being determined by the 'lowest' level. For the macro non-reductive physicalist, *if* there are levels, then there is nothing to prevent some given level from being entirely autonomous in respect of how its properties are determined. Levels can be 'closed',

---

[2] And their properties and relations, of course.

as it were (although they need not be — such physicalists commonly accept the reductive explanation of heat in terms of molecular kinetic energy, for example); and the experiential/mental may well be such a level. Arrange the microphysical into a brain, the view goes, and experience certainly 'emerges'. That is to say, if there is a brain, experience appears. But this is not to say that experience emerges *from* the small and non-experiential, nor that it is *determined by* the small and non-experiential. Perhaps experience is a distinct and autonomous feature or level of existence, which through natural law comes into being under specific conditions.[3]

*But surely such a position is **dualism**? If the experiential is a distinct level of reality floating (even: supervening) upon the small physical, then it is certainly not physical, right? The claim that it is not upwardly determined by the microphysical, in contrast to heat, just goes to show that this is right.*

In fact, macro non-reductive physicalism comfortably qualifies as physicalism, by Strawson's own lights. For he holds the *object conception* of physicalism, rejecting the popular physics-based definition.[4] He stresses that 'physical' is a natural kind term, defined as qualifying whatever stuff actually makes up tables, chairs and experiences. If experience makes its entrance when non-experiential matter is arranged as a brain, it can still count as physical on Strawson's view.[5]

What kind of criticism is it of this position to complain that there is nothing about the non-experiential ultimates composing a brain in virtue of which experience emerges? It is no criticism at all. For the macro non-reductive physicalist, there need be nothing about ultimates in virtue of which experience emerges from them. For she rejects the claim that experience emerges *from* them, as it were. She rejects smallism.

It might be thought that Strawson has a notion of *bruteness* with which to object here. He tells us that emergence cannot be brute, nature is not this way. It is hard to draw out an *argument* for this claim, all the harder because *bruteness* seems to straddle epistemological and metaphysical categories, but still there are things we can say in reply.

---

[3] This mention of natural law is important: the macro non-reductive physicalist is free to hold that experience *supervenes* on the material make-up of the brain. But supervenience being a notoriously weak relation, this can fall short of saying that the state of the non-experiential small things is *responsible for* or *determines* experience. It may just be the natural way we find the world that this is so.

[4] I follow Stoljar's account of this distinction. See his entry on physicalism in the *Stanford Encyclopaedia of Philosophy* (http://plato.stanford.edu/entries/physicalism/).

[5] There seems a threat here that 'physical' might then range over a disjunction of quite different kinds. More on this later.

Perhaps, *perhaps* smallistic *emergence-from* can't be brute.[6] The way that liquid phenomena emerge from non-liquid phenomena makes pleasing sense. But on the present view we do not have emergence-from. The way things are is held to be such that, as a matter of nature, a certain level of organization of physical ultimates is accompanied by experience.

This accompaniment certainly counts as brute. Yet this bruteness is not a legitimate target for Strawson when one bears in mind his own view. He appears to hold *micro property dualism*.[7] He thinks that there are ultimates, smallistically responsible for the facts about levels above them, at least some of which have both experiential *and* non-experiential properties. Given the difference between experiential and non-experiential modes of being — the fundamental gulf in ontology for Strawson — the accompaniment of one property type by the other at the lowest level cannot be other than brute: there can be no reason in the experiential nature of an ultimate why it is also non-experiential, and conversely. But then if brute accompaniment is acceptable at the lowest level, it is hard to see why it should be objectionable at the macro-level. Brute accompaniment, as such, seems no problem for Strawson. It is brute *emergence from* that is the problem. There is of course a way to object to macro brute accompaniment: to endorse smallism (even from a micro-level where brute accompaniment obtains). Then the accompaniment at a macro-level wouldn't be brute, it would follow from the machinations of lower levels. But smallism is precisely the doctrine whose assumed truth we are questioning.

I end the discussion of smallism by noting an argument Strawson does give that might be thought to support it. He considers a view on which non-experiential ultimates *give rise to* the experiential, 'something wholly ontologically distinct' (Strawson, 2006, p. 23). Perhaps this comes close to the macro non-reductive physicalism described here. Strawson objects not on the grounds that such a giving-rise-to looks like the ushering in of dualism, but that it means the production of 'brand new physical stuff (mass/energy)' (ibid.), and that neither physicalist nor emergentist will want to ditch the principles of conservation. If this argument were successful, it would count against a higher-level

---

[6] Though this claim is criticized below.

[7] I use 'micro' and 'macro' in the names for positions to specify the level at which relevant properties are held to obtain. The macro non-reductive physicalist thinks that experience is a macro-physical phenomenon, not reducible to the micro-physical. The micro property dualist thinks that micro-level things have two fundamentally different kinds of property. A macro property dualist might think that macro-things (e.g. people) are both mental and physical, but that at other levels the duality doesn't apply. More on such distinctions and the positions that follow from them can be found in Coleman and Willemarck (forthcoming).

physical product being anything other than the sum of lower-level goings on, which would in turn constitute a defence of smallism.

But the argument is too narrow in scope. Strawson's decision to focus on the generation of novel physical *stuff* is puzzling, since the most natural way to understand the addition of experience to ontology is as the addition of a new *property*. It certainly seems true that if the macro non-reductive physicalist were committed to new physical *matter* accompanying the assembly of ultimates into a brain, this would violate conservation. But this physicalist can hold instead that what comes along is a new physical[8] property, undetermined by the ultimates, though perhaps supervenient upon them.

Another principle dear to many physicalists *is* violated, however, on the assumption that consciousness has causal power. For macro non-reductive physicalism then involves genuine *downward causation*. Thus the *causal completeness of physics* — the claim that every physical happening has an origin describable by physics — is broken; given that (micro)physics could not capture the novel macro-physical property at work. This, however, does not threaten the causal completeness of the *physical*. Granted the object conception of the physical on which emergent experiential properties are wholly physical, consciousness exercising its power would do nothing to break a purely physical story of causation. So the real question for the macro non-reductive physicalist is whether the completeness of *physics* is held as dear as the completeness of the *physical*. If yes, then she indeed has a problem. But Strawson says nothing to defend this former doctrine, and appeal to the obvious means of motivating it — smallism — would again beg the question.

We may conclude that to rule out macro non-reductive physicalism, which would successfully combine the non-experientiality of physical ultimates with experiential physicality in Strawson's sense, smallism is relied upon at all turns. But such a substantive thesis cannot just be assumed, it needs motivating.

3. Perspicuity

Strawson's second assumption is *perspicuity*: *the perspicuity of upward determination*. This is an epistemological thesis, which only comes into play *given* the upward determination thesis of smallism. Perspicuity is the view that where some item at a higher level is determined by goings-on at a lower level, this determination is in principle visible a priori. Another way to put it — closer to Strawson's way of

---

[8] On the object conception.

thinking — is that where higher-level Y is determined by lower-level X, it will *make sense* why this should be so. He would claim an almost metaphysical weight for this *making sense*: it is the notion that there must exist features of X in virtue of which Y comes about, and that these are in principle discoverable, if only by God.

Perspicuity is assumed, but it is made explicit, as smallism is not; and it is wielded at superficially the same target, Strawson's 'emergence'. He claims that 'if it really is true that Y is emergent from X... all features of Y trace intelligibly back to X' (Strawson, 2006, p. 18). It is clear that the macro non-reductive physicalist is immune to the use of perspicuity: since she denies upward determination of experience, she need take no view on whether such upward determination must be perspicuous. The real target of perspicuity, therefore, must be the smallist. Specifically: Strawson wishes to attack physicalists who believe that the microphysical upwardly determines experience and that the microphysical is non-experiential. This is the real combination of views he objects to. The concrete claim against the holder of this combination is that if it is unintelligible how the experiential could emerge from the non-experiential microphysical, then it cannot so emerge. Yet, given smallism, experience *does* emerge from the microphysical. So the microphysical cannot be non-experiential.

But, again, a position seems to be passed over here. For there are physicalists who think that the microphysical is non-experiential and upwardly determines experience, but who appear invulnerable to Strawson's argument as it stands. The key is that the notion of 'intelligibility' appealed to *is* epistemological; and what a knower can know depends on the powers of that knower. The *a posteriori physicalist* holds that, for one reason or another, *we* cannot see how experience is a purely physically generated phenomenon. This is consistent with the claim that matters to Strawson, that God can understand the upward determination of experience by the non-experiential physical. For example, philosophers following Loar[9] have argued that there is something special about our 'phenomenal' concepts — those we use to refer to experience in respect of its qualitative character — that prevents us from connecting them a priori with physical concepts. No story told in physical terms will ever compel us to see what is described in experiential terms, for the two vocabularies just do not meet in conceptual space. If this is so, then even if God can see how experience is microphysically generated, *we* cannot. Yet this epistemological failing — a contingently human one perhaps — need not count against smallist physicalist

---

[9] See Loar (1997). See also Papineau (2002), for an example of a follower.

metaphysics. It is as much as to say that our powers of reasoning are limited. And who is there who doesn't agree with that?

As against this view, the argumentative examples Strawson supplies do not help:

1. *Liquidity* is the model, for him. If experience emerges from the non-experiential, this needs to be as perspicuous as the way that liquid emerges from the non-liquid. *The a posteriori physicalist replies*: this emergence *is* perspicuous, in the sense that there are indeed features of the non-experiential ultimates in virtue of which experience comes about. God could see this. *We* cannot, for our phenomenal concepts don't connect a priori to physical ones. So perspicuity on the model of the emergence of liquidity is denied us.

2. Strawson claims that it is unintelligible how *extension* could come from unextended ultimates, or *space* from non-spatial ultimates. And sure enough, it is plausible that such emergences are in fact impossible. So, it seems that where emergence is unintelligible, it is impossible. So much for experiential emergence from the non-experiential. *Reply*: it seems likely that if a case of emergence is in fact impossible, like those cited, it will rightly appear so. But the entailment needed is *from* unintelligibility *to* impossibility, and this is not supported by the examples featured. Such a move from epistemology to modality — as seen in, for example, the *zombie argument* against physicalism — is highly controversial and rejected by many. A popular way of denying the move, in connection with the putative physicality of experience, is to endorse the view that phenomenal concepts don't connect a priori with physical ones. In this case experiential emergence will indeed be unintelligible, but the source of the unintelligibility won't be anything metaphysically grand.

3. To support his intuition that the experiential cannot emerge from the non-experiential, Strawson points out that this is just as strong as the intuition that the non-experiential cannot emerge from the experiential. This latter claim seems obviously true, so why not the former? *Reply*: this argument from parity is disingenuous. Emergence is a matter of higher-level phenomena coming from the organization of lower-level phenomena. So it no more makes sense to talk of lower-level non-experientiality *emerging from* higher-level experientiality — given the emergence of experience from non-experience — than it does to talk of non-liquidity emerging from liquidity. It is clear that to get a lower-level phenomenon from a higher-level one, you must *take something away*; there is no question of *emergence from*. The move is towards a decrease in worldly richness, not the increase we get by going up a level. From this perspective it is clear that liquidity *can*

yield non-liquidity: just move the liquid's molecules far enough apart that they can't slide over one another. In the same sense, *if* it's true that experientiality emerges via some complex interaction of non-experiential ultimates, then taking these ultimates out of that complex interaction will seem a sure way of deriving non-experientiality from experientiality. For Strawson's opponent, this is just what happens when we die, perhaps.

But still, surely, Strawson has hold of a powerful intuition: if the ultimates are non-experiential, then how can they generate experientiality? This claim, setting aside the notion of intelligibility (to us), takes the form of a metaphysical challenge, and it is a stiff one, to be sure. Strawson points out that liquidity reduces to 'P' phenomena: 'shape-size-mass-charge-etc. phenomena' (Strawson, 2006, p. 13). Thus, in a sense, when liquidity arises there are nothing but P phenomena: 'Analysed in terms of P properties, liquid bodies of water and H_2O molecules have exactly the same sorts of properties' (ibid., p. 22). But, he says, this cannot be the case with non-experiential/experiential phenomena 'for it is built into our starting point ... that they do not have the same sorts of properties at all in this sense' (ibid.). Thus it seems that experience must really be something extra that — given smallism — needs to be plugged in at the lowest level.

But again the a posteriori physicalist is well placed to respond. If he is right, then what goes for liquidity goes for experience as well: analysed in terms of P properties, experiential and non-experiential matter have exactly the same sorts of properties. It is begging the big question to claim that non-experiential and experiential phenomena do not have the same sorts of properties in this sense. For the a posteriori physicalist there is nothing more to experience than P phenomena, it is just that its emergence from these is not perspicuous to us, due to the tricks played by phenomenal concepts.

If perspicuity is true, then the a posteriori physicalist is mistaken. But this physicalist has a good reason why perspicuity does not hold; at least, not in the experiential case. To really press the smallist physicalist believer in non-experiential ultimates, this *phenomenal concepts strategy* needs arguing against.

Overall, we see that Strawson's 'emergence' obscures two positions: on the non-smallist version of emergence — macro non-reductive physicalism — a new property needn't emerge *from* the base, it just comes along; on the a posteriori physicalist version, though the property emerges *from*, we can't see how. To challenge these two (popular) positions, *smallism* and *perspicuity* require defence. I believe that these two assumptions can be satisfactorily defended, but I cannot do this here.[10] Instead, I will conclude by challenging Strawson's view of what follows if the assumptions hold.

4. Panexperientialism

Given smallism, and *given* perspicuity, Strawson seems correct that there must be experientiality in micro-ontology. Otherwise macro-experience could not emerge from ultimates. To the extent that his panpsychism comes to this claim, I agree with it. Other claims that Strawson makes about it can however be criticized.

The official position is that at least *some* ultimates must have experiential as well as non-experiential properties. Where only some have this dual nature, this is *micropsychism*, and the view where all ultimates are experiential is *panpsychism*. It is worth lingering on the move Strawson makes from micropsychism to his preferred panpsychism. His belief in the neatness of ontology surfaces when he declares that it is improbable to find the 'radical heterogeneity at the very bottom of things' (Strawson, 2006, p. 25) that micropsychism involves. Needless to say, such a belief doesn't supply the promised *entailment* of panpsychism, even if we grant the entailment of *micro*psychism. And there is perhaps good reason for Strawson to avoid putting the 'pan' in 'panpsychism'.

For he holds that an instance of experience implies the presence of a *subject* of experience. Consequently, every experiential ultimate counts as a subject for Strawson. And on panpsychism, *every* ultimate is a subject. I think that if we stick anywhere close to the normal conception of a subject — examples: human person, fairly evolved animal, *maybe* insect — such a wide and deep spread of them throughout ontology is to be resisted at all costs. Then, given the weakness of the move from micropsychism to panpsychism, it seems a strong point in favour of the former that it at least limits the number of subjects, on Strawson's terms, as compared with the latter.

But micropsychism is difficult to adhere to, for a different reason. To many it is plausible that brain-making ultimates are *refundable*. It goes with our conception of, for example, sub-atomic particles that, *if* they can constitute brains, then any suitably arranged set of them will do. Maybe this is the grain of truth in Strawson's intuition of deep-down homogeneity: we don't feel that there can be some privileged set of ultimates out of which, and out of which alone, a conscious brain can be assembled. Any ultimates will do. This is the thought that

---

[10] For some defence, especially of perspicuity, see Coleman (2006).

underwrites Nagel's stunning list of possible brain-ingredients: 'books, bricks, gold, peanut butter, a grand piano' (Nagel, 1986, p. 28). If this is correct, and given the argument for low-level experientiality, it will follow that all ultimates must have experiential nature. So panpsychism is, for this reason, unavoidable.

What to do now? The best solution, surely, is to ditch the thesis about subjects. If experience must permeate the lowest level, let us deny that mere experience suffices for subjecthood. What, then, does suffice? This is not the place to go deeply into this issue, but it is intuitive that being a subject implies being placed to interact in certain ways with the environment: to receive information from it, perhaps to act upon it. Certainly to have representations, of one's insides and outsides. And much else besides that all speaks of *complexity*. I suggest that genuine subjecthood requires something like a brain: a sophisticated organ of representation which enables a rich, multi-faceted relationship with what is around the subject. Such a way of relating to the environment is not, in the least, open to an experiential *ultimate*. So we can discount such an ultimate as a subject. Rather, on the view thus entertained, experiential ultimates can upwardly determine subjects when arranged in the correct way.

Two other problems noted by Strawson are also immediately avoided by this move. The first is that something needs to be said on his view about whether any subjects exist at levels of ontology *between* beings like ourselves and the ultimates: are any less complex parcels of matter also entire subjects, like cells? It is hard to see what materials Strawson's position has to draw subject-lines round ultimates, and also round animals, but at no place in between, as he wishes to. So this is quite a challenge. By restricting subjecthood to brained beings, we are spared it.

The second problem is James's objection that putting little subjects together doesn't give a big subject: subjects are, intuitively, inviolable *individuals*; they cannot be pooled. Interestingly, Strawson parses this difficulty as that of explaining how the macro-experiential derives from the micro-experiential. This latter is certainly a problem, but it is one that is hard enough without the added positing of subjects throughout micro-ontology! To avoid James's *subject-combination problem*, we just refrain from endorsing Strawson's small subjects. Yet still the difficulty remains of saying how macro-experience is upwardly determined by micro-experience. This is the *combination problem* proper, and it is a central obligation to answer it. Recall that, given *perspicuity*, if it is not intelligible how macro-experience emerges from micro-experience, then it does not so emerge. More on this in a moment.

We've denied that experience implies an experiencer, a subject. How is this denial tolerable? For many the implication has the status of a necessary truth. Brief reply:[11] We are apt to confuse experience with *human* experience. Human experience — consider an instance of pain — is phenomenally and representationally rich. It is even arguable that a pain makes essential reference to a subject that has it; this is so on any view where the content of pain includes *damage to the sufferer*. But on our view — restricting subjects to the macro-level and complex creatures — the rich intentional content of such states is a product of their being embedded in a brain. The idea is that phenomenology is basic, but can be harnessed by the system of a cognitive engine to yield states that in addition to feel have *representation*, or: *feel as representation*. If feel can exist without representation — as this view implies — then it is tolerable to say that there can be experience without subjecthood. This is the situation at the lowest level, the story goes. The experiential properties of the ultimates will be the contentless raw-feels of the qualia freaks. But that is not to say that a *human being* can have any such property, as intentionalists seem correct to deny.[12]

It is no objection to this view to say that we cannot *conceive* of raw-feels. As we have been taught, one needs to have a phenomenal state, or one close enough to it, to be able to imagine it. If our phenomenal states are shot through with intentionality, we will not be capable of imagining phenomenal states that are not so shot through. But only an unacceptable tendency to generalize from our limited case — against a known law of phenomenology — would permit the inference from there to representation, and subjecthood, in micro-ontology. One of the distinguishing features of *panexperientialism*, as opposed to panpsychism, is that it posits experience throughout micro-ontology, but not subjects; not little psyches.

Back to the combination problem. Perspicuity makes it pressing that macro-experience should intelligibly emerge from micro-experience. But how does it? The mind boggles at this question, which augurs trouble for the claims we wish to hold together. Two points: First, we

---

[11] Again, more is to be found in Coleman (2006).

[12] A clue lies here to an increasingly popular position not examined: some philosophers, enchanted by the notion that experience is through-and-through intentional, see there a pathway for a physical (non-experiential) analysis of consciousness. Such theorists agree with us on smallism and perspicuity, but differ in holding that non-experiential analysis is possible. If content (in general) can be naturalized — as many believe it can, with time — then experience ought to follow shortly after. If someone takes this view seriously, they will find another lacuna in Strawson's argument. For my part, I take it that this *a priori physicalism* is deeply implausible — see the usual thought experiments.

are not now faced with the conceptual mountain of how experience could come from non-experience. By endorsing panexperientialism, we acknowledge the truth that consciousness cannot come from the non-conscious: thus the problem of emergence — the broader combination problem — is made that much less serious. Second, this leaves the issue of the *manner* in which micro-experience yields macro-experience untouched, but we have already made gestures in this direction. If the major distinction between the micro-phenomenal and the macro-phenomenal is that the latter has rich representational capacities, then perhaps we need look no further than ongoing attempts to naturalize content. Take your favourite candidate for an analysis of aboutness: some brain-world causal relation. Maybe when suitable unities of intrinsically conscious ultimates are placed in this relation, they represent. Starting out from conscious bedrock, perhaps the feeling that naturalizing intentionality leaves something out can now also be overcome. This is, of course, sketchy. The combination *problem* is not so called for nothing. It plausibly represents the major theoretical 'I owe you' of the panexperientialist/panpsychist. But that there is work to be done does not imply the falsity of a view, and there are avenues to be explored.

The other major part of panpsychism that is better done without is its property dualism. This is disguised by the object conception of physicalism: since both micro-experiential and micro non-experiential features qualify the stuff of chairs, tables and us, both property-types count as physical. Nonetheless, the dualism is apparent when one bears in mind that Strawson's ultimates have experiential *and* non-experiential properties, and that the experiential/non-experiential divide is for him the most fundamental in nature. The major weakness of the dualism is that it evolves a new problem of causation. On the object conception, experiential as much as non-experiential properties are 'physical', even if the former might not figure in *physics*.[13] But this does not avoid the essential difficulty that goes back to Descartes: how does the mental impact causally on the non-mental, when these are taken to be radically different aspects of ontology? I take it that it is a weakness of Strawson's object conception of physicalism that it conceals this problem, and the problem itself is quite daunting. On panexperientialism, the doctrine of physicalism can be better expressed, and the interaction problem avoided.

---

[13] Although, this could depend on how the reference of concepts in physics is fixed. More on this point, and the position it leads to, in a moment.

I follow Russell in holding that the concepts of physics only express the extrinsic natures of the items they refer to (see Russell, 1927). The question of their intrinsic natures is left unanswered by the theory, with its purely formal description of micro-ontology. If there must be experientiality at the lowest level, then let it stand as the intrinsic character of the things physics mentions. The essence of the physical then is experiential. Now we can define physicalism as the doctrine that everything that exists is either part of, or upwardly determined by, physics' catalogue of items. First, the thesis that the physical is experiential is preserved: in fact the *essence* of the physical is now experiential. Second, physicalism is not defined in such a way that it permits the physical to be radically disjunctive: there is micro-property monism. Third, the interaction problem is avoided. Causal commerce on this view is *all* down to the doings of intrinsically experiential existents: causality as described by physics, as currently conceived, captures the *structure* of these goings on, but leaves out the real loci of causal power. On the novel conception of physics enjoined by panexperientialism, physics' terms pick out these causal loci — the experiential properties — with reference to the relational structure they support; and since causality involves the interaction ('*intra*-action') of properties of only one type, the causation problem disappears. A final boon is that physical causal completeness, *and* the completeness of physics, are both neatly entailed.

In conclusion: given smallism, and given perspicuity, we should be panexperientialists.

References

Coleman, S. (2006), *In Spaceships They Won't Understand: The Knowledge Argument, Past, Present and Future* (PhD Thesis, University of London).

Coleman, S. and Willemarck, F. (forthcoming), 'How many positions on the mind/body problem?'.

Loar, B. (1997), 'Phenomenal states (revised version)', in *The Nature of Consciousness*, ed. N. Block, O. Flanagan and G. Guzeldere (Cambridge, MA: The MIT Press).

Nagel, T. (1986), *The View from Nowhere* (New York: Oxford University Press).

Papineau, D. (2002), *Thinking About Consciousness* (Oxford: Oxford University Press).

Russell, B. (1927), *The Analysis of Matter* (London: Routledge).

Strawson, G. (2006), 'Realistic monism: Why physicalism entails panpsychism', *Journal of Consciousness Studies*, **13** (10–11), pp. 3–31.

Philip Goff
Experiences Don't Sum

According to Galen Strawson, there could be no such thing as 'brute emergence'. If we allow that certain x's can emerge from certain y's in a way that is unintelligible, even to God, then we allow for anything: for something to emerge from nothing, for the concrete to emerge from the abstract. To suppose that experiential phenomena could emerge from wholly non-experiential phenomena would be to commit ourselves to just such a brute emergence, to enlist in the 'Humpty Dumpty army' for life, with little chance of honourable discharge. It is this revulsion for the notion of brute emergence which leads Strawson to hold that the only viable form of physicalism is panpsychism, the view that the ultimate constituents of the physical world (which I will follow Strawson in calling 'ultimates') are essentially experience involving. Unfortunately, panpsychism is also committed to a kind of brute emergence which is arguably just as unintelligible as the emergence of the experiential from the non-experiential: the emergence of novel 'macroexperiential phenomena' from 'microexperiential phenomena'.

Any *realistic* version of panpsychism must hold that certain macroscopic physical entities, at least human beings or parts of them, have conscious experience, conscious experience which is presumably very different from the conscious experience of ultimates. On the assumption that these experience-involving macroscopic entities are wholly constituted of physical ultimates — there are no souls — we must suppose that the experiential being of macroscopic physical entities is wholly constituted by the experiential being of physical ultimates. Strawson consents to all this. Somehow thousands of experience-involving ultimates come together in my brain to constitute the 'big' experience-involving thing that is the subject of my experience.

But it is perfectly unintelligible how this could be. William James puts the point very vividly, in a passage referred to by Strawson (Strawson, 2006, fn 48):

> Take a hundred of them [feelings], shuffle them and pack them as close together as you can (whatever that may mean); still each remains the same feeling it always was, shut in its own skin, windowless, ignorant of what the other feelings are and mean. There would be a hundred-and-first feeling there, if, when a group or series of such feelings were set up, a consciousness *belonging to the group as such* should emerge. And this 101st feeling would be a totally new fact; the 100 feelings might, by a curious physical law, be a signal for its *creation*, when they came together; but they would have no substantial identity with it, nor it with them, and one could never deduce the one from the others, or (in any intelligible sense) say that they *evolved* it. (James, 1983, p. 162)

Suppose that my brain is composed of a billion ultimates which have no experiential being. Strawson claims that if this were the case it would be unintelligible why the arrangement of these ultimates in my brain should give rise to experience. But now let us suppose that each of the billion ultimates that compose my brain is a subject of experience: that there is something that it is like to be each of the ultimates of which my brain is composed. Imagine that each of the ultimates in my brain feels slightly pained. It is unintelligible why the arrangement of these ultimates in my brain should give rise to some *new* subject of experience, over and above the billion slightly pained subjects of experience we already have. The emergence of novel macroexperiential properties from the coming together of microexperiential properties is as brute and miraculous as the emergence of experiential properties from non-experiential properties. Strawson's panpsychism is itself committed to the very kind of brute emergence which it was set up to avoid.

Epistemological Limitations

It is not intelligible to us how our 'macro conscious experience' might be constituted of the 'micro conscious experience' of billions of micro subjects of experience. But why should we not think that this is a reflection of our epistemological limitations, rather than the metaphysical reality? Perhaps when God attends to my conscious experience, he conceives of it as something that is constituted of the conscious experience of billions of micro subjects of experience, even though this is not revealed to me in introspection. The problem with this response is that Strawson's overall project is committed to our having, though introspection, a kind of transparent understanding of the essential nature of our conscious experience which is inconsistent, or at the very least in tension, with this proposal. I will explain this in a little more detail.

There is an important sense in which Strawson is a Cartesian and an important sense in which Strawson is not a Cartesian. Descartes held that we are able to achieve a transparent understanding of the essential nature of physical stuff (i.e. its being extended) — let us call this a commitment to the *transparency of the physical* — and that we have a transparent understanding of the essential nature of our own mental states, which we may call a commitment to the *transparency of the mental*. It was his commitment to both the transparency of the physical and the transparency of the mental, and his ability to conceive of his mind and body as separate, which convinced Descartes that his mind and body must indeed be distinct.

Strawson is a non-Cartesian in that he denies the transparency of the physical. It is his denial of the transparency of the physical that allows him to identify mind and brain: 'we have no good reason to think that we know anything about the physical that gives us any reason to find any problem in the idea that experiential phenomena are physical' (Strawson, 2006, p. 4). It is because our physical concepts do not afford us a transparent understanding of the (complete) essential nature of physical stuff that it is coherent to suppose that physical stuff might turn out, as a matter of empirical fact, to be mental stuff.

But Strawson's entire discussion is premised on a commitment to the transparency of the mental, or at least of *our own conscious experience*. In this paper, Strawson describes conscious experience as 'the fundamental given natural fact' (p. 4). In a previous paper, Strawson puts his commitment to our having a transparent understanding of the nature of our conscious experience more explicitly:

> ... we have direct acquaintance with — know — fundamental features of the mental nature of (physical) reality just in having experience in the way we do, in a way that has no parallel in the case of any non-mental features of (physical) reality. We do not have to stand back from experiences and take them as objects of knowledge by means of some further mental operation, in order for there to be acquaintance and knowing of this sort: the having is the knowing ... we are acquainted with reality *as it is in itself*, in certain respects, in having experience as we do. (Strawson, 2003, p. 54)

In his commitment to our having a transparent understanding of the nature of our conscious experiences Strawson is a card-carrying Cartesian; and this commitment is not an unimportant background assumption. None of Strawson's worries in this paper have any force if we do not have, through introspection, a transparent grasp of the essential nature of our conscious experience. If we are not, in introspectively attending to the properties of our conscious experience, afforded a transparent grasp of the essential nature of those experiential properties, then those very properties might turn out to be physicSal (which is the name Strawson gives to those properties which physics does afford us a transparent grasp of) or functional properties under a different guise.

This is the claim of the orthodox a posteriori materialist: experiential properties, although *conceptually* distinct from physicSal or functional properties, turn out, as a matter of empirical fact, to be identical with such properties.[1] Brian Loar, an a posteriori materialist, expresses his conception of physicalism thus: 'Physicalism is the thesis that, however odd it may seem, that quale (which I am now conceiving phenomenally) might, for all we know, be a physical-functional property' (Loar, 2003, p. 121). The orthodox a posteriori materialist can agree with Strawson that there is an *epistemological* gap between physicSal/functional properties and experiential properties, and that, in this sense, the emergence of experiential properties from physicSal/functional properties is unintelligible, but will deny that this epistemological gap has *metaphysical* significance. As a matter of empirical fact, those very physicSal/functional properties which are conceptually distinct from experiential properties, turn out to be identical with experiential properties.

The conceptual gap between experiential and physicSal/functional properties only begins to have metaphysical significance if we suppose that introspection affords us a grasp of the essential nature of our experiential properties. For this seems to place limits on what those properties could *turn out*, as a matter of empirical fact, to be. The Cartesian commitment to our having a transparent understanding of the

---

[1] I say 'orthodox' a posteriori materialist because in a sense Strawson himself is a kind of a posteriori materialist, in that he wants to claim that physical stuff turns out, as a matter of empirical fact, to be essentially experience involving. But Strawson's a posteriori materialism is a non-standard kind of materialism in that it is reliant on physical stuff having an irreducible phenomenal essence which is hidden from empirical investigation. In contrast, the orthodox a posteriori materialism of philosophers like Levine (1983), Loar (1990, 2003), Papineau (1993, 2002), Tye (1995), Lycan (1996), Hill (1997), Hill and McLaughlin (1998), Block and Stalnaker (1999), Perry (2001), identifies phenomenal properties with the kind of ordinary physicSal or functional properties which are straightforwardly empirically discernible. Other examples of the kind of non-standard a posteriori materialism Strawson advocates include Russell (1927), Eddington (1928), Maxwell (1979), Lockwood (1989), Stoljar (2001). See Papineau (2002, p. 85, fn 5), Loar (2003, pp. 114–15) for statements differentiating their position from this kind of non-standard materialism.

essential nature of our conscious experience is crucial for Strawson's argument.

But then this commitment to my having, through introspection, a transparent understanding of the essential nature of my conscious experience is sharply in tension, if not inconsistent, with my conscious experience turning out to be, in and of itself, quite different from how it appears to be in introspection: i.e. turning out to be constituted of the experiential being of billions of micro subjects of experience. Strawson claims that in introspecting one's conscious experience, one perceives that metaphysical reality '*as it is in itself*' (Strawson, 2003, p. 54). It is this that allows us to know that this reality is not constituted of physicSal or functional properties. How then could this metaphysical reality turn out to be in any way different, as it is in itself, from how it appears to us in introspection? Just as a commitment to the transparency of the experiential gives the epistemological gap between experiential phenomena and non-experiential phenomena metaphysical significance, so it gives the epistemological gap between microexperiential phenomena and macroexperiential phenomena metaphysical significance.

Phenomenal Parts and Wholes

Even if we could make sense of the idea of the experiential being of several distinct subjects of experience coming together to constitute the experiential being of some higher-order subject of experience, or at least suppose that there is some explanation beyond our ken, it can be shown that it is contradictory to suppose, in the way we surely must do if panpsychism is viable, that the experiential being of a higher-level subject of experience is significantly qualitatively different from the experiential being of the lower-level subjects of experience of which it is constituted.

Consider a physical ultimate that feels slightly pained, call it LITTLE PAIN 1. Consider ten such slightly pained ultimates, LITTLE PAIN 1, LITTLE PAIN 2, etc., coming together to constitute a severely pained macroscopic thing, call it BIG PAIN. The pained-ness of each of the ultimates comes together to constitute the pained-ness of BIG PAIN: an entity that feels ten times the pain of each LITTLE PAIN. The severe pained-ness of BIG PAIN is wholly constituted by the slight pained-ness of all the LITTLE PAINS.

Assuming the coherence of this, the experiential being of each LITTLE PAIN is part of the experiential being of BIG PAIN; the experiential being of the BIG PAIN is a whole which contains nothing other than the experiential being of all the LITTLE PAINS. But it is a conceptual truth, as James rightly points out shortly after the passage above, that 'as a psychic existent *feels*, so it must *be*' (James, 1983, p. 165). It follows that for LITTLE PAIN 1 to be part of BIG PAIN is for *what it feels like to be LITTLE PAIN 1* to be part of *what it feels like to be BIG PAIN*. But what it feels like to be LITTLE PAIN 1 is not part of what it feels like to be BIG PAIN. LITTLE PAIN 1 feels slightly pained, BIG PAIN does not. The phenomenal character of LITTLE PAIN 1's experience, i.e. feeling slightly pained, is no part of the phenomenal character of BIG PAIN's experience, i.e. feeling severely pained.

In the same way, the experiential being of BIG PAIN is supposed to be wholly constituted by the experiential being of all the LITTLE PAINS. But to suppose that what it feels like to be BIG PAIN is wholly constituted by what it feels like to be all the LITTLE PAINS (if this comes to anything at all) must be to suppose that BIG PAIN feels how all the LITTLE PAINS feel and feels nothing else. But, by stipulation, this is not right. BIG PAIN feels a certain way that all the LITTLE PAINS do not: that is, severely pained.

Whatever sense we can make of experiences summing together, it is contradictory to suppose that the experiential being of lots of little experiencing things can come together to wholly constitute the *novel* experiential being of some big experiencing thing. Even the experience of a severely pained subject of experience is sufficiently different from the experience of slightly pained subjects of experience as to make it incoherent to suppose that the former could be formed from the latter. For the experiential being of some little experiencing thing 'LITTLE' to be part of the experiential being of some big experiencing thing 'BIG' is for what it is like to be LITTLE to be a part of what it is like to be BIG. But it follows from this that BIG feels how LITTLE does (even if it also feels other things). Correspondingly, for the experiential being of some BIG to be wholly constituted by the experiential being of LITTLE 1, LITTLE 2, LITTLE 3 … (again assuming this makes any sense at all) can be nothing other than for BIG to feel how it feels to be all those LITTLES and to feel nothing else. Even if it is intelligible how experiential states can *sum* together, it is contradictory to suppose that they could sum together to form some novel conscious state.

If my experiential being were constituted by the experiential being of billions of experience-involving ultimates, then what it is like to be each of those ultimates would be part of what it is like to be me. I would literally feel how each of those ultimates feels, somehow all at the same time. Assuming that my experiential being is *wholly* constituted by the experiential being of a billion experience-involving ultimates, then what it is like to be me can be nothing other than what it is like to be each of those billion ultimates (somehow experienced all at the same time).

But this surely cannot be right. My experience is of a three dimensional world of people, cars, buildings, etc. The phenomenal character of my experience is surely very different from the phenomenal character of something that feels as a billion ultimates feel.

Strawson's Response

Strawson is aware of the intuitive difficulties of making sense of the emergence of 'macroexperiential phenomena' from 'microexperiential phenomena'. He says the following:

> Human experience or sea snail experience (if any) is an emergent property of structures of ultimates whose individual experientiality no more resembles human or sea snail experientiality then an electron resembles a molecule, a neuron, a brain or a human being... This is not to advance our detailed understanding in any way. Nor is it to say we can ever hope to achieve, in the experiential case, the sort of feeling of understanding that we achieve in the liquid case. The present proposal is made at a very high level of generality (which is not a virtue); it merely recommends a general framework of thought in which there need be no more sense of a radically unintelligible transition in the case of experientiality than there is in the case of liquidity. (Strawson, 2006, pp. 27–8)

Strawson seems to suppose that the fact that macroexperientiality and microexperientiality are both the same *kind* of thing, i.e. they are both experiential phenomena, implies that the emergence of the former from the latter does not constitute a 'radically unintelligible transition'. But it is at least not clear how the fact that microexperientiality and macroexperientiality are the same kind of thing makes the emergence of the latter from the former any more intelligible. Subjects of experience are just not the kind of things that could intelligibly join together to form 'bigger' subjects of experience, any more than non-experiential things are the kind of things that can intelligibly come together to form subjects of experience.[2]

---

[2] Perhaps what it is like to be the ultimates that compose my brain is very different from what it is like to be me. But it is difficult to see how this difference could have significance for the intelligibility of their experiential being summing. I take it that it is something about the nature of *experiential being as such*, rather than the nature of any *specific experiential being*, in virtue of which experiential phenomena are capable of summing, if indeed they are capable of summing. I take it that our having a transparent grasp of our own experiential being implies that we have a transparent grasp of experiential being as such. If there is some feature f of the nature of experiential being as such in virtue of which

Nor does giving this 'general framework of thought' help make the notion of the emergence of macroexperientiality from microexperientiality intelligible. The above passage seems not to explain this emergence, but rather to express a faith that *it must happen somehow*. How does this differ from the faith of the non-panpsychist physicalist that *somehow* experience must emerge from the wholly non-experiential? According to Strawson, the hypothesis that experiential being emerges from non-experiential being comes to *feel* intelligible to many physicalists only because their other commitments make it such that *it must be the case*, it being the only option in logical space which preserves all their prior commitments: 'the notion of brute emergence comes to feel substantial to them by a kind of reflected, holographical energy. It has to be there, given these unquestioned premises, so it is felt to be real' (Strawson, 2006, pp. 18–19). Surely we could say the same of Strawson's hypothesis that macroexperiential being emerges from microexperiential being: something we can make no intelligible sense of, but which is needed to keep the theory consistent.

Strawson goes on to suggest that working on our general metaphysics will help make this picture clearer: 'The object/process/property/state/event cluster of distinctions is unexceptionable in everyday life but it is hopelessly superficial from the point of view of science and metaphysics, and one needs to acquire a vivid sense that this is so' (Strawson, 2006, p. 28). But it is difficult to see how considerations of *general* metaphysics could help with the problems I have been describing. The unintelligibility of the emergence of macroexperientiality from microexperientiality is a reflection of the *specific* nature of experiential phenomena, rather than the general ontological categories they fall into. At the very least, Strawson is obliged to explain how reflection on general metaphysical concerns could help here.

Conclusion

In order to avoid the brute emergence of conscious experience in the physical world, Strawson supposes that the fundamental constituents of matter are subjects of experience. But it is unintelligible how 'little' subject of experience can sum together to form 'big' subjects of experience. Because of this, panpsychism does nothing to explain, in a way that does not appeal to brute emergence, the conscious experience of people and animals.

---

experiential phenomena are capable of summing, then, in having a transparent grasp of the nature of experiential being as such, we ought to have a transparent grasp of f.

References

Block, N. and Stalnaker, R. (1999), 'Conceptual analysis, dualism and the explanatory gap', *Philosophical Review*, **108**, pp. 1–46.

Chalmers, D.J. (ed. 2002), *Philosophy of Mind: Classical and Contemporary Readings* (Oxford, New York: Oxford University Press).

Eddington, A. (1928), *The Nature of the Physical World* (New York: Macmillan).

Hill, C. (1997), 'Imaginability, conceivability, possibility and the mind-body problem', *Philosophical Studies*, **87**.

Hill, C. and McLaughlin, B. (1998), 'There are fewer things in reality than are dreamt of in Chalmers's philosophy', *Philosophy and Phenomenological Research*, **59**.

James, W. (1983), *The Principles of Psychology* (Cambridge, MA, and London: Harvard University Press).

Levine, J. (1983), 'Materialism and qualia: the explanatory gap', *Pacific Philosophical Quarterly*, **64**, pp. 354–61.

Loar, Brian (1990), 'Phenomenal states', in *Philosophical Perspectives*, ed. J. Tomberlin, Vol. 4: repr. in Chalmers (ed. 2002).

Loar, Brian (2003), 'Qualia, properties, modality', in *Philosophical Issues*, **13**.

Lockwood, M. (1989), *Mind, Brain and the Quantum* (Oxford: Oxford University Press).

Lycan, W.G. (1996), *Consciousness and Experience* (MIT Press).

Maxwell, G. (1979), 'Rigid designators and mind-brain identity', *Minnesota Studies in the Philosophy of Science*, **9**, pp. 365–403, repr. in Chalmers (ed. 2002).

Papineau, D. (1993), 'Physicalism, consciousness, and the antipathetic fallacy', *Australasian Journal of Philosophy*, **71**.

Papineau, D. (2002), *Thinking about Consciousness* (Oxford: Clarendon Press).

Perry, J. (2001), *Knowledge, Possibility and Consciousness* (MIT Press).

Russell, B. (1927), *The Analysis of Matter* (London: Routledge).

Stoljar, D. (2001), 'Two conceptions of the physical', *Philosophy and Phenomenological Research*, **62**, pp. 53–81.

Strawson, G. (2003), 'Real materialism', in *Chomsky and his Critics*, ed. Louise Antony and Norbert Hornstein (Oxford: Blackwell), pp. 49–88.

Strawson, G. (2006), 'Realistic monism: Why physicalism entails panpsychism', *Journal of Consciousness Studies*, **13** (10–11), pp. 3–31.

Tye, M. (1995), *Ten Problems of Consciousness: A Representational Theory of the Phenomenal Mind* (MIT Press).

Frank Jackson

Galen Strawson on Panpsychism

We make powerful motor cars by suitably assembling items that are not themselves powerful, but we do not do this by 'adding in the power' at the very end of the assembly line; nor, if it comes to that, do we add portions of power along the way. Powerful motor cars are nothing over and above complex arrangements or aggregations of items that are not themselves powerful. The example illustrates the way aggregations can have interesting properties that the items aggregated lack. What can we say of a general kind about what can be made from what by nothing over and above aggregation? I think that this is the key issue that Galen Strawson (2006) puts so forcefully on the table.

One thing we can say more or less straight off is that a certain supervenience thesis will hold. Suppose that we have a stock of ingredients that are all of kind K, where K might cover a highly diverse set — all items with the properties mentioned in one or another physical science, as it might be, and where K will in general include modes of arrangement as well as properties of the arranged — being an electron moving in a certain force field, for example. Suppose that we have two aggregations, A and B, which are, K thing for K thing, and K mode of arrangement for K mode of arrangement, identical, and are nothing but such aggregations. There are no gratuitous additions. If A has some non-K property, then so does B, and conversely. The non-K properties of mere aggregations supervene on the K properties of those aggregations. Why so? To deny the supervenience thesis would seem tantamount to denying that we are dealing with 'mere' aggregations, to denying the 'nothing but' part of the specification of the aggregations.

If we think of the emergence of properties that Strawson discusses as properties that come from mere aggregation, then we can put it this way: emergent properties supervene on that which they emerge from. Moreover, one way to view his insistence that emergence be 'intelligible' and not 'brute', is as insisting, to put the matter in terms that figure in the current debate over physicalism (in the usual sense of that term, not his) (see, e.g., Jackson, 2003), that any properties that emerge do so a priori and not a posteriori.

Strawson gives a number of examples of emergence and they are all, in my view, cases where the properties that emerge are a priori determined by the nature of the aggregations — the nature, that is, of that which is aggregated and how it is aggregated — on which they emerge. Take being liquid. He is right that 'we have ... no difficulty with the idea that liquid phenomena ... are emergent properties of wholly non-liquid phenomenon' (Strawson, 2006, p. 15). The reason is, I would say, that when we have enough detail about the non-liquid phenomenon, we have something that leads a priori to liquidity. The non-liquid phenomenon will include the way molecules move past each other, the way certain bodies of molecules tend to group together but in ways that allow easy changes in overall shape, the way force fields including especially gravity move the molecules around while allowing a certain degree of cohesion (without the cohesion, we'd have a gas), and so on. After a certain amount of information of this kind, what else is there to say but something like 'That counts as being liquid', and when that happens we have something a priori. Similar points apply to his example of life*. The smooth reduction he talks about seems to me to be another way of saying that life* follows a priori from enough information of the right kind about constituents that are not themselves alive*.

I think Strawson is right to reject brute emergence; that is to say, on my way of construing matters, I think he is right to insist that cases of emergence are always cases of a priori supervenience. But how much does this help the cause of *panpsychism*? One way to resist panpsychism while agreeing with his rejection of brute emergence is to hold that consciousness a priori supervenes on items that are wholly non-conscious. This is in fact the view I hold. Strawson clearly thinks it is very implausible (as I once did; it was my reason for not being a physicalist). But there is another way of resisting panpsychism. It is to insist that consciousness does not emerge in the relevant sense. Items which are conscious are more than mere aggregations of the non-conscious.

Strawson does not reject this position outright but finds it implausible. I think this is because he does not consider the most attractive version of the position. The most attractive version holds that there are fundamental laws of nature that go from certain complex arrangements of the non-conscious to consciousness. These arrangements are more than mere arrangements of the non-conscious because they have the extra property of causing consciousness. We know that different arrangements of the non-conscious may possess very different kinds of conscious experience. There are very many different kinds of conscious experience and they go along with differences in regard to, for example, how things are around some human body, the neurological differences between dogs and humans, the nature of an injury, blood sugar levels and so on. The panpsychist needs certain kinds of composition laws — laws that take the bits of consciousness in everything and tell us about the rules of additivity, as we might put it. Surely it is at least as plausible that the laws be ones that don't talk of additivity but instead talk of the *generation* of the various kinds of conscious experience. Panpsychists like to argue that something as special as consciousness could not plausibly come from aggregating items that separately lack consciousness. But once we acknowledge that consciousness comes in very many forms, we have to allow a role for the generation of these various types of consciousness by items that lack the types in question. But if we can get quite new types of consciousness, why not consciousness *per se*?

References

Jackson, F. (2003), 'From H_2O to water: the relevance to a priori passage', in *Real Metaphysics*, papers for D.H. Mellor, ed. Hallvard Lillehammer and Gonzalo Roderiguez-Pereyra (London: Routledge), pp. 84–97.

Strawson, G. (2006), 'Realistic monism: Why physicalism entails panpsychism', *Journal of Consciousness Studies*, **13** (10–11), pp. 3–31.

William G. Lycan

Resisting ?-ism

Professor Strawson's paper is refreshing in content as well as refreshingly intemperate. It is salutary to be reminded that even the Type Identity Theory does not entail physicalism as that doctrine is usually understood (since c-fiber firings are not *by definition* purely physical).[1] And it's fun to consider versions of panpsychism.

I can see why Strawson finds his position hard to classify (p. 7), and I sympathize. In my title I have cast my own vote for '?-ism' on the grounds that any familiar label would be either misleading or unwieldy.[2]

My main purpose here is to assess Strawson's case for panpsychism and then to offer some objections to panpsychism, but first I want to answer an interesting and serious charge he makes against me.

Have I Looking-Glassed 'Experience'?

In his dialectical footnote 6 (p. 5), Strawson accuses me, along with Dennett, Tye, Dretske and Rey, of having used the term 'experience' or equivalents 'in such a way that whatever . . . [we mean] by it, it excludes what the term means'. Rather than having offered theories of experience as each of us claims to have done, we have covertly denied the existence of experience; we are really eliminativists.

If Strawson is right, it would not be the first time I had misunderstood one of my own views. (I hate that.) But I am entirely unconvinced by his argument. Its assumption is true: We, his targets, do 'still dream of giving a reductive analysis of the experiential in nonexperiential terms'. In fact, I have actually given one that I allow

---

[1] The point has also been made by Gregg Rosenberg (2004, p. 59).

[2] In *Mental Reality* Strawson called himself a materialist (1994, p. 105), but inevitably had to add qualifications, and the qualifications were the distinctive and important features of his view.

myself to suspect is correct.[3] 'This, however, amounts to denying the existence of experience, because the nature of (real) experience can no more be specified in wholly non-experiential terms than the nature of the (real) non-experiential can be specified in wholly experiential terms.' (Strawson goes on to hypothesize a fruitless dialogue between us alleged deniers and the experiential realists, featuring tedious mutual accusations of question-begging, but that part of the footnote is not, I take it, part of his argument for classifying us as deniers in the first place.) So, the argument goes: (1) The nature of (real) experience cannot be specified in wholly non-experiential terms. Therefore, (2) no one who claims to have specified the nature of experience in wholly non-experiential terms could be using the term 'experience' to mean *experience* (whatever Ersatz thing s/he may mean by it). Evidently (3) such a theorist believes that every real relevant phenomenon has been explained, and experience is relevant; so (4) the theorist must not believe that experience is real, which means s/he is a denier.

My objection is to the move from (1) to (2). Arguably, (2) would follow if (1) were analytic. But (1) is not analytic; it is a highly contentious philosophical claim. And my colleagues and I are deniers of *it*. We may be wrong to reject (1), but it is (1), not the reality of experience, that we are rejecting.

Strawson's Case for Panpsychism

1. The question of evidence is a substantive and interesting one. Strawson provokes us by saying that 'there is *absolutely no evidence whatever*' against panpsychism (p. 20, italics very much original). I find myself agreeing with that, if only on a very strict reading of 'evidence': there is nothing I can exhibit to *show* decisively that a muon or a quark is not a locus of experience. But neither is there any scientific evidence for panpsychism; there is no scientific reason, as opposed to philosophical argument, for believing it.[4]

I do not see that as an even standoff; I think the burden lies with the panpsychist. For what is at issue is a specifically scientific claim about

---

[3] It isn't, of course. Feyerabend's meta-induction may or may not work against scientific theories, but it has great force against philosophy. Every previous philosophical theory has been false.

[4] The premise would be contested by interpreters of quantum mechanics who maintain that in order for there to be determinate physical fact in a spatiotemporal region, there is 'consciousness' in that region. Some of those interpreters may actually mean *consciousness* by 'consciousness'. Such an interpretation might attract adherents; but remember that any interpretation of quantum mechanics is just that, an interpretation of the quantum facts, not itself a fact, and the interpretation of quantum mechanics is considerably affected by philosophical considerations.

the nature of the most fundamental posits of physics. If quarks or photons have mental properties, that is for the microphysicist to find out. If microphysics has no need of the panpsychist hypothesis, we should apply Occam's Razor in the direction opposite to Strawson's.

Strawson will protest that I have cheated: we already know science could not show that subatomic particles have experiential features, but that is because, as Eddington says,[5] science can reveal only relational properties. The particles' intrinsic or nonrelational character must after all be left to philosophers. Strawson himself has given two philosophical arguments, the Russell-Eddington line (pp. 9–12) and the appeal to his principle of total dependence (pp. 13–19).[6]

2. I cannot say I have ever been much moved by Russell-Eddington. First, what grounds the assumption that the ultimate constituents of the physical world must have intrinsic properties at all? Perhaps the nature of a subatomic particle is exhausted by the totality of its relations to other things. (Notice that that suggestion is perfectly consistent with the obvious fact that a relation must have relata.)

Second, even if subatomic particles do have intrinsic natures and the experiential properties featured in sensation and perception are the only intrinsic properties that we know directly, that gives us little reason to suppose that the particles' natures are experiential properties. Eddington calls it 'silly' to resist the latter idea and adds, 'If we must embed our schedule of indicator readings in some kind of background, at least let us accept the only hint we have received as to the significance of the background — namely, that it has a nature capable of manifesting itself as mental activity' (Eddington, 1928, p. 260). The only hint, perhaps, but not a promising one. We do know our experiential states from the inside, but we also know that they are states of macroscopic, yea hulking, complex beings who get about their complex environments in an amazingly well-adapted and intelligent way by dint of those mental states, and that the mental activity is closely mediated by gazillions-per-second of electrochemical operations throughout the brain. In my view, each of the latter features makes them poor candidates for being ascribed to muons or neutrinos.

---

[5] It is interesting that Strawson (2006, note 17, p. 9) records the occasion on which he came across *The Nature of the Physical World* (Eddington, 1928). I too vividly recall my first encounter with the book, in a guest house at Franklin & Marshall College in 1982; it kept me up most of the night. This testifies to the power of Sir Arthur's writing.

[6] The latter argument is similar to and I daresay inspired by the one considered by Thomas Nagel in 'Panpsychism' (Nagel, 1979); cf. also Sellars' 'grain' argument, especially as formulated in Sellars (1971). But the total-dependence argument differs importantly in that unlike Nagel's as reconstructed by me, it does not presuppose property dualism.

Third, I do not grant Russell's, Eddington's and Strawson's assumption that experiential properties are nonrelational in the first place. Elsewhere I have defended the claim that they are all relational (Lycan, 1996, ch. 6; Lycan, 2001).

3. Turning to the principle of total dependence (as stated on p. 14), I accept the principle but reject its application to the case at hand. Strawson maintains that one cannot hold both NE and RP, because the notion of emergence required by that conjoint position is 'incoherent' (p. 12). Some part or aspect of experience will have to hail from somewhere other than the physical stuff conceived as wholly and utterly non-experiential. Strawson makes it clear (p. 15) that this is an ontological claim, not epistemological only.

Why will some aspect of experience have to be superadded? That there is an 'explanatory gap', in Joe Levine's famous phrase, I not only admit but insist (Levine, 1993; Lycan, 1996, ch. 3). But that gap is epistemological; no ontological gap may be inferred from it. Strawson here gives no further reason, except to demand 'an analogy of the right size or momentousness' (p. 15). In Lycan (1996) I have offered an ontological theory of all the main phenomena of consciousness — state consciousness, qualia properly so called, subjectivity, 'what it's like' and El Gappo itself — according to which each of those ontologically reduces to the underlying NE-style physical stuff. So (again) I would need to hear why the total dependence of those things on wholly non-experiential phenomena 'boggles the human mind' (p. 15) (though undoubtedly it continues to boggle some human minds).

What of the demand for an analogy? First, why do we need an analogy if we have actual theories? (Mine is not the only one.)[7] But, second, the analogies that Strawson goes on to consider are, as he says, pretty outrageous. Pseudo-Boscovich's thesis (p. 15) is *mathematically* absurd. So, as I understand it, is the emergence of spatiality from the nonspatial (p. 17). Even if experiential phenomena had to emerge from the nonexperiential in an objectionable way, that emergence would not be outright absurd as in those cases.

But as before, I maintain that the experiential does not violate the principle of total dependence in the first place. I grant that there is a difference, indeed a dramatic one, between the cases of experience and liquidity: Levine's Gap. But that gap is epistemological only.

---

[7] I guess my second favourite would be that of David Papineau in Papineau (2002).

Arguments against Panpsychism

1. One problem for the panpsychist is to avoid epiphenomenalism. Epiphenomenalism may be all right for macroscopic property dualists, but it would be very bad for a theorist who contends that elementary particles are loci of experience. Because of the causal closure of physics as we know it (i.e. without the assumption that microphysics involves mind), the panpsychist's tiny mental properties could play no added causal role. Since every nonrandom physical event has a sufficient purely physical cause, there is no work for the experiential properties to do. They are brought into existence by philosophers only to accomplish nothing at all.

(Yet the panpsychist could hold that some or all physical events are causally overdetermined, each one having both a sufficient physical cause and a sufficient or contributing mental cause. Perhaps there is a kind of mental-physical parallel in the world's causal structure. This response would be ad hoc and, in my view, grievously implausible, but if there is reason to accept panpsychism and no more devastating objection than the present one, the panpsychist could live with it.)

2. A further difficulty, anticipated by Strawson on p. 26, is raised by the total-dependence argument itself, that of saying 'how macroexperientiality arises from microexperientiality'. The mental properties we all know about are of course properties of complex organisms such as ourselves. If the total-dependence argument is sound, those properties must be a function of the mental properties inhering in their subjects' ultimate components. But what sort of function?

Suppose I am looking out of my kitchen window, and simultaneously seeing a rabbit in my back yard, hearing my wife's cat yowling that he wants to behead the rabbit, feeling the touch of my fingertips on a bottle of salad dressing, smelling the spaghetti sauce in the pot, suffering an ache in my right shoulder, and imagining in anticipation a very tall frosty beer. In what way could such a mental aggregate consist of or be determined by or otherwise 'arise from' a swarm of smaller mentations? Is it that some of my ultimate components are experiencing some of those very same mental states, and when enough of them do, I myself do? Or are the mental states of my components little, primitive states that somehow together add up to macroscopic states such as the ones I am in? Either alternative is hard to imagine, as is any further alternative.

(That objection need not bother the panpsychist who makes no appeal to the total-dependence argument. For such a person is free to reject the idea that my mentation is a function of the mental states of my components, and can suppose they are unrelated.)

Incidentally, I can't see why Strawson is so confident that 'things like stones and tables' are not subjects of experience (p. 26). To be sure, there are excellent reasons for such confidence: stones and tables do not have sense organs; they do not have nervous systems; they do not behave intelligently in response to environmental contingencies. But those reasons apply to ultimate particles as well — which is itself also a good argument against panpsychism.

3. Finally, if every ultimate particle has mental properties, what sorts of mental properties in particular do the particles have? It seems ludicrous to think that a photon has either sensory experiences or intentional states. (It does not even have mass.) How could it see, hear or smell anything? And if it has experiential properties, then presumably it also has rudimentary propositional attitudes. What would be the contents of its beliefs or desires? Perhaps it wishes it were a **u** quark.

As both Strawson and Dave Chalmers remind us,[8] to say that the universe is suffused with mind is not to say that it is suffused with minds like ours. So perhaps my semi-sarcastic query about sensory states and propositional attitudes is unfair. But the serious question remains: What sort of experiences *could* a photon have? Strawson rightly points out (p. 27) that there may be sensory modalities in other species whose qualitative deliverances we cannot imagine; but those are, in his own word, *sensory* modalities, and photons do not have sense organs.

4. Pretty bad. But I want to assure Strawson that even if physicalism really does 'entail' panpsychism, I will accept panpsychism rather than going eliminativist.

References

Chalmers, D. (1996), *The Conscious Mind* (Oxford: Oxford University Press).
Eddington, A. (1928), *The Nature of the Physical World* (New York: Macmillan).
Levine, (1993), 'On leaving out what it's like', in *Consciousness*, ed. M. Davies and G. Humphreys (Oxford: Basil Blackwell).
Lycan, W.G. (1996), *Consciousness and Experience* (Cambridge, MA: Bradford Books/MIT Press).
Lycan, W.G. (2001), 'The case for phenomenal externalism', in *Philosophical Perspectives, Vol. 15: Metaphysics* (Atascadero, CA: Ridgeview Publishing).

---

[8] Chalmers, 1996, pp. 298–9. In Chalmers' view, very simple systems would have very simple phenomenology or protophenomenology, so simple that they would fall short of having what we usually think of as minds.

Nagel, T. (1979), 'Panpsychism', in *Mortal Questions* (Cambridge: Cambridge University Press).
Papineau, D. (2002), *Thinking About Consciousness* (Oxford: Oxford University Press).
Rosenberg, G. (2004), *A Place for Consciousness* (Oxford: Oxford University Press).
Strawson, G. (1994), *Mental Reality* (Cambridge, MA: Bradford Books/MIT Press).
Strawson, G. (2006), 'Realistic monism: Why physicalism entails panpsychism', *Journal of Consciousness Studies*, **13** (10–11), pp. 3–31.
Sellars, (1971), 'Seeing, sense impressions, and sensa: a reply to Cornman', *Review of Metaphysics*, **24**, pp. 391–447.

Fiona Macpherson

Property Dualism and the Merits of Solutions to the Mind–Body Problem

A Reply to Strawson

1. Introduction

This paper is divided into two main sections. The first articulates what I believe Strawson's position to be. I contrast Strawson's usage of 'physicalism' with the mainstream use. I then explain why I think that Strawson's position is one of property dualism and substance monism. In doing this, I outline his view and Locke's view on the nature of substance. I argue that they are similar in many respects and thus it is no surprise that Strawson actually holds a view on the mind much like one plausible interpretation of Locke's position. Strawson's use of terminology cloaks this fact and he does not himself explicitly recognize it in his paper. In the second section, I outline some of Strawson's assumptions that he uses in arguing for his position. I comment on the plausibility of his position concerning the relation of the mind to the body compared with mainstream physicalism and various forms of dualism. Before embarking on the two main sections, in the remainder of this introduction, I very briefly sketch Strawson's view.

Strawson claims that he is a physicalist and panpsychist. These two views are not obvious bedfellows, indeed, as typically conceived, they are incompatible positions. However, Strawson's use of the term 'physicalism' is not the mainstream one. Strawson, clearly, recognizes this and takes some pains to distinguish how his conception of physicalism differs from mainstream physicalism. Strawsonian physicalism is the position that there are both non-abstract 'experiential' phenomena (by this Strawson means conscious mental phenomena, including both experience, traditionally conceived, and conscious thought) and non-experiential phenomena and that, 'there is, in some fundamental sense, only one kind of stuff in the universe' (Strawson, 2006, p. 7). He thinks that the experiential cannot be explained in principle by the non-experiential. The former does not reduce to the latter, and it does not emerge from the latter in any explicable way (Strawson, 2006, p. 12). Strawson urges a panpsychist view, which he claims to be the view that 'all physical ultimates are experiential' (Strawson, 2006, p. 25).

2. What Strawson's Position Really Is

2.1 Strawsonian Physicalism and Mainstream Physicalism

Strawsonian physicalism is the claim that all real concrete phenomena in our universe are physical. (Concrete phenomena are contrasted with abstract ones such as numbers and concepts.) Further, Strawsonian 'real physicalism' is the view that, in addition to the previous claim, conscious experience and conscious thought are concrete existents that require an explanation.[1] This is because, according to Strawson, the existence of consciousness 'is more certain than the existence of anything else' (Strawson, 2006, p. 3). Thus, Strawson wishes to defend a position that, in his terminology, would be stated as 'experiential phenomena are physical'.[2]

In the conventional usage, 'physicalism' is taken to be a position that embraces Strawson's claim that all real concrete phenomena in our universe are physical. However, a definition is usually given of what 'physical' means that is at odds with Strawson. Crane and Mellor tell us that in the eighteenth century the physical was defined a priori, requiring it to be 'solid, inert, impenetrable and conserved, and to interact deterministically and only on contact' (Crane and Mellor, 1990, p. 186). However, the posits of modern physics have few of these properties, yet are still taken to be clearly and paradigmatically physical. Thus, mainstream physicalists today usually define 'physical' as that posited by fundamental science.

However, this latter notion needs to be made more precise. What exactly is 'fundamental science'? Fundamental science would describe the basic particles and forces and the laws governing them. What would make such things basic is that all other phenomena can be

---
[1] From now on, when I talk about something's existing, I will mean concretely existing, unless I explicitly specify otherwise.
[2] I will follow Strawson's usage in this paper and use 'experiential' to refer to any conscious phenomena.

reduced to them. That is, all other phenomena can be explained and predicted, with no remainder, by fundamental science, perhaps together with suitable bridge laws (laws that specify the relations between the fundamental and non-fundamental phenomena).[3,4]

Can we be more precise still? One answer that could be given is that present day physics is the fundamental science that explains everything. However, this answer is very likely false. The physics of today is not complete. It probably does not list all the basic particles, forces and laws which together can explain everything else. Thus, we should expect fundamental science in the future to alter from its present state. Therefore, a seemingly better answer that could be given to the question of what is physical is that it is the posits of true and complete fundamental science.

However, here a large problem looms. If the answer were simply left at that then mainstream physicalism would have become a vacuous doctrine. It would state that what exists is the physical and the physical is that which is needed to explain everything that exists. There would be no limit as to what sort of thing can count as physical. To see this, one need simply note that if completed and true fundamental science had to posit a fundamental experiential force or particle then the physicalist position under consideration would have to say that that experiential particle or force was physical. Mainstream physicalists wish to resist this trivialization of 'physicalism' and deny that among the posits of true and complete fundamental science will be experiential particles or forces or the like. (Similarly, people who wish to be mainstream physicalists concerning morality, aesthetics or some other area, would hold that moral entities, or aesthetic entities, or entities in that other area, will not be among the posits of completed and true fundamental science.) This point is well-made by Crane and Mellor, and Papineau, who write:

---

[3] Of course there is a large dispute today about whether everything can be reduced to physics, or to physics plus other sciences, and whether there will be a 'unity of science'. It is not my intention to suggest here that there must be such a reduction or such a unity. I simply wish to convey that a completed science must explain everything, and thus must posit what is needed to do so.

[4] What I have outlined here is reductive physicalism. Those who believe in non-reductive physicalism hold that there can be a looser relation than reduction between the posits of fundamental science and higher-level phenomena that, nonetheless, still warrants holding the higher-level phenomena to be physical. Instead of there being bridge laws that show higher-level phenomena to be identical with their lower-level counterparts (as is the case in reduction), it is held that the lower-level phenomena merely constitute or give rise to the higher-level phenomena. That is to say, a weaker relation than identity, such as supervenience, is taken to hold between the different levels, and the obtaining of this relation establishes that the higher-level is physical. Whether or not non-reductive mainstream physicalism is plausible is the subject of much debate in the literature.

> One may debate the exact boundary of physical science: but unless some human sciences, of which psychology will be our exemplar, lie beyond its pale, physicalism, as a doctrine about the mind will be vacuous.[5]

> ... it is not crucial that you know exactly what a complete physics would include. Much more important is to know what it will not include. Suppose, for example, that you have an initial idea of what you mean by 'mental' (the sentient, say, or the intentional, or perhaps just whatever events occur specifically in the heads of intelligent beings). And suppose now that you understand 'physical' as simply meaning 'non-mental', that is, as standing for those properties which can be identified without using this specifically mental terminology. Then, provided we can be confident that the 'physical' in this sense is complete, that is, that every non-mental effect is fully determined by *non-mental* antecedents, then we can conclude that all mental states must be identical with something non-mental (otherwise mental states couldn't have non-mental effects). This understanding of 'physical' as 'non-mental' might seem a lot weaker than most pre-theoretical understandings, but note that it is just what we need for philosophical purposes, since it still generates the worthwhile conclusion that the mental must be identical with the non-mental; given, that is, that we are entitled to assume that the non-mental is complete.[6]

Thus, mainstream physicalists hold that, applied to the posits of completed and true fundamental science, 'mental' and 'physical' are incompatible or opposing terms. However, this does not stop the main claim of mainstream physicalists about the mind being that the experiential, or the mental more generally, is the physical; by this they mean that the mental is a higher-level phenomenon that can be explained in a reductive (or non-reductive[7]) way by non-mental, physical fundamental entities. (Note that eliminativists have a mainstream physicalist ontology, however, rather than being physicalists about the mind, they think that it does not exist because it cannot be explained by non-mental, physical fundamental entities.)

The above is well established in the literature and I don't take it that I am saying anything that will be new to Strawson. He is careful to distinguish his physicalism from the mainstream variety, or at least something like it. However, explicating mainstream physicalism in this

---

[5] Crane and Mellor (1990), p. 186. They argue that the debate about the nature of the mind should not be conducted in terms of whether it is physical or not due to the kind of problems that I indicate here, which they go into in greater detail.

[6] Papineau (2000), pp. 183–4. The emphasis is his. He, unlike Crane and Mellor, is a proponent of mainstream physicalism. However, that people with different views agree on how mainstream physicalism has to be construed is instructive.

[7] See footnote 4.

way allows me to highlight certain features of Strawson's physicalism. Recall that Strawson's physicalism was the claim that all real concrete phenomena in our universe are physical. One can now see that, as Strawson does not go on to say what 'physical' means, it may look as if Strawson's physicalism is open to just the same charge of vacuousness that is outlined above.

Is Strawson's physicalism vacuous? To see the answer, note that Strawson claims that he does not define the physical as being 'concrete-reality-what-ever-it-is' (Strawson, 2006, p. 8). If he did that then he would be open to a blatant charge of vacuousness. Rather, he holds that 'physical stuff' is a natural kind term whose reference is fixed as being that kind of 'stuff' that comprises the concrete phenomena that we actually find in our world. This move has three consequences of note. First, Strawson is just assuming that there is one type of 'stuff'. Thus, he is simply assuming monism. Strawson does not present an argument that there are not two (or more) distinctive natural kinds of 'stuff'. Thus he has not provided us with an argument for dismissing dualism about 'stuff'. For all that has been said, dualism could be true. (I will come back to this point later in this paper.) Second, it renders Strawson's physicalism a little less trivial than we might have thought because he is explicitly stating that the 'stuff' is of one (natural) kind. It is not that there are many kinds of 'stuff' each of which deserves the epithet 'physical'. However, third, the claim that this one kind of 'stuff' is physical does look to be a vacuous one. However, Strawson himself in various passages admits this, stating that he would be happy to concede that his view is just a monist view, so long that it is understood that this one 'stuff' can be both physical (in the mainstream use of the term which excludes the experiential) and experiential (which, recall, in Strawson's terminology, refers to all conscious mental phenomena).

2.2 The Metaphysics of Locke and Strawson and their Commitment to Substance Monism

Suppose we adopt neutral terminology (neutral between the mainstream view of physicalism and Strawson's view of physicalism) and say that Strawson is clearly making the claim that he is a monist. I now wish to assess whether the view Strawson opines really is monist. I will do so by considering, first, the question of whether Strawson is a monist about substance and, second, the question whether he is a monist about properties.[8] After establishing what Strawson's position really is, I will go on, in section three, to comment on the nature of the arguments that he gives for it and the plausibility of his position.

At the end of the previous section, I used the term 'stuff' to explicate Strawson's monism, which is the term that Strawson himself uses. On one conception of substance, substances are just kinds of stuff.[9] Thus, it might be reasonable to think that Strawson is at least being a monist about substance. However, investigation of these matters thoroughly leads us deep into metaphysical territory, only hinted at in Strawson's paper. On page eight, in referring to the possibility that a concrete thing 'involves experiential being, even if it also involves non-experiential being', we are directed to Locke's views on substance to understand how this could be so.[10] Later in the paper, on page twenty-eight, amongst cryptic remarks concerning how one might flesh-out Strawson's view, we are referred to Strawson's views on the nature of substance and properties outlined in another paper (namely, Strawson, 2003). So what are these views and how do they help us address the question of whether Strawson is a substance monist?

Locke held that we have an idea of 'substance in general' and an idea of 'particular sorts of substance' (Locke, II, xxiii, 2 and II, xxiii, 3). Our idea of a substance in general is the idea of something that exists to support properties, which, he says, we cannot imagine existing on their own. It is unclear whether Locke thinks that there is substance in general, or whether he thinks that we merely have such an idea but that the idea is confused.[11] It seems certainly true that Locke thinks that if substance exists then we can know very little about it — perhaps only that it is that in which properties inhere. The general idea here is that if substance is the thing in which properties inhere then substance itself must be property-less. If that is the case then there

---

[8] If one is a dualist about substance then one will be a dualist about properties. Each different substance will have fundamentally different types of properties. However, if one is a monist substance, then one can either be monistic about properties or one can be dualist about properties.

[9] See Robinson (2004). The 'stuff' conception of substance is contrasted with the 'thing' conception. To illustrate: Descartes, a dualist, thought that all physical matter was part of the one physical substance and so thought of it as a stuff. At the same time, he thought each person was a different mental substance, which conforms more with the thing idea of substance.

[10] We are referred to Locke (1690). All references to Locke in the rest of this paper will be to this work and will take the usual form, so that 'III, iii, 3' should be read as 'Part Three, Chapter Three, Section Three'.

[11] There is debate in the secondary literature about which view is right and some people simply acknowledge the unclarity, see Robinson (2004).

would be little to know of its nature. This problem is at least part of why Locke entertains the thought that the idea of substance dubious.

Given Locke's views on substance in general, it is no surprise that he goes on to say what he does about particular sorts of substance. He says that we have the idea of physical substance as being that in which the properties that affect our senses inhere — ideas of these properties are got from 'without'. We also have ideas of properties got from 'within' — ideas concerning the workings of the mind, particularly thought. We tend to think that conscious mental properties could not belong to physical substance so we posit a mental substance in which those properties inhere.[12] From these and other remarks of Locke's, it is sometimes taken that Locke is a substance dualist concerning the mind-body relation (see Aaron, 1955, p. 143; and Woolhouse, 1983, p. 180). However, based on what he goes on to say about these notions, it is far from obvious that he is.[13] Locke holds that a substance with physical properties alone could not produce thought:

> yet matter, incogitative matter and motion, whatever changes it might produce of figure and bulk, could never produce thought (Locke, V, x, 10).

Yet he claims that God could add to physical substance, which has physical properties, mental properties:

> I see no contradiction in it, that the first Eternal thinking Being, or Omnipotent Spirit, should, if he pleased, give to certain systems of created senseless matter, put together as he thinks fit, some degrees of sense, perception, and thought.[14]

This shows that Locke is agnostic about whether there are two distinctive substances underlying the mind-body relationship — the physical and mental — or whether there is one substance that can have both mental and physical attributes. He thinks that there are no good grounds on which to choose between these two positions.[15] However, what he does seem to hold is that *if* there is only one substance then it has *distinctive* mental and physical properties. The physical properties that inhere in a substance cannot produce the mental ones. God has to

---

[12] Locke, II, xxxiii, 5. Note that Locke's understanding of 'physical' is unlike that of Strawson. As will become clearer below, it corresponds more closely to Strawson's 'non-experiential'.

[13] See Bermúdez (1996). He argues, contra Ayers (1991), p. 44, that Locke was not clearly a substance dualist abut the (human) mind-body relation. (He notes that it is clear that Locke thought that God was an immaterial substance.)

[14] Locke, V, iii, 6. Other passages in this section indicate the same.

[15] He does, however, think that the substance dualist account is slightly more plausible. See IV, iii, 6.

add mental properties to a substance that has physical properties in order for it to have mental attributes. This view, in standard terminology, is one of substance monism together with property dualism. The mental cannot be explained in terms of the physical because the two are of fundamentally different kinds.

In summary, Locke is rather circumspect about substance. Either he thinks that it does not exist or he thinks that our idea of it is confused or at the very least exceedingly limited. To the extent that he endorses the notion, Locke is agnostic between substance dualism and substance monism about the mind-body relation. According to Locke, if substance dualism is true then there is a distinctive mental substance that has only mental properties and a distinctive physical substance that has only physical properties. If substance monism is true then, according to Locke, any mental attributes that such a substance has will be had in virtue of its having distinctive mental properties — distinct from any physical properties that it may have. This is because he thinks no combination of physical properties can produce mentality.

How does this relate to Strawson's views? When we look to Strawson (2003), which we are instructed to do on page 28 of the target paper, we find that Strawson, even more than Locke, is explicitly sceptical about the notion of substance:

> 'Bare particulars' — objects thought of as things that do of course *have* properties but are *in themselves* entirely independent of properties — are incoherent. To be is necessarily to be somehow or other . . . The claim is not that there can be concrete instantiations of properties without concrete objects. It is that objects (just) are collections of concretely instantiated properties . . . When Kant says that 'in their relation to substance, accidents [or properties] are not really subordinated to it, but are the mode of existing of the substance itself' I think that he gets the matter exactly right . . . the distinction between the actual being of a thing or object or particular, considered at any given time, and its actual properties, at that time, is merely a conceptual distinction (like the distinction between triangularity and trilaterality) rather than a real (ontological) distinction.[16]

This shows us that, for Strawson, there are, ontologically, only collections of properties. There is no independently existing substance. Strawson is either sceptical or deflationist about substance. On the one hand, taking a sceptical reading of the above, one might be tempted to say that he is neither a dualist nor a monist about substance as he thinks there really is no substance. On the other hand, a

---

[16] Strawson (2003), pp. 299–300. All parentheses are Strawson's own. The quotation within the paragraph is from Kant (1781–7/1889) A414/B441.

deflationary reading seems more appropriate as he does not seem to deny that there is substance (conceived of only as the right kind of collections of properties) and in the target paper, Strawson (2006), insists on the claim that there is only one type of 'stuff'. This latter claim about the number of types of 'stuff' appears to show that Strawson is professing to be a substance monist. This claim to monism would be backed up if Strawson was a monist about properties. If all the properties are of the one type then collections of such properties will be of the one type also. But what if Strawson turns out to be a dualist about properties, as I will argue is the case below? In that case, so long as there is no bar to any object or substance coming to have any type of property then all objects or substances are of the same type — the type in which any sort of property can inhere. It would only be reasonable to attribute substance dualism to Strawson on this reading if he also held that properties of different types could not form collections of a sort that constitute substances, but there is every reason to think that this is precisely not what Strawson's view is, as his claim to panpsychism attests.[17] Thus, regardless of whether Strawson turns out to be a property monist or a property dualist, the best interpretation of his position is that he is a substance monist, but that claim should be taken as follows: substances, or objects, simply are the right kind of collections of properties and such collections can consist of both experiential and non-experiential properties.[18]

To sum up this section, either Strawson should be thought of as thinking that there is no substance or, more plausibly I think, he should be seen as deflationist about substance. If the latter is accepted then Strawson's position is clearly a substance monist view. As I have said before, I will address his arguments for this view and the plausibility of this position in Section 3. Before doing so, I turn now to the question of whether Strawson's position is a property monist position.

2.3 *Property Dualism*

There are two sorts of property dualist. One sort is compatible with substance dualism. On this view, the two different types of properties cannot exist in the same type of substance, or cannot exist in collections of the right kind that constitute a substance. (Exactly how one will put

---

[17] Further elucidation of this point occurs at the beginning of the next section.

[18] Strawson goes on to argue, in the second half of his paper, for panpsychism — which would suggest that he thinks that the right kind of collections will always involve experiential properties. Note that it is unclear whether Strawson would countenance objects that had only experiential properties. While I think he might not, I can't see anything in Strawson (2006) that rules it out.

this claim will depend on whether one holds a deflationary view of substance or not.) The other sort is compatible with substance monism. On this view, the two different sorts of properties can exist in the same type of substance or can exist in collections of the right kind that constitute a substance. If, despite the fact he does not claim to be one, Strawson was to turn out to be a property dualist, he would clearly be a property dualist of the latter kind.[19,20]

Recall that I concluded that Strawson is a substance monist. If someone is a substance monist how do you decide whether they are a property monist or dualist? What exactly is a property dualist? Rosenthal says:

> human beings are physical substances but have mental properties, and those properties are not physical. This view is known as property dualism, or the dual-aspect theory.[21]

We know that Strawson would not agree to the idea that humans are physical substances, when 'physical' is used in the mainstream way to exclude the mental. However, I think that it is in the spirit of Rosenthal's definition that a view would still be clearly property dualist if it claimed that human beings are composed of one type of substance but have conscious mental properties that are not mainstream physical properties (as well as mainstream physical properties). Given this definition, Strawson is a property dualist. He holds that experiential properties (conscious, mental properties) are not mainstream physical properties. This simply is property dualism in mainstream terminology.

In addition, when we look to other slightly different definitions of property dualism, we see that one can take Strawson's arguments from the core of his paper, which argue for the conclusion that the non-experiential cannot in principle explain the experiential, as being vehement arguments for property dualism. To see this, recall that Locke held that mental properties could not be explained by physical

---

[19] Philosophical terminology here is slippery. Occasionally property dualism is taken to be only the latter view: substance monism and property dualism. However, I will not adhere to this usage.

[20] Strawson makes the remark that 'one needs to grasp fully the point that "property dualism", applied to intrinsic, non-relational properties, is strictly incoherent insofar as it purports to be genuinely distinct from substance dualism, because there is nothing more to a thing's being than its intrinsic, non-relational properties' (Strawson, 2006, p. 28). I have to admit that I don't see why this is true, given that there are the two versions of property dualism that I have just outlined.

[21] Rosenthal (1998). Note that this conception of property dualism is incompatible with substance dualism. This is, I think, a mistake, as I articulated in the paragraph above. Nonetheless, Rosenthal's definition is useful as it articulates the form that pertains to Strawson.

properties. If a physical object or substance had mental properties, then they had to be put there by God in addition to the physical ones. In standard terminology, this renders Locke a property dualist. Bermúdez, who reads Locke in this way, says:

> The crucial tenet of property dualism is that in principle we will not be able to explain mental properties in terms of physical properties, or vice versa. (Bermúdez, 1996, p. 233)

He is not alone in having this conception. For example, Calef (2005) says:

> Property dualists argue that mental states are irreducible attributes of brain states.

What applies to Locke here, applies to Strawson. Strawson spends the bulk of his paper defending the position that experiential properties cannot in principle be explained in terms of non-experiential properties. He thinks that the experiential cannot be reduced to the non-experiential and then argues at length that, in principle, the experiential cannot be wholly and fully explained by the non-experiential in some emergentist way.[22] Thus, by the lights of these central definitions of property dualism, Strawson is a property dualist, and the arguments that are at the heart of his paper are precisely arguments for that position.

Lastly, to back up this conclusion, note that Strawson's position also conforms to the following definition:

> Fundamental property dualism regards conscious mental properties as basic constituents of reality on a par with fundamental physical properties such as electromagnetic charge. (Van Gulick, 2004)

Strawson clearly holds this position, as can be seen from the following quotations (remembering what his non-standard usage of the term 'physical' is):

> Assuming, then, that there is a plurality of physical ultimates, some of them at least must be intrinsically experiential, intrinsically experience-involving. (Strawson, 2006, p. 24)

> Once upon a time there was relatively unorganized matter, with both experiential and non-experiential *fundamental* features. It organized into increasingly complex forms, both experiential and non-experiential. (Strawson, 2006, p. 27, the emphasis is mine)

---

[22] Rather than talk of what is explicable in principle, Strawson (2006, p. 15) talks of the notion of what is intelligible or explicable to God, which is equivalent to the former and which he holds, rightly, is not an epistemological notion.

Thus, by this third definition too, Strawson is a property dualist. (Moreover, we can see from the quote above that he is genuinely dualist, as opposed to a property monist of the kind who thinks that the one kind of fundamental properties are experiential.[23])

How does this charge of property dualism affect Strawson's claim that he is a panpsychist? Suppose one thought that panpsychism is the claim that all the fundamental constituents of reality are experiential. If, on the one hand, 'the fundamental constituents of reality' can refer to the fundamental properties then Strawson is not a panpsychist as, as I have been arguing, he thinks that there are both fundamental experiential and fundamental non-experiential properties. (Indeed, a person who thought that there were only fundamental experiential properties would be an idealist of some type.) If, on the other hand, 'the fundamental constituents of reality' only refers to fundamental objects, substances, or collections of properties of the right kind that comprise objects, then because Strawson holds that these things always involve at least one fundamental experiential property he can reasonably be classified as some type of panpsychist.

Before finishing this section, I wish briefly to comment on two ways Strawson might reply to this charge. First, Strawson explicitly states that he wishes to eschew, as far as possible, the subject/predicate form and the substance/property distinction. Thus he might claim that my insistence that he is a property dualist forces him to recognize the substance/property distinction that he denies. However, I spelled out in detail the position Strawson takes on this distinction. I have not ignored his position on this. He wishes to be deflationist about the notion of substance and claims that substances simply are the right kind of collections of properties. I have taken pains to show that, even understanding his metaphysics, the best and correct classification of Strawson's position is one of property dualism.

Second, because Strawson would state his own position as being one in which all the properties are Strawsonian physicalist properties, I think he might try to defend his position as being monist about properties. But recall also that Strawson's physicalism claim could be broken in two. There was the monist claim and the claim that the monism deserved the epithet 'physicalism'. However, recall that Strawson

---

[23] This claim is backed up further by the following quotation from Strawson (2006): 'you can't get experiential phenomena from P phenomena, i.e. shape-size-mass-charge-etc. phenomena, or, more carefully now — for we can no longer assume that P phenomena as defined really are wholly non-experiential phenomena — from *non-experiential* features of shape-size-mass-charge etc. phenomena' (p. 24). Strawson again suggests that there are both experiential and non-experiential features or properties. Thus his view is not that there are only experiential properties, which supports my charge of property dualism.

explicitly said that the monism claim was an assumption. I do not think that he is entitled to this assumption as regards properties on his view. I have been arguing that there are very good reasons to think that in fact the properties that Strawson posits are of fundamentally different types. This is because he clearly thinks that the experiential properties are not mainstream, non-experiential physical properties and are not reducible to the non-experiential ones, and that both experiential and non-experiential properties are fundamental features of the 'ultimates'. These are defining features of property dualism.

In conclusion, thus far I have attempted to spell out what Strawson's position is in standard terminology. I have outlined his usage of 'physical' and 'physicalism' and compared them to the standard. I noted Strawson's view of the nature of substance, which was seen to be rather similar to Locke's. Finally, I claimed that Strawson's position, like a view Locke finds plausible, is property dualism, combined with a deflationist monism about substance. My claim is not simply that Strawson's use of 'physicalism' is not the mainstream — a fact that Strawson acknowledges. It is that Strawsonian physicalism involves two claims: a monist claim and a claim that the monism has a right to the name 'physicalism'. I argue against the first by claiming that, concerning properties, there is good reason to think that he is in fact dualist. Because Strawson does not go into detail concerning the underlying metaphysics, the reasons to think he is really a property dualist are masked.

I noted along the way that Locke remained agnostic about whether substance dualism or substance monism was true, whereas Strawson simply assumed that substance monism was correct. In the next section of the paper, I wish to build on this last remark and comment on some of Strawson's other assumptions and his arguments and position.

3. Comments on Strawson's Arguments and Position

I wish to identify some assumptions that Strawson makes in his paper. As I have already claimed, I think that Strawson assumes that there is only one type of substance. Strawson might think that he is entitled to make this assumption because he is deflationist about substance. Recall he thought that substances were just to be identified with groups of properties conjoined in some manner. However, even if one holds this view of substance, there are still two available views on how many types of substances there are. One view would be that any mixture of experiential and non-experiential properties can form the kind of collection that is an object or substance. In this case, one could maintain that there is only one type of substance and that it can have either experiential or non-experiential properties, or both. The second view would be that there can be suitable collections of experiential properties and suitable collections of non-experiential properties, but that there cannot be suitable collections of mixed experiential and non-experiential properties. This view would amount to substance dualism. There would be some objects or substances composed solely from experiential properties and some from solely non-experiential properties, but none composed from both. Thus, even a deflationist can claim that there are distinctively experiential objects or substances and distinctively non-experiential ones. As we saw above, Locke, who tended towards similar views on the nature of substance to Strawson, was keen to leave open this possibility.

A second assumption arises after Strawson claims to have established that the experiential does not reduce to the non-experiential and that it cannot emerge from the non-experiential. Strawson goes on to try to establish 'micropsychism'. (In addition he goes on to try to establish 'panpsychism' but I will not deal with this further move here.) Micropsychism is the claim that 'some ultimates are intrinsically experience-involving'; panpsychism is the claim that they all are (Strawson, 2006, p. 25). Suppose that the former claims have been established and thus that we are agreed that there are distinctive mental and non-mental properties. How does the claim that micropsychism must be true arise? Strawson says:

> So if experience like ours (or mouse experience, or sea snail experience) emerges from something that is not experience like ours (or mouse experience, or sea snail experience), then that something must already be experiential in some sense or other. (Strawson, 2006, p. 24)

He concludes that some of the 'ultimates' must be experiential. But why should we suppose that our experience emerges from anything? Why not suppose that the property of having an experience, or the property of having an experience of a particular kind, is a fundamental property that can, together with other (experiential and non-experiential) properties combine to form an object or substance: a human being or a human mind or a subject of experience more generally? Why must the experiential property I have when I see something red emerge from other more fundamental experiential properties? After all, according to Strawson, the fundamental experiential properties from which my experiential properties arise are experiential properties of 'ultimates' — the things of which I am, and everything is, composed. As Strawson himself notes, the idea that there are such properties is exceedingly problematic: this must mean that the 'ultimates' themselves have experiences — are subjects of experience. Moreover, we have no good idea about how the experiential properties of ultimates would combine to produce the kind of experiential properties that we are ourselves familiar with. Doesn't that view posit such extremely problematic notions that we should give it up?

Strawson's answer comes only in the following passage:

> Given that everything concrete is physical, *and that everything physical is constituted out of physical ultimates*, and that experience is part of concrete reality, it seems the only reasonable position (Strawson, 2006, p. 25, the emphasis is mine)

There are several ways in which one could take the italicized passage. Strawson must at least intend that, together with the other claims in the sentence, it makes likely the truth of micropsychism. But taken in that way, it appears to involve much the same assumption that was present in the previous quotation, namely that the property of having an experience of such and such a type (a property that humans have) cannot be a fundamental property. It must be reducible to, or emerge (in a completely explicable way), from more basic mental properties. But it is not clear why we should accept such an assumption. Why not be a property dualist and think that one of the fundamental properties is the property of having experience of the kind with which we are familiar? One could hold that that property is not reducible to, or does not emerge from, other properties — experiential or non-experiential. One could hold that that property can attach to bundles of other properties to create creatures with experience.

Strawson might object here that such an experiential property would look as if it appeared by magic. It would look as if such a property appeared when you got the right sort of non-experiential complexity. It would look as if it emerged in a problematic sense from something non-experiential. However, I would make two points in reply. First, if the option under consideration were true then it might look to us as if such a property must be related to non-experiential properties and that it emerged from non-experiential properties. But, however it would look to us, or whatever we would be tempted to conclude if the world were that way, that would not be the truth of the matter. That is not the position being outlined. Rather, the position being outlined is that the non-experiential is not responsible for the experiential and is not something from which the experiential emerges.

My second point is the main point that I wish to make in this section. While I am sympathetic to Strawson's claims that no brand of mainstream physicalism explains the mental, it seems to me that all versions of solutions to the mind-body problem posit rather large mysteries. Strawson does not persuade me that his version posits less of a mystery than other versions, including the alternative form of property dualism that I have just outlined.

Strawson claims that mainstream physicalism posits something mysterious: the experiential arising from the non-experiential. I think Strawson is right that no brand of mainstream physicalism has explained away this mystery with success. (Note, however, that there are many different ways in which physicalists try to do so. In addition to there being many types of emergentism that Strawson does not consider there are more straightforward types of physicalist reductionism that try to explain away the appearance of an explanatory gap by claiming that it is merely a conceptual gap. While I am not persuaded by such arguments, it would have been good if Strawson had addressed them.) Besides physicalism, there are, roughly, three brands of dualism: (i) substance and property dualism; (ii) substance monism and property dualism of a non-micropsychic kind; and (iii) substance monism and property dualism of a micropsychic kind. Each of these positions posits something unexplained. Position (i) brutely posits mental substance and properties. In its favour, it does not have to explain how the one substance can have properties of both kinds. As I have argued in Section 1 above, Strawson does nothing to show that his position (iii) is less mysterious or any more motivated than (i). (Recall too that Locke held that substance dualism was no more problematic than mere property dualism.) Position (ii) brutely posits fundamental experiential properties of the kind that humans have (macro experiential properties). It has an advantage over (i) in that it also does not posit mental substance. It has an advantage over (iii) in not positing further (micro) experiential properties, which are the properties of 'ultimates' and are not those experiential properties that we are familiar with. It has the disadvantage of not explaining why the macro experiential properties only seem to attach to non-experiential matter arranged in certain ways. As I have tried to show in this section, I believe that (ii) is no more mysterious and no less motivated than (iii). Position (iii) brutely posits micro experiential properties that are the properties of ultimates. These properties are not like the macro experiential properties that we know of but, in some unknown way, the macro experiential properties we do know emerge from them.

In short, Strawson's position has the advantage that it:

recommends a general framework of thought in which there need be no more sense of a radically unintelligible transition in the case of experientiality that there is in the case of liquidity. (Strawson, 2006, p. 28)

Yet, it has a great number of disadvantages. Other mysteries or problematic features of the account that are just as great, if not greater, replace the mystery that is solved:

(1) There are a large number of subjects of experience and these include the fundamental particles.
(2) If fundamental particles are the subjects of experience then is anything composed from them such a subject? If not, what is the principle that makes creatures like us such subjects (at least during our wakeful and dreaming periods), and other conglomerations not?
(3) The position says little about how macro experience (and different types of such experience) arises from micro experience, other than that it emerges (as does liquidity) in a wholly dependent way.
(4) Do the fundamental micro experiential properties have independent causal powers? Do they abide by laws? If they interact can they do so with non-experiential properties too?

Strawson explicitly mentions some of these problems for his account, yet I think he fails to make a fair assessment of his position relative to the others that I have outlined here. In my view, his position is worse off than the other forms of dualism I outlined and it is nearly every bit as problematic as physicalism.

4. Conclusion

Strawson articulates a view on the mind-body problem that goes against the mainstream physicalist view held by many today. I think that he is right to bring it to our attention. If, as I believe, mainstream physicalism is problematic, then more attention ought to be given to spelling out alternatives in order to assess them for plausibility.

In the first part of this paper I argued that, despite Strawson's claims to the contrary, his position is in fact one version of substance monism together with property dualism. In the second part of this paper, I argued that by looking at the assumptions that Strawson makes when arguing for his position and by looking at the problems of such a position we can see that his variety of dualism is no better off than other varieties, and seems indeed to have more explanatory work to do. I also claimed that Strawson is right to point out that mainstream physicalism is not, at least as we understand it at present, a good solution to the mind-body problem. There is an explanatory gap that has not been crossed. However, Strawson's position brutely posits experiential properties that are unlike ours, but from which ours emerge. This position also has an explanatory gap that, while not as wide as that which the physicalists posit, nonetheless has, at one side of the gulf, brute unfamiliar experiential properties.[24]

References

Aaron, R.I. (1955), *John Locke* (Oxford: Clarendon Press, second edition).
Ayers, M. (1991), *Locke*, Vol. 2 (London: Routledge and Kegan Paul).
Bermúdez, J.L. (1996), 'Locke, property dualism and metaphysical dualism', *British Journal of the History of Philosophy*, **4**, pp. 223–45.
Calef, S. (2005), 'Dualism and mind', in *The Internet Encyclopedia of Philosophy*, ed. J. Fieser and B. Dowden, retrieved 26 January 2006 from http://www.iep.utm.edu/d/dualism.htm
Crane, T. and Mellor, D.H. (1990), 'There is no question of physicalism', *Mind*, **99**, pp. 185–206.
Kant, I. (1781–7/1889), *Critique of Pure Reason*, ed. E. Adickes (Berlin: Mayer and Müller).
Locke (1690), *An Essay Concerning Human Understanding*, retrieved 26 January 2006 from http://www.class.uidaho.edu/mickelsen/locke310.htm
Papineau, D. (2000), 'The rise of physicalism', in *The Proper Ambition of Science*, ed. J. Wolff and M.W.F. Stone (London: Routledge).
Robinson, H. (2004), 'Substance', *The Stanford Encyclopedia of Philosophy* (Winter 2004), ed. E.N. Zalta, retrieved 26 January 2006, from http://plato.stanford.edu/archives/win2004/ entries/substance
Rosenthal, D.M. (1998), 'Dualism', in *Routledge Encyclopedia of Philosophy*, ed. E. Craig (London: Routledge), retrieved 26 January 2006 from http://www.rep.routledge.com/article/V011
Strawson, G. (2003), 'What is the relation between an experience, the subject of experience, and the content of experience?', *Philosophical Issues*, **13**, pp. 279–315.
Strawson, G. (2006), 'Realistic monism: Why physicalism entails panpsychism', *Journal of Consciousness Studies*, **13** (10–11), pp. 3–31.
Van Gulick, R. (2004), 'Consciousness', *The Stanford Encyclopedia of Philosophy*, ed. E.N. Zalta, retrieved 26 January 2006 from http://plato.stanford.edu/archives/fall2004/entries/consciousness
Woolhouse, R.S. (1980), *Locke* (Brighton: Harvester Press).

[24] Thanks to Daniel Stoljar and David Chalmers for helpful discussion and comments on this paper.

Colin McGinn

Hard Questions

Comments on Galen Strawson

I find myself in agreement with almost all of Galen's paper (Strawson, 2006) — except, that is, for his three main claims. These I take to be: that he has provided a substantive and useful definition of 'physicalism'; that physicalism entails panpsychism; and that panpsychism is a necessary and viable doctrine. But I find much to applaud in the incidentals Galen brings in to defend these three claims, particularly his eloquent and uncompromising rejection of the idea of brute emergence, as well as his dissatisfaction with standard forms of physicalism. I certainly find his paper far more on target than most of the stuff I read on this topic.

Physicalism

Galen treats standard physicalists (his 'physicSalists') as blinkered victims of scientism, compared to his more 'open-minded' alternative; but actually there is method to their madness and it is worth recalling why. They are convinced of two (connected) things: that experiential facts supervene on non-experiential facts (henceforth E and non-E facts), and that the causal powers of E facts, in respect of non-E facts at least, are specifiable in entirely non-E terms. They think that the best explanation of these two things is that E facts *reduce* to non-E facts — that the 'real essence' of an experience is given by the kinds of descriptions used to specify the supervenience base and causal powers of experiences, i.e. 'physical' descriptions. (Their problem is that they seem to miss out the very essence of what an experience is.) Galen wants to inherit their rhetoric but avoid their omissions; he wants a physicalism that is 'realistic'. But what is his attitude towards the twin motivations for the old-style physicalism? Does he want to retain these motivations? If so, how does his version preserve them? Is it even compatible with them? The problem is to see how they can hold and yet E facts be totally different in kind from non-E facts. How does his nonreductive brand of physicalism square with the intimate connections that appear to obtain between the E and non-E realms?

We can see what motivated the old-style physicalists, and also appreciate that their doctrine is far from trivial; but what motivates Galen's revamped version and is it substantive? His view is that experience 'just is' physical, quite independently of its connections with non-E facts; it is *intrinsically* physical, even before we ask how it might be related to non-E facts about the brain (say, supervenient on them). But what *is* this notion of the 'physical'? The danger is that it is just an honorific synonym for 'concrete': everything concrete is physical, we are told, but what is it to be physical, other than to be concrete? Galen doesn't say spatial, or subject to the laws of physics, or even causally connected to non-E facts. All he says is that he is treating 'physical' like a natural kind term, whose extension is fixed by ostensive samples — tables, gravity, molecules and experiences. Being physical, presumably, is what all these have in common. But *what* do they have in common — beyond being 'concrete' (as opposed to abstract)? A radical dualist will say that Galen's samples of the 'physical' are wildly heterogeneous, that he is just lumping metaphysically unlike things together. We need to be told in what *respect* experiences are like molecules before we can assess whether the class constitutes a genuine natural kind. Of course, if Galen is determined to classify these things together under the label 'physical', we can't fault his stipulation (though we may find it eccentric); but we can protest that he has not advanced the subject of ontology. He simply *wants to call* experiences physical — just as I may want to call ocean waves spiritual. The complaint in both cases is that very different things have been brought together under the same label, in flagrant violation of common usage.

The point can be put another way. Consider a world of pure experience, with no non-E facts in it at all, or a world with both sorts of fact completely cut off from each other, causally and metaphysically. So far as I can see, Galen will be just as insistent that these worlds are purely physical worlds, since experiences 'just are' physical, independently of any connection to non-E facts. That is, disembodied experiences are in their intrinsic nature 'physical', even though they bear no relation to physical facts of the usual kind. What is gained by this piece of linguistic legislation? Surely an opponent will simply observe that, if so, there are two very different kinds of 'physical' fact in the world, and want to know how they are related (is one reducible to the other, say?). This is just playing with words. I might as well announce that I have become an idealist because I use the word 'mental' to cover a class of things including tables, molecules and experiences! (Is Galen perhaps a physicalist *and* an idealist?)

Galen should drop the word 'physical', accept that it has no useful definition and move on. Nor should he cleave to the more neutral word 'monism': short of a *theory* of the nature of experience that unites it with non-E facts (like old-style physicalism), it is vacuous. Even a full-blooded Cartesian dualist is a monist in the sense that he believes that mind and matter are both concrete (non-abstract) substances, though of course insisting that they differ in other ways. E facts and non-E facts have different *essences*, as Galen seems to accept; so why not respect this fact by declining to classify them together, except in trivial ways (as by saying they are both 'concrete')? By his methods we could extend the reach of physicalism still further, by declaring that 'physical' is a natural kind term for such things as bodies, minds and numbers!

Entailment

Galen's subtitle is 'why physicalism entails panpsychism'. Of course, old-style physicalism has no such entailment, unless we simply stipulate that whatever in the brain gives rise to experience is 'mental', even things like neural firings and chemical transfers (the converse of Galen's move with 'physical'). It is only physicalism, as Galen construes it, which is supposed to have this entailment, since E facts can only in the end emerge from E facts, not from non-E facts. But it is surely wrong to suggest that Galen's physicalism by itself leads to panpsychism; it is only in conjunction with an emergence thesis that we can get this entailment. For consider the worlds I mentioned in the previous section — the disembodied and dualistic worlds. In these worlds experience is still 'physical' by Galen's standards, but there is no emergence of the experiential in these worlds, and hence no demonstration of panpsychism. It isn't his *physicalism* that entails panpsychism; it's his commitment to emergence. You could be a non-physicalist about experience and still be a panpsychist (indeed that is the usual combination of views), so long as you take experience to emerge from constituents of the brain; and you could agree with Galen's physicalism and deny that there is any such emergence, and hence not be committed to panpsychism. You might hold that experience is a fundamental feature of the universe, not emerging from anything else, as basic as space and time, that just becomes *attached* to brains when they reach the right level of complexity, and at the same time insist, like Galen, that experiences 'just are' physical (you might even have a substantive notion of the physical to back up this claim, as that experiences are 'really' spatial entities).

Of course, Galen might reply that the *best* version of physicalism, with emergence as part of the package, entails panpsychism, but the fact remains that the physicalism by itself has no such entailment; so Galen's title, striking as it is, is misleading. Even *his* very idiosyncratic type of physicalism has no such entailment, let alone more orthodox types. What instead is true is that if you agree that experiences are irreducible to non-E facts, and if you accept emergence, then you end up with panpsychism — which is not a very surprising outcome given the assumptions. In other words: emergence from the ultimates, plus irreducibility, implies panpsychism — big deal! Whether experiences themselves are declared physical plays no role whatever in that argument.

Panpsychism

Panpsychism is surely one of the loveliest and most tempting views of reality ever devised; and it is not without its respectable motivations either. There are good arguments for it, and it would be wonderful if it were true — theoretically, aesthetically, humanly. Any reflective person must feel the pull of panpsychism once in a while. It's almost as good as pantheism! The trouble is that it's a complete myth, a comforting piece of utter balderdash. Sorry Galen, I'm just not down with it (and isn't there something vaguely hippyish, i.e. stoned, about the doctrine?).

Panpsychism has two massive things to be said in its favour. The first is that it would apparently solve the problem of emergence, which really is the heart of the mind-body problem, since it finds in the originating basis of consciousness precisely consciousness itself — we get like from like, not from unlike. The second is that physics does indeed leave the intrinsic nature of matter unspecified, as Eddington and Russell urged, so that panpsychism can step in to plug the ontological gap — the intrinsic nature of the ultimates is mental! I think the reason for this gap is less that physics is mathematical and structural as that it is *operationalist*: it tells us what the ultimates *do*, particularly their dynamic properties, but it doesn't tell us what they *are*. Physics is a kind of functionalist theory of material reality. But I won't dwell on this; the key point is that panpsychism offers to fill an epistemological and ontological gap — and with something so dear to us, mentality itself. All we have to do is accept that the ultimate entities of physics have experiences (*gulp*) and everything will turn out fine.

Here is a list of problems, some of which I brought up as long ago as *The Character of Mind* (1982). Do the E properties of elementary particles (or molecules or cells) contribute to their causal powers? If so, how come physics (and chemistry and biology) never have to take account of their contribution? The trajectories and interactions of these entities ought to be affected by their experiential properties, if these properties have causal powers (of roughly the kind they have when instantiated in human minds), and yet the relevant sciences can get on without recording them. They make no difference to how the things having them behave, considered in the relevant sciences. But if they are agreed not to have any causal powers — and so are entirely epiphenomenal — how can they blossom into properties that *do* have such powers once they take up residence inside brains? Unless, that is, we conclude that regular mental properties must be epiphenomenal *since* their panpsychist basis evidently is — contrary to the appearances. For example, suppose we suggest that a particular particle has an experience of red (presumably a hallucination not a veridical perception — unless we suppose that such particles can literally *see* things). Does it act as if it has this experience or not? Apparently not, so the experience lacks causal powers; but then how does it manage to constitute an experience with causal powers once it is inserted into the right part of a person's brain?

Are the E properties of particles supervenient on their non-E properties or not. If not, then there can be particles that are exactly alike in all non-E respects yet differing, perhaps dramatically, in their E properties — as it might be, identically vibrating strings in a 9D space that are hallucinating totally different colours. But, again, how come these differences don't show up at the microscopic level? If, on the other hand, there is supervenience here, then we have a total dependence of the E properties on the non-E properties; but this cannot presumably be just a miracle (for the kinds of reasons Galen gives), so there has to be some sort of account of it that makes it intelligible — but then we have our original problem all over again. *Why* is there this kind of dependence between the experiential and the non-experiential? And isn't this exactly the sort of picture that nonreductive supervenience theorists (like Davidson) envisage for the human mind; if so, why not accept it for those higher minds and cut out the panpsychism? If there is supervenience of the E properties on the non-E properties of particles, it will be hard to avoid accepting that there is *emergence* there — that combining the non-E properties in that way gives rise to the E properties. But that is the problem we started out with.

What kinds of E properties do particles have? Galen at one point accepts that they could have ordinary vivid experiences of red, as well as alien subjectivities — certainly not dimmer or fainter or blurrier than our usual experiences. But to provide an adequate basis for the full panoply of human phenomenology they are going to have to be rich and wide-ranging: not just sensory states but also emotional states, conative states and cognitive states — willing and thinking and feeling as well as sensing. How will this work? Will each type of particle possess a wide range of experiences, including emotions and thoughts, or will particles specialize in certain types of experience — electrons doing sensory, protons handling the emotional, neutrinos taking care of the cognitive? Either position seems totally arbitrary and empirically unconstrained. This is a game without rules and without consequences. Is it really to be supposed that a particle can enjoy these kinds of experiences — say, feeling depressed at its monotonous life of orbiting a nucleus but occasionally cheered up by its experience of musical notes?

Here the persistent panpsychist might retreat to *watered-down* phenomenology, perhaps imagining faint and blurry qualia, along the lines possibly of those in the nascent mind of a foetus. Galen does not take this weaselly line, to his credit; it obviously makes no difference to the general issue and merely registers the natural unease that the honest-to-goodness panpsychist provokes. Even the faint and blurry is phenomenology too much for the humble electron. The problem is that we can solve the emergence problem only if we credit the ultimates with a rich enough phenomenology to form an adequate basis for a full-bodied human mind, or else we have to suppose input from outside to pump up the volume (and hence relinquish emergence). So we simply can't scale it back when we come to the basic elements, on pain of resurrecting the old emergence problem. There is really no alternative but to accept that particles have minds in much the same way we (and other animals) do. And please don't say that the particles are only required to be *potentially* experience-endowed for panpsychism to be true, since this is common ground for *any* view of the relation between experience and the wider world — *of course* matter must have the potential to generate mind, since it patently does (unless we are radical dualists). The whole question is, in virtue of what sort of property — and the honest panpsychist at least has a nontrivial answer, viz. experiential properties. The potentiality move simply says that particles produce minds when combined into brains, and hence have that potential; but that is not a *theory* at all, just the datum we are trying to explain.

Panpsychism raises what might be called the *derivation problem*: how are higher-level experiences derived from lower-level ones? Here I think Galen is too sanguine, inviting us to consider how much variety in the spatial world can be derived from exiguous materials at the elementary level. The reason for this fecundity is that there are so many possibilities of combination of simpler elements, so we can get a lot of different things by spatially arranging a smallish number of physical primitives. But there is no analogous notion of combination for qualia — there is no analogue for spatial arrangement (you can't put qualia end-to-end). We cannot therefore envisage a small number of experiential primitives yielding a rich variety of phenomenologies; we have to postulate richness all the way down, more or less. An easy way to see this is to note that you can't derive one sort of experience from another: you can't get pains from experiences of colour, or emotions from thoughts, or thoughts from acts of will. There are a large number of phenomenal primitives. Accordingly, we cannot formulate panpsychism in terms of a small number of phenomenal primitives — say, one for each type of elementary particle — and hope to derive the rest. We have to postulate richness at the basis. It would be impossible, say, to begin with simply an array of faint experiences of shades of grey and then hope to derive all of human phenomenology! For the same reason, we cannot suppose that the particles have an alien phenomenology, perhaps more suitable to their limited and peculiar 'form of life' (rattling around a nucleus, subject to powerful electromagnetic forces, in imminent danger of annihilation), because there is no coherent way to derive from such an alien form of experience the kinds of familiar experiences that we enjoy. To suppose otherwise is to fall victim to the kind of magical thinking that the brute emergentist indulges in; there can be no miraculous transformation of one type of experience into some other quite distinct type — as it might be, yellow experiences into the sound of a trumpet (and if anyone mentions synaesthesia at this point I will scream).

Then there is the question of the need for a brain. We normally suppose that one of these is pretty useful when it comes to having a mind, indeed a *sine qua non* (even if it's made of silicon); we suppose that, at a minimum, a physical object has to exhibit the right degree of complexity before it can make a mind. But the panpsychist is having none of it: you get to have a mind well before even organic cells come on the market, before molecules indeed. Actually, you get mentality — experience — at the point of the Big Bang, fifteen billion years before brains are minted. So brains are a kind of contingency, a kind of pointless luxury when it comes to possessing mental states. It becomes puzzling why we have them at all, and why they are so big and fragile; atoms don't need them, so why do we? And this puzzle only becomes more severe when we remind ourselves that the panpsychist has to believe in full-throttle pre-cerebral mentality — genuine experiences of red and pangs of hunger and spasms of lust. As Eddington puts it, the mental world that we are acquainted with in introspection is a window onto the world of the physical universe, and the two are qualitatively alike: introspection tells us what matter is like from the inside, whether it is in our brain or not. But then the brain isn't necessary for the kind of experiential property it reveals to us; it is only necessary for the revealing to occur. *What* is revealed by introspection is spread over the entire physical universe. In fact, it would not be stretching a point to say that *all* bits of matter — from strings, to quarks, to atoms, to molecules, to cells, to organs, to animals — are themselves brains. There can be brains without brains! But if so, why bother with brains?

One last point: Galen says he has got used to crediting particles with experiences, so impressed is he with the problem-resolving power of this move; but why stop there — why not credit *space* with experiences? That is, if experience is everywhere that matter is, why not say that it is also everywhere that space is — empty space included? That region of space between the earth and moon, for example — it pullulates with experience. Since nothing is *required* of bits of matter for them to have experiences (no neurons or functional complexity), why not extend to space the courtesy of recognizing *its* mentality? After all, most of the brain — like all lumps of matter — is mainly empty space, and maybe this space itself contributes to the mind (the right density of matter is needed if human mentality is to take off). We know from physics that matter and space are deeply interwoven, so it is unlikely that such a fundamental property of matter as mindedness would not spill over to space, which is the medium in which matter has its being. So I fearlessly propose *extended panpsychism*: experience exists at every point in the spatial universe, whether occupied by matter or not. You may think me extravagant, but you must surely concede the explanatory power of my hypothesis, and it has a wonderful simplicity and symmetry to recommend it. So I invite my more conservative panpsychist colleagues to join me in extending panpsychism to the limit; and if they will not, I would like to hear their objections — my rule being that they must not recapitulate the objections to conservative panpsychism that I have just been citing.

Emergence

Perceptive readers will gather that I don't hold with panpsychism as a solution to the problem of emergence. They will also surmise that I am not an old-style physicalist. They will then wonder what solution I have to offer to the emergence problem, given that I can't stomach brute emergence. My answer is that I don't have the solution; the problem is hard. It may look as if we have precious few options: if experience is irreducible to non-E facts, and yet it emerges from the physical components of the brain, how can panpsychism *not* be true? The only way to avoid it is to deny irreducibility or deny emergence, it seems.

The only way I can see out of this is to hold out the hope of a third level of description (whether or not it is humanly accessible). Brains have properties beyond those of experience and those of basic physics and biology. These properties may then mediate between the other two sorts of property, offering some sort of unification of all three levels. Functional properties have the right form to provide such mediation, but functionalism as a theory of the mental is wrong. The idea of structural phenomenology is also in the right logical ballpark, but it also is problematic. (I will not repeat here what I have said about this in other places.) My point is just that we should be open to the idea that the simple dichotomy that Galen is working with — of E properties and non-E properties (mass, charge, momentum, force, neuron) — might be too simple; in particular, the properties of basic physics are not exhaustive, even before we get to experiences themselves. As I remarked earlier, I agree with friends Arthur, Bertie and Galen that physics does not tell us the intrinsic nature of matter, only its operationally definable aspects; so there is room for the idea that there is deep ignorance at the heart of our knowledge of matter. I also think that we don't really know the inner nature of consciousness itself. In these areas of ignorance the solution to the problem of emergence probably lies. All I have argued here is that panpsychism is not the way to solve the problem.

If we want an analogy for the situation we face — and looking for one is a risky undertaking for a problem so singular — we could do worse than consider the problematic relationship between the quantum world and the so-called classical world. Somehow the classical world of macroscopic objects in space, subject to Newtonian or Einsteinian principles, emerges from the world of quantum fluctuations. Gravity, in particular, has no place in the quantum world, but it is the essence of the classical world. The two worlds seem incommensurate and yet we have every reason to suppose that the one constitutes the other; the classical world must somehow result from the principles that govern its constituents. This is the emergence problem in physics, and physicists are well aware of it; they want a unifying theory that shows how the entities and forces at the classical level emerge from those at the quantum level. The problem is they don't have such a theory. It would obviously be folly for them to simply *inject* the classical world into the quantum world ('panclassicism'); what they need is a framework for reconciling and uniting the two (which is what string theory is supposed to do). From the microscopic point of view, the world looks one way; from the macroscopic point of view it looks another way — yet the one emerges from the other. From the point of view of introspection, the world looks one way; from the point of view of brain science it looks another way — yet the one emerges from the other. In both cases we know that there *is* emergence, but in neither case do we understand how it is possible.

Reference

Strawson, G. (2006), 'Realistic monism: Why physicalism entails panpsychism', *Journal of Consciousness Studies*, **13** (10–11), pp. 3–31.

David Papineau

Comments on Galen Strawson
'Realistic Monism:
Why Physicalism Entails Panpsychism'

1. Straightforward Physicalism

Galen Strawson (2006) thinks it is 'obviously' false that 'the terms of physics can fully capture the nature or essence of experience' (p. 4). He also describes this view as 'crazy' (p. 7). I think that he has been carried away by first impressions. It is certainly true that 'physicSalism', as he dubs this view, is strongly counterintuitive. But at the same time there are compelling arguments in its favour. I think that these arguments are sound and that the contrary intuitions are misbegotten.

In the first two sections of my remarks I would like to spend a little time defending physicSalism, or 'straightforward' physicalism, as I shall call it ('S' for 'straightforward', if you like). I realize that the main topic of Strawson's paper is panpsychism rather than his rejection of straightforward physicalism. But the latter is relevant as his arguments for panpsychism depend on his rejection of straightforward physicalism, in ways I shall explain below.

Let me first explain how I understand straightforward physicalism. I don't suppose that current science has yet established, for any specific phenomenal kind M and any specific physical kind P, that M = P. Our understanding of the brain is as yet too fragmentary. Indeed it is possible that our access to such identities will require significant advances in brain science. Even so, I see no reason to think that such advances need to take us beyond physical theories of the same general sort as we already have. So I do suppose that every phenomenal kind M is identical to *some* P that is generally similar to the kinds currently recognized by the physical sciences.[1] Moreover, I suppose that when we have established such M = P identities, then we will therewith have 'fully captured the nature or essence of experience' in physical terms, in that the relevant physical term will refer to nothing other than the phenomenal kind M.

I shall not here rehearse the arguments in favour of this straightforward physicalism. These are well-known, and derive from familiar causal-explanatory considerations (cf. Papineau, 2002, ch. 1). Rather, the question we need to ask is what countervailing reasons might nevertheless block these arguments. Strawson is by no means alone among contemporary thinkers in judging that the familiar causal-explanatory arguments are somehow defeated by the special nature of consciousness, and that therefore we must find some alternative to straightforward physicalism (such as epiphenomenalism or Strawson's own 'real physicalism'). But what exactly is it about consciousness that is supposed to prevent the causal-explanatory arguments from establishing their straightforward conclusion?

One possible thought is that it is simply a conceptual contradiction to suppose that a conscious state might be one and the same as a physical state. Doesn't our concept of a physical state definitionally exclude any conscious element, and vice versa? Isn't there something contradictory in supposing that a being with only physical properties could have conscious experience?

But this doesn't strike me as contradictory at all. As far as our concepts go, it seems to me completely open what it would feel like to be a purely physical being with firing C-fibres, say. What would you expect that to feel like? Why suppose it must feel like nothing? That's one possibility, but conceptually it seems just as possible that it would feel like being light-headed, or indeed feel like being in pain. Our concepts are not going to decide this issue.

Perhaps the problem is not that our concepts of physical states conceptually exclude any conscious element, but that they *don't conceptually guarantee* any.[2] We certainly can't tell straight off that a physical being with C-fibres firing will be in pain. (As is now conventional, let me use C-fibres as a place-holder for whichever physical state is in fact correlated with pain.) At best, the equation of pains with C-fibre firings will be the result of empirical investigation.

---

[1] In these remarks I will use 'physical' broadly, and not just to refer to kinds studied in physics departments. So it should be understood as including physically realized role states along with strictly physical states.

[2] This lack of a conceptual guarantee is the 'standard argument' against physicSalism that Strawson cites in footnote 4, referring to Strawson (1994), pp. 62–5.

Still, why is that any reason to deny that pains are one and the same as C-fibre firings? Many true identities are only established a posteriori, such as that table salt is sodium chloride, or that light is electromagnetic radiation, or that Cary Grant is Archie Leach. But that is no reason to deny, say, that 'the term *sodium chloride* fully captures the nature and essence of table salt'. So why not similarly allow that the term *C-fibres firing* fully captures the nature and essence of pain?

Maybe the underlying thought is that identities and other necessities only appear a posteriori when we are thinking about them in a second-hand way that does not *acquaint* us directly with the entities involved. When we think about NaCl as *table salt*, we are thinking about it indirectly, as *the stuff, whatever it is, that is left when sea water evaporates, is good for flavouring food, etc.* That is why we can't tell straight off that it is NaCl. But our concept of *pain* surely doesn't similarly pick out its referent indirectly. So if pain = C-fibre firings were true, it should be a priori, given that we think about pain (and C-fibres) directly, and not at second-hand. But of course pain = C-fibre firings is not a priori. So it can't be true. Or so at least this story goes.

This line of thought has been gaining currency over the past decade (Chalmers, 1996; Jackson, 1998). But the notion of direct acquaintance on which it rests strikes me as highly suspicious. It assumes some mode of thought where objects become completely transparent to the mind and all their essential properties are thereby laid bare. It is hard not to see this as inspired by some misplaced visual model, in which we are able to peer in at some immaculately illuminated scene. I myself doubt that there is any such mode of thought. No doubt there are ways of thinking of things that make certain essential properties a priori knowable. But I take such a priori knowledge to derive from (possibly implicit) compositionality in the relevant modes of thinking, and so not to be associated with the most basic ways in which thought makes contact with reality.

When it comes to these basic points of contact, I find it hard to take seriously any alternative to the assumption that our atomic concepts are related to reality by facts external to our a priori grasp, such as causal or historical facts. From this externalist perspective, there is no reason to suppose that when any two atomic concepts latch onto the same entity this identity will be a priori transparent. For example, suppose that the facts which give my *Cicero* and *Tully* concepts their semantic value are external to the mind, as supposed by an orthodox Kripkean causal theory of reference. Clearly this leaves it open that I could possess both concepts and yet not know that they co-refer.

Reference here will be 'direct' in that it does not depend on any descriptive or other complexity in the relevant concepts, but this kind of directness is no guarantee at all that relations of identity and other essential properties of the referents will be transparent to me. So, in the case at hand, both *pain* and *C-fibres* might similarly refer directly to the same entity, yet this not be transparent to someone who possesses these concepts (cf. Levine, 2001, ch. 2; Papineau, 2007, sect. 4).

2. Contrary Intuitions

Some of you might be getting impatient. Do we really need an *argument* to show that pains are not the same as C-fibre firings? Isn't it *obvious* that conscious states are distinct from straightforward physical states?

I agree entirely that this *seems* obvious. There is no doubt that we are all subject to a strong 'intuition of distinctness'. How can feelings — the colours, the smells, the excitement — possibly be the same as grey mushy brain states?

However, it is philosophically very unwise to rest one's case on unanalysed intuitions. Intuitions can be quite mistaken, and they need to be examined very carefully before arguments can be grounded on them.

It seems obvious that the earth is stationary. It seems incontrovertible that time moves. And so on. If intuitions like these were allowed to stand, many important theories would be strangled at birth.

Of course argument has to stop somewhere. If every assumption always has to be scrutinized and assessed, then we will be off on an interminable regress. Some claims must be taken as given in any argumentative context. But this does not justify resting a case on an intuition when that intuition runs counter to an otherwise well-supported theory. Given such a theory-intuition clash, the intuition has the status of a disputed premise. So we need to subject it to further assessment, hoping either to show that it follows from further agreed assumptions, or alternatively to explain why it arises even though it is false. (Consider, for example, how Galileo showed why the earth appears to be stationary even though it is moving, and B-series theorists aim to show why we have the impression of a moving present even though there is no such thing.)

I take the intuition of mind-brain distinctness to raise just this kind of issue. On the one hand we have straightforward physicalism, which is strongly backed by causal-explanatory considerations. On the other hand we have the brute intuition that mind and brain must be distinct.

Given this, we need to know whether there is any independent support for this intuition or whether it stems from some fallacious source.

This is why it is necessary to consider whether there are any good *arguments* that might back up the intuition of mind-brain distinctness, as I did in the previous section. My verdict there was that this intuition of distinctness lacks independent argumentative support. Perhaps, then, we should take seriously the possibility that it is ungrounded.

I myself think this is the right diagnosis. In my view, the intuition of distinctness is a confusion engendered by the peculiar way in which we think about consciousness. Consider what happens when we think about some conscious state *as* a conscious state. For example, we think about what it is *like* to feel pain or to see something as red. Such thoughts are characteristically accompanied by some version of the conscious state being thought about. Thus for example, we might think about the experience of seeing something red while we are actually having that experience; alternatively, we might think about what it is like to see red while we are imaginatively recreating the experience. I shall call thoughts of these kinds 'phenomenal thoughts'.[3] Their characteristic feature is that the conscious *referent* itself is involved in the *vehicle* of thought. We can think of this as the 'use-mention feature' of phenomenal thoughts: phenomenal thoughts use the very states that they mention

Now, this use-mention feature carries much potential for confusion. In fact, I think that there are a number of *different* ways in which it can seduce thinkers into an intuition of dualism. Here I want to focus on a line of thought that seems to me to have some affinity with Strawson's 'real physicalism'.[4]

When we reflect introspectively on phenomenal thoughts, we become aware that the thought is accompanied by the very experience it refers to, or an imaginative recreation thereof. (The conscious feeling is *right there* in our thinking.) This can lend credibility to the idea that phenomenal thinking gives us a special acquaintance with reality — here at least the object of thought are 'given' to us. For, to repeat, in phenomenal thought the conscious referent seems to be present in the thinking itself, without any veil between subject and

---

[3] Sometimes this kind of thought is said to involve a distinctive species of 'phenomenal concept' (cf. Loar, 1990; Papineau, 2002, ch. 2). But this characterization is not essential to the point I wish to make. It is enough that, in a familiar range of cases, thoughts about conscious states are accompanied by versions of those same states, whether or not we view these thoughts as involving special phenomenal concepts.

[4] For other ways in which the use-mention feature of phenomenal thoughts might generate an intuition of distinctness, see Papineau (2002), ch. 6; and Melnyk (2003).

object. So we suppose that the nature of conscious states must be completely transparent to phenomenal thinking — nothing essential will be omitted. From which it follows that conscious states cannot be identical to physical states, at least not to straightforward physical states. For it is certainly not immediately apparent to phenomenal thought that pains are identical to C-fibres firing or any other straightforward physical state.

I think that this whole line of thought is misguided. There is no such thing as unmediated reference, even in the phenomenal realm. Phenomenal thoughts do not have any magical power to reach out and grasp their objects transparently. Phenomenal thoughts, just like any other intentional states, gain their referential powers from causal and historical relations, and those referential relations can leave many essential features of the referents opaque.

Still, even if the idea of epistemologically transparent phenomenal reference is misguided, we can still see how it might seduce us into the intuition that the mind must be distinct from the brain. For conscious introspection does indeed show us that conscious states are present in our phenomenal thinking, in a way that the referents of other thoughts are not; and it is a natural enough step to infer from this that phenomenal thinking, unlike other thoughts, somehow casts a pure light on conscious reality as it is in itself. This latter step may be fallacious, but this needn't stop it being plausible enough to account for the intuition that consciousness, as revealed to us in phenomenal thought, can't be identical with anything physical.

If this diagnosis is right, then I take it that the opposition to straightforward physicalism falls away. On the one hand we have the causal-explanatory arguments weighing strongly in favour of straightforward physicalism. On the other, we have the brute intuition that straightforward physicalism cannot be true. But if this intuition can be explained away, in the sense of showing how it would still arise even if straightforward physicalism were true, then the causal-explanatory arguments stand unopposed. Those who feel that they should still stand by their anti-physicalist intuition, even in the face of this explanation, would do well to consider the fate of other anti-theoretical intuitions, like the intuition that earth stands still.

3. Real Physicalism and Panpsychism

Those who believe that phenomenal thought offers unmediated acquaintance with conscious reality don't necessarily reject all varieties of physicalism. True, they will reject, along with Strawson, any physicalism that supposes that straightforward terms like *C-fibres* 'can fully capture the nature or essence of experience'. But that leaves room for a form of physicalism ('real physicalism') in which physical states are identified in some superior way which makes their identity with conscious states manifest.

It is not immediately obvious, however, exactly how the position which then results differs from straightforward physicalism. Even though straightforward physicalists do not share the motivations of real physicalists, they are likely to agree with them on the following two propositions.

(1) The terms of current physics, like *C-fibres*, or indeed *quark*, pick out their referents in ways that fail to make all their essential features transparent. That is why claims like *C-fibres* = *pain* cannot be known a priori. Moreover, any terminology developed by future physical theories is likely to be similar in this respect.

(2) An identification of some real physical state with a given conscious experience will only be transparent to the mind if we conceive of that state in phenomenal terms — that is, *as* that conscious experience.

In short, straightforward physicalists are likely to agree with real physicalists that scientific talk of relevant brain states picks out states which are in fact essentially conscious, but does not a priori display those states as conscious.

True, straightforward physicalists will attach far less significance to this epistemological opacity than real physicalists, for straightforward physicalists do not accept *any* transparent modes of thought, and in particular do not accept that phenomenal thought is superior in this respect. So from their perspective there is nothing epistemologically second-class about scientific terminology — for no other mode of thought makes all essential properties transparent either. (That is why I am happy to say that ordinary scientific terms 'can fully capture the nature or essence of experience'. I don't recognize any way in which the mind 'captures' something, apart from simply referring to it.)

Still, despite this difference of perspective, straightforward physicalists will agree with real physicalists that the metaphysics of mind involves states that are inherently conscious, but that are not displayed as conscious by straightforward physical terminology.

This raises the question of what makes real physicalism a distinctive metaphysical position. The answer, I take it, is to do with panpsychism. Real physicalism, but not straightforward physicalism, implies panpsychism.

Strawson holds that consciousness cannot be constituted out of materials that are not conscious, in the way that liquidity can be constituted out of materials that are not liquid. So he concludes that complex conscious states, like states of human awareness, must have simple components that are themselves experiential in nature.

From my point of view, this micropsychism (and the panpsychism that Strawson quickly infers from it) is unmotivated, for reasons I will spell out in a moment. But first I would like briefly to point to two internal difficulties that seem to face Strawson's view.

First, it is not clear that Strawson's micropsychism can play the role it is invoked to fill, of explaining how complex conscious states can be constituted out of simpler components. Strawson says:

> It is at this point, when we consider the difference between macroexperiential and microexperiential phenomena, that the notion of emergence begins to recover some respectability in its application to the case of experience. (Strawson, 2006, p. 27)

Somehow, the experientiality of the microcomponents is supposed to help explain the emergence of the macroexperiential states. But how exactly will this work? Explanations of normal physical macrophenomena like liquidity hinge on some understanding of the macrophenomenon's *causal role* — we can show how liquidity arises from certain microarrangements by showing how those arrangements ensure that liquidity's causal role is filled. But when we think phenomenally about macroexperiential events like seeing something green, or listening to a high-pitched noise, we do not conceive these macroexperiences in terms of any causal role. This makes it quite obscure exactly how their emergence is to be explained in terms of microcomponents whose simple experiential nature is quite different. Why should certain arrangements of these simple experiential components give rise to the experience of seeing something green? This would seem just as 'brute' as the straightforward physicalist explanations that Strawson is inclined to reject.

Second, the idea that complex conscious states are composed of simpler experiential components would seem to be in tension with the idea that phenomenal thinking gives us unmediated epistemological access to the nature of conscious experience. If the experience of seeing something green is a complex state, composed of experiential simples, then this complexity is presumably an essential property of the macroexperience. Yet phenomenal thinking about the experience of green does not reveal it to be some kind of structured complex, in the way it ought to if it is to lay bare all the essential features of experience.

I said that Strawson's panpsychism is unmotivated. Let me now explain why. Strawson needs panpsychism because he takes there to be an unbridgeable divide between the non-experiential and experiential realms. As he sees it, the whole cloth of experience cannot possibly be woven from threads which are themselves non-experiential. So Strawson is driven to posit that the simples from which complex mental states are built are themselves experiential in nature.

However, this whole line of thought is driven by his initial rejection of straightforward physicalism. To see this, consider how the issue of emergence will appear if we do not abandon straightforward physicalism. According to straightforward physicalism, the general phenomenal category of *being conscious*, like more specific phenomenal categories, will be identified with some broadly physical category (such as activity in a global workspace, or representation that controls action, or oscillations at 40 hertz in the sensory cortex . . .). Now, presumably there will be no difficulty in explaining how such a complex physical category can be constituted out of non-experiential physical simples. This would be just the kind of explanation that Strawson allows can account for the emergence of liquidity from non-liquidity. But this will now amount to an explanation of consciousness in terms of non-experiential physical simples. For if consciousness is known to be identical to some complex physical category, and we can explain this category in terms of non-experiential simples, then we will therewith have explained consciousness.

It might seem as if straightforward physicalism still owes some further explanation that can't be given in straightforwardly physical terms. *Why* is it like something to have a brain state that is active in a global workspace, say? Why are just those brain states conscious, and others not? Surely this is just the same mystery as motivates Strawson; and surely it cannot be answered simply by showing how global workspace activity can be realized by some complicated arrangement of non-experiential physical simples.

However, the demand for further explanation depends on the intuition that conscious mind and physical brain are distinct. If we did not have this intuition, we would not feel there was anything more to explain. Perhaps the point is easiest to see with some specific phenomenal category, like pain, say. Suppose, for the sake of the argument, that pain is identified with C-fibre firings. Do we still need to explain why C-fibre firings 'give rise to' pain? I say not. If C-fibre firings *are* pains, then there is no remaining question of *why* pains are found where C-fibres are firing. Given that's what C-fibre firings are, there's no possibility of their being otherwise. To ask for further explanation illegitimately presupposes that the pain is distinct from the brain state.

Sometimes people accept this point for pains and other specific phenomenal categories but resist it for the general category of consciousness as such. Why should it be just *that* range of physical states, and not some other, that 'gives rise to' feelings? But the straightforward physicalist should simply say the same thing again. If activity in a global workspace *is* consciousness, there's no issue of explaining *why* it is — it couldn't have been otherwise. The request for explanation arises only as long as we remain in the grip of the intuition that being conscious must be extra to any straightforward physical property.

So I see nothing wrong with explaining the 'emergence' of consciousness from non-experiential physical simples. If the property of being conscious is identical to some straightforward physical property, there can be no barrier to such an explanation. We are only driven towards panpsychism if we posit a radical divide between the experiential and non-experiential realms. Straightforward physicalism rejects any such divide. Those who listen to argument, and ignore brute intuition, can thus steer clear of panpsychism.

References

Chalmers, D. (1996), *The Conscious Mind* (Oxford: Oxford University Press).
Jackson, F. (1998), *From Metaphysics to Ethics* (Oxford: Oxford University Press).
Levine, J. (2001), *Purple Haze* (Oxford: Oxford University Press).
Loar, B. (1990), 'Phenomenal states', in *Philosophical Perspectives*, 4, ed. J. Tomberlin.
Melnyk, A. (2003), 'Papineau on the intuition of distinctness', *SWIF Forum on Thinking about Consciousness*, http://lgxserver.uniba.it/lei/mind/forums/ 004_0003.htm
Papineau, D. (2002), *Thinking about Consciousness* (Oxford: Oxford University Press).
Papineau, D. (2006), 'Phenomenal and perceptual concepts', in *Phenomenal Concepts*, ed. T. Alter and S. Walter (Oxford: Oxford University Press).
Strawson, G. (1994), *Mental Reality* (Cambridge, MA: MIT Press).
Strawson, G. (2006), 'Realistic monism: Why physicalism entails panpsychism', *Journal of Consciousness Studies*, **13** (10–11), pp. 3–31.

Georges Rey

Better to Study Human Than World Psychology

Commentary on Galen Strawson's 'Realistic Monism: Why Physicalism Entails Panpsychism'

Strawson argues that 'the only reasonable position' regarding the inexplicability of experience in non-experiential terms is that experience is a fundamental feature of the world: 'All physical stuff is energy, and all energy . . . is an experience-involving phenomenon' (Strawson, 2006, p. 25). Now, I take it we haven't the slightest reason *independent* of this argument for any such fantastic conclusion. No one has produced the slightest evidence that anything but certain animals (and maybe certain machines) have experiences. Indeed, on the face of it, the conclusion seems about as plausible as a claim that all energy involves lactation, or that quarks are a kind of mammal. Perhaps I don't *know* that these latter claims are false, but there's obviously not a shred of reason to take them seriously. Strawson's claims also seem to me indistinguishable from dualism, as I understand it, which, after all, just is the claim that the mental is not explicable by the non-mental — i.e. not explicable at all — a view that I join many in finding improbable, and in any case intellectually unsatisfying.

The considerations Strawson raises on behalf of his conclusion are, of course, familiar enough from philosophy of the last sixty years or so. Wittgenstein (1953/68) captured some of them when he described 'turn[ing] my attention in a particular way on to my own consciousness and, astonished, say[ing] to myself: THIS is supposed to be produced by a process in the brain!' (§412).[1] But, unlike many who are

---

[1] Although I'm leery of Wittgenstein's approach to the philosophy of mind generally, I do think his treatment of consciousness and qualia, here and in §§271–308, are a useful antidote to much of the recent fascination with those phenomena in the philosophical

similarly astonished, Wittgenstein goes on to notice how odd it is to turn one's attention in this way. There and elsewhere he advises us to think of the multiplicity of ways we ordinarily talk about mental states, and to try to resist the picture that 'consciousness' and 'experience' name some 'inner thing'. Although I don't share Wittgenstein's confidence that this advice will suffice, I do share his suspicions of philosophers' too facile reifications in this domain, and of their disregard of details these reifications can invite. They make me wary of Strawson's (2006, p. 5 fn. 6) reliance on Louis Armstrong's 'If you gotta ask, you ain't never gonna get to know' as a reply to someone's wondering what exactly is being discussed. Is it the 'qualia', the purported properties of certain sensory experience? Or is it some property that is shared by all and only the various states we sometimes call 'conscious': not only sensations, but thoughts, ideas, beliefs, desires, emotions, moods, memories, tunes going round in one's head, 'senses' one has about what one wants to say? But why think there is any one property that we're aware of in all these cases? Speaking for myself, I'm as unaware of any such property as Hume is of a soul accompanying all his impressions and ideas.

In addition to attending to the richness of ordinary thought and talk, it's also worth attending to the details of actual research. Too often, the research is treated as if it were no more than psycho-physical correlations, as in those passages of his (Strawson, 1994, pp. 62–5) to which Strawson (2006, p. 4 fn. 3) refers us, where he focuses on the difficulty of verifying 'correlation statements' relating experiential and non-experiential states, and of the uselessness of a 'psychoscope' (1994, pp. 64–5). 'So long as we still possessed only the resources of current science,' he complains, 'we would still have no account of how experience is possible' (p. 65). Well, that might well be true if the only resources were correlations. But science doesn't traffic in correlations alone: there are also its *systematic explanations*. I remember being mystified when I was young by the correlations between football games and the images on television. The mystery was not removed by merely establishing more such correlations. Rather, I had to learn a *theory* of light and electricity, and something about how it applied to the intricate details of a TV. Just so: if you want to begin to remove some of the mystery of mind, you should look at detailed theories of the mechanisms that underlie, say, visual perception, mental imagery, memory, introspection. It's certainly not unilluminating to learn how

---

literature, particularly if it is read against a background of a causal/computational theory of mind, rather than the quasi-behaviourism against which Wittgenstein himself cast it; see Rey (2004) for discussion.

various illusions are due to the visual system automatically imposing a 3D interpretation on 2D figures, that vibrating colours are due to opponent processing, or that vivid mental imaging may involve simply running through procedures compiled from memory (see, e.g., Pylyshyn, 2003). As Strawson presumably knows (see 2006, p. 5 fn. 6), various recent authors have proposed ways that qualitative experience might be captured by certain sorts of representations, perhaps ones that are accessible in certain ways, e.g. as second-order thoughts (as in Rosenthal, 1986/2005), or as the contents of specific buffers (as in Dennett, 1978). Although I also have defended one such approach (Rey, 1997, ch. 11), I am under no illusion that any of them — or any of the other research — has fully succeeded. But I also don't think they should be dismissed or completely ignored, as I was surprised to find them to be by Strawson. One wants to know *precisely* where they go wrong, what *exactly* they fail to explain.[2]

There are plenty of mysteries about the mind. But it is important to note that they concern not only its relation to *body*, but simply the mind *itself*. *Pace* Strawson and Eddington's confidence about how 'the mental activity of the part of the world constituting ourselves occasions no surprise; it is known to us by direct self-knowledge' (Strawson, 2006, p. 29), we don't have anything remotely like a serious understanding of our mental lives. Introspection is all well and good in providing us with some sort of data to go on, but it provides little insight into any *explanation* of it. Worse, as Nisbett and Wilson (1977) famously documented, where it purports to do so, it often reflects not genuine insight, but merely popular belief. Of course, we do have some understanding here and there, but a moment's reflection reveals that we have no adequate explanations of such basic activities as perception, thinking, reasoning, language, decision making, motor control, much less creativity, scientific insight, morally responsible action, or any of the various, quite special phenomena associated with conscious experience. Our position is rather like that of precocious children, who know how to play fancy games on their computers but haven't the foggiest idea of how they work, in particular, of the ways in which transducers, compilers, operating systems, 'virtual machines' and specific programs are responsible for

---

[2] Strawson is not alone in disregarding detail; as I complain in Rey (1993), McGinn (1991) exhibits a similar failing. A sceptical discussion that *does* go into admirable detail is Levine's (2001), which I think raises a specific problem about the richness of particularly colour experiences that is a problem for all current approaches, a problem that would persist even if problems of consciousness and subjectivity were otherwise solved; see my (forthcoming-b) for discussion.

the images produced on their monitors.[3] It is this massive ignorance of the mind that gives the lie to the idealism with which Strawson flirts (2006, p. 5 fn. 6; p. 29). We really haven't a serious idea about how 'ideas' explain much of what goes on *in* the mind, let alone how they begin to offer a rival explanation to standard physics of what goes on *outside* it![4]

Strawson would, I suspect, be unfazed by the need for details here, since he is convinced that *any* attempt to explain experience by the non-experiential is doomed to failure, evidently by merely the very *nature* of the experiential. Reduction in this domain is elimination (Strawson, 2006, p. 5 fn. 6; p. 7). Thus, he regards efforts by Dennett, Dretske, Tye, Lycan and myself to try to account for various features of experience as

> 'looking-glassing' the term 'consciousness' — i.e. 'us[ing] a term in such a way that whatever one means by it, it excludes what the term means' ... It seems that they still dream of giving a reductive analysis of the experiential in non-experiential terms. *This, however, amounts to denying the existence of experience*, because the nature of (real) experience can be no more specified in non-experiential terms than the nature of the (real) non-experiential can be specified in wholly experiential terms' — (Strawson, 2006, p. 5 fn. 6, italics mine).

Of course, most defenders of a reductive strategy might find this view more than a little question-begging — but, ironically enough, not myself. I suspect Strawson is on to something quite right and important here: that there is something intrinsic to our concepts of 'experience' that does oddly seem to render explanation hopeless, much as there's something about our concepts of the divine and 'supernatural' that make them similarly resist explanation (once a 'magic' trick is explained, it's no longer 'magic'). However, as these comparisons are intended to suggest, this is a reason to suspect there is simply something wrong with the concepts. Indeed, I and a few others are prepared to conclude for this reason that there really are no phenomena answering to our usual concepts of qualia, consciousness and experience.

---

[3] I develop this point in greater detail in Rey (2002) and (forthcoming-a). Note, again, the issues are not about the *physics* of these systems, but about their logical and sometimes intentional organization.

[4] Note that the passages from Eddington that Strawson quotes on behalf of idealism are not very clear about either physics or the mind, given that 'atoms' are regarded as 'schedules of pointer readings' (p. 10). What Eddington seems to be defending is not serious idealism, but some sort of operationalism about theoretic terms, which is, mercifully, a different (and, I would have thought, a long dead) dispute.

Strawson, of course, has no patience with such a view, and worries in our case about 'the capacity of human minds to be gripped by theory, by faith ... the deepest woo-woo of the human mind' (Strawson, 2006, p. 5). But I don't see why he, of all people, should be so abusive, given his own (1986) similar — and quite penetrating — criticism of our ordinary notion of free-will. Indeed, the general point seems so plausible: once one notices just how little we know of our minds (again, just *psychologically*), surely it is a serious possibility that we misconceive them, just as we have historically often misconceived phenomena in most any other domain of which we've been ignorant, i.e. virtually *every* domain, particularly ones, like the mind, in which there's been enormous social and religious investments. In any case, it's more than a little unfair to claim that eliminativism is simply some sort of article of 'faith'. Wittgenstein's quasi-eliminativism ('pain is not a something, but not a nothing either' [1953/68, §307]) was certainly not motivated by any sort of worship of physics, but rather by reflections on what he took to be the 'grammar' of ordinary talk. Quine (1960, ch. 2) was motivated by scepticism about intensional semantics. Richard Rorty (1979) was sceptical of the 'glassy essence' conception of the mind that he argued was a peculiar inheritance of Cartesian and perhaps certain religious traditions. Churchland (1981) argued that psychology is just a 'degenerate research program', and Feyerabend (1963/71) and Quine (1960, p. 264) were impressed with the fact that there is no causal break in the purely physical explanation of human bodily motion.

I, myself, am motivated not by a *general* eliminativism, but only by an eliminativism about phenomena for which there is no non-question-begging data, i.e. data whose description presupposes the truth of a claim being reasonably disputed. I argue (Rey, 1997, pp. 88–134, 307–10; forthcoming-b) that there *is* in fact such data to be had for intentional states, but not for qualia, consciousness — or a soul or free-will — at least as these are standardly conceived.

'To which,' Strawson replies:

> the experiential realist's correct reply is: 'It's question-begging for you to say there must be an account of it that's non-question begging in your terms.' Such an exchange shows that we have reached the end of argument, a point illustrated by the fact that reductive idealists can make exactly the same 'you have no non-question-begging account' objection to reductive physicalists that reductive physicalists make to realists about experience. — (Strawson, 2006, p. 5 fn. 6).

Well, perhaps; but I don't like to give up so easily. I really don't see what question is being begged in asking for non-question-begging evidence — this seems like a rational demand in most any dispute[5] — I certainly don't think there are no question-begging objections to idealism: again, no idealism begins to rival physics. In any event, it's a healthy and venerable philosophical tradition to doubt what everyone takes for granted, no matter how central and indubitable it may be, as in the case of doubts about induction, knowledge, even non-contradiction.

There is, of course, a familiar 'transcendental' strategy against eliminativism, according to which the very denial of the mental presupposes the mental, of which Strawson can be regarded as providing a special case:[6]

> Experiential phenomena... cannot be mere appearance, if only because all appearance depends on their existence. (Strawson, 2006, p. 17)

But current psychological theory plainly gives the lie to this claim: there are plenty of 'appearance' representations in, say, the visual and phonological systems that are not part of anyone's ordinary conscious experience, e.g. representations of the generalized cones by which vision theorists such as Marr (1982) and Beiderman (1987) propose we are able to quickly identify ordinary objects, or the abstract phonological features by which Halle (2002) and others propose we identify morphemes.[7] And there's the well-established phenomena of 'blindsight', where people lacking visual experience can nevertheless guess above chance about the appearance properties (e.g. colour) in their visual field.

Towards the end of his article, Strawson calls into question the 'object/process/property/state/event cluster of distinctions' (Strawson, 2006, p. 28), and, indeed, speculates that 'discursive thought is not adequate to the nature of reality' (p. 28). Well, perhaps he is right about that. But if he's willing to entertain doubts at *that* fundamental level about our thought, why isn't he willing to entertain comparatively more modest ones about some of our ordinary, probably culturally conditioned conceptions of our minds? In any case, for the reasons I and others have

---

[5] More exactly, in which there are at least serious *prima facie* reasons for accepting either of two alternative claims (this perhaps rules out extreme examples such as 'nothing exists' or 'every claim is true'). I submit that the various figures I have cited on behalf of eliminativism provide plenty of such reasons.

[6] There is no need to address the general case here. I address it in Rey (1997, pp. 78–82).

[7] To be sure, Strawson (1994) claims that 'no non-experiential phenomena are intrinsically mentally contentful' (p. 164). He admits 'it is only an intuition' (p. 167), but it is an intuition that would seem to be contradicted by vast tracts of current psychological research, from Freud to the vision and language theorists I've mentioned — as well as by any number of historical writers, from the ancient Greeks to Nietzsche (see Whyte, 1960, and Rey, 1997, pp. 256–8, for further discussion).

mentioned, the psychology of those conceptions seems to me far more worth serious consideration than do otherwise unsupported and extravagant panpsychological claims about the world.

References

Beiderman, I. (1987), 'Recognition-by-components: a theory of human understanding', *Psychological Review*, **94**, pp. 115–47.
Churchland, P. (1981/90), 'Eliminative materialism and prospositional attitudes', in *Mind and Cognition*, ed. W. Lycan (Oxford: Blackwell).
Dennett, D. (1978), 'Towards a cognitive theory of consciousness', in D. Dennett, *Brainstorms* (Cambridge, MA: MIT Press).
Feyerabend, P. (1963/71), 'Mental events and the brain', in D. Rosenthal, *Materialism and the Mind/Body Problem* (Oxford: Oxford University Press), pp. 172–3.
Halle, M. (2002), *From Memory to Speech and Back: Papers on Phonetics and Phonology 1954–2002* (New York: de Gruyter).
Levine, J. (2001), *Purple Haze* (Oxford: Oxford University Press).
Marr, D. (1982), *Vision* (San Francisco: W.H. Freeman & Co.).
McGinn, C. (1991), *The Problem of Consciousness* (Oxford: Blackwell).
Nisbett, R. and Wilson, T. (1977), 'On telling more than we can know', *Psychological Review*, **84** (3), pp. 231–59.
Pylyshyn, Z. (2003), *Seeing and Visualizing:It's Not What You Think* (Cambridge, MA: MIT Press).
Quine, W. (1960), *Word and Object* (Cambridge, MA: MIT Press).
Rey, G. (1993), Review of Colin McGinn, *The Prob!em of Consciousness*, *Philosophical Review*, **102** (2), pp. 274–8.
Rey, G. (1997), *Contemporary Philosophy of Mind* (Oxford: Blackwell).
Rey, G. (2002), 'Physicalism and psychology: a plea for substantive philosophy of mind', in *Physicalism and Its Discontents*, ed. Carl Gillet and Barry Loewer (Cambridge: Cambridge University Press).
Rey, G. (2004), 'Why Wittgenstein ought to have been a computationalist (and what a computationalist can learn from Wittgenstein)', *Croation Journal of Philosophy*.
Rey, G. (forthcoming-a), 'Psychology is not normative', *Blackwell Debates in Philosophy of Mind*, ed. J. Cohen and B. McLaughlin.
Rey, G. (forthcoming-b), 'Phenomenal content and the richness and determinacy of colour experience', *Journal of Consciousness Studies*.
Rorty, R. (1979), *Philosophy and the Mirror of Nature* (Princeton: Princeton University Press).
Rosenthal, D. (1986), 'Two concepts of consciousness', *Philosophical Studies*, **49** (3), pp. 329–59.
Strawson, G. (1986), *Freedom and Belief* (Oxford: Oxford University Press).
Strawson, G. (1994), *Mental Reality* (Cambridge, MA: MIT Press).
Strawson, G. (2006), 'Realistic monism: Why physicalism entails panpsychism', *Journal of Consciousness Studies*, **13** (10–11), pp.3–31.
Whyte, L. (1960), *The Unconscious Before Freud* (New York: Basic Books).
Wittgenstein, L. (1953/68), *Philosophical Investigations*, 3rd edn., trans. G.E.M. Anscombe (Oxford: Blackwell).

David M. Rosenthal

Experience and the Physical

Strawson's challenging and provocative defence of panpsychism[1] begins by sensibly insisting that physicalism, properly understood, must unflinchingly countenance the occurrence of conscious experiences. No view, he urges, will count as 'real physicalism' (p. 4) if it seeks to get around or soften that commitment, as versions of so-called physicalism sometimes do.

Real physicalism (hereinafter physicalism *tout court*) must accordingly reject any stark opposition of mental and physical, which is not only invoked by many followers of Descartes, but even countenanced by many recent physicalists. Conscious experiences, Strawson persuasively urges, are a special case of the physical, just as cows are animals.

Panpsychism enters the picture because, despite the physical nature of conscious experiences, Strawson maintains that we cannot describe or explain the experiential using the terms of physics or neurophysiology. Since the experiential is nonetheless physical, Strawson concludes that the physical ultimates must, in addition to whatever properties physics and neurophysiology reveal, have experiential properties as well.

Strawson argues that we cannot avoid this conclusion by maintaining that combinations of physical ultimates constitute or give rise to experiential properties. Physical ultimates cannot give rise to conscious experience unless those ultimates are themselves in some way 'intrinsically experiential' (p. 22). The experiential cannot emerge from nonexperiential ultimates in the way that macroscopic liquidity is standardly held to emerge from molecular properties. Strawson concludes that at least some physical ultimates must be 'intrinsically experience-involving' (p. 22); and, since it's reasonable to see the physical ultimates as homogeneous in nature, we should assume that they all have experiential properties.

---

[1] Strawson (2006). Except as otherwise noted, all references are to this article, and all emphasis in quotations is in the original.

In Section I, I will raise doubts both about Strawson's claim that we cannot describe the experiential in nonexperiential terms and his argument for that claim. In Section II, then, I will suggest some concerns about Strawson's argument against emergence and his consequent ascription of experiential properties to the physical ultimates. I will close in Section III with some general remarks about physicalism and subjectivity.

I. Subjectivity and the Physical

The pressure for panpsychism stems from combining physicalism with Strawson's claim that we cannot fully describe conscious experience using the terms of 'physics and neurophysiology or any non-revolutionary extensions of them'. It may well seem that the only way to reconcile this claim with physicalism is to adopt Strawson's conclusion 'that there is a lot more to neurons than physics and neurophysiology record (or can record)' (p. 7).

Strawson sensibly understands the physical not solely by reference to physics, but rather in terms of the clear paradigms of the things we know to be physical. 'The physical is whatever general kind of thing we are considering when we consider things like tables and chairs and [assuming physicalism] experiential phenomena. It includes everything that concretely exists in the universe' (p. 8). Physicalism does not imply 'that the terms of physics can fully capture the nature or essence of experience' (p. 4).

Still, one might well see Strawson's denial that physics or neurophysiology can describe conscious experience not as supporting panpsychism, but rather as telling against physicalism. Indeed, many would see the rejection of physicalism as far the more sensible of those alternatives. But independently of that there is reason to contest Strawson's claim that we cannot describe the experiential in the terms used in 'physics and neurophysiology or any non-revolutionary extensions of them'.

To sustain that claim, Strawson refers us to his splendid book, *Mental Reality* (Strawson, 1994, pp. 62–5). There he urges that 'experiential phenomena outrun the resources of human language' in that, even if you and I describe our experiences in exactly the same way, 'still we cannot know that we are similar (identical) in respect of our experiential properties' (p. 62). Even when exactly the same terms apply to conscious experiences, what it's like for each of us to have those experiences may well differ from one individual to another.

Strawson concludes that no terms available to us can capture what conscious experiences are like for any particular individual. No matter how fully we describe a conscious experience, there remains on Strawson's view 'a real and unanswerable question about whether the experience is the same or different for any two of us' (Strawson, 1994, p. 63). Conscious experiences are, he concludes, outside the reach of 'human science' (p. 62).

Many would agree that quality inversion is possible that cannot be detected by any third-person means.[2] But there are compelling reasons to doubt that it is. Fundamental to that view is the claim that we know about the qualitative character of experiences only by appeal to the way each individual is conscious of those experiences. It may seem tempting to hold that we cannot determine whether conscious qualitative properties are the same from one individual to another because what is conscious to one individual is not conscious to anybody else. But that would matter only if our knowledge about conscious qualitative properties derived exclusively from the way each individual is conscious of them. If we can know about conscious mental qualities in some way that's independent of the way one is conscious of them, that would enable us to tell whether qualitative character is the same from one individual to another.

There is reason, moreover, to think that we do know about conscious mental qualities apart from any first-person access to them. Our knowledge about mental qualities could depend solely on such first-person access only if such qualities never occur without being conscious. If mental qualities do sometimes fail to be conscious, we can, in those cases at least, determine which qualities occur without appeal to first-person access. And presumably such third-person access would also apply when the relevant qualities are conscious. We would have information about mental qualities that's independent of how, and even whether, the relevant individual is in any way conscious of it.

There is compelling empirical evidence that qualitative states do occur without being conscious. Priming experiments show that nonconscious perceptual states have qualitative character, since such states prime for subsequent perceptual recognition in ways that reflect differences in mental quality, for example, differences in colour.[3] Since nonconscious states prime differentially for various conscious

---

[2] Some, such as Sydney Shoemaker, concede that, in the human case, asymmetries in the similarities and differences among mental qualities preclude undetectable quality inversion, but urge that creatures are conceivable in which such inversion occurs. (Shoemaker, 1982; reprinted in Shoemaker, 1984, p. 336; and Shoemaker, 1996, p. 150.) I'll argue in Section III that such asymmetries are unavoidable.

[3] See, e.g., Marcel (1983), and Breitmeyer, Ro and Singhal (2004). For a useful review of various kinds of evidence for the occurrence of nonconscious qualitative character more generally, see Kim and Blake (2005).

qualities, we can determine the qualitative character of those nonconscious states by reference to which conscious qualitative states they prime. We can accordingly determine qualitative character apart from the way the relevant individual is conscious of it. This also allows for useful theoretical ways to identify and taxonomize mental qualities independently of the way individuals are conscious of them, and hence precludes undetectable quality inversion; I'll return to that in Section III.

One might simply insist that no state can have qualitative character without being conscious, and that nonconscious states achieve their priming effects by way of some nonmental, physiological properties. But there is no independent reason to interpret the empirical results in that way, independent, that is, of the unsubstantiated claim that consciousness is our only avenue to information about qualitative character. The natural interpretation of these results is that nonconscious mental qualities prime for conscious versions of the very same qualities.

Strawson does seem to hold that we know about mental qualities only from the way an individual is conscious of them.[4] So he might deny that any account that does enable us to determine whether two individuals have the same mental qualities can capture the true nature of experiential states, and hence that it is relevant to his 'real physicalism'. But it's again unclear what independent reason there might be for that denial.

If we cannot tell whether two individuals have states with the same mental qualities, Strawson would be right that there can be no 'human science' of conscious experience. We would be unable to account for conscious qualitative character in terms of a nonrevolutionary extension physics or neurophysiology. But if we can tell whether qualitative character is the same from one individual to another, an objective account of qualitative character may well be possible. And then we may well be able to use the terms in a nonradical extension of neurophysiology in developing such an account.

---

[4] Thus his appeal, following Ned Block, to the Louis Armstrong style answer to what qualitative character is: 'If you gotta ask, you ain't never gonna get to know' (n. 6), which in effect denies the possibility of any informative account of qualitative character. But the Louis Armstrong reply is apt only if one builds into the very notion of qualitative character that we can know about it solely by the way an individual is conscious of it.

It is striking that those who hold that no more informative account of qualitative character can be given are just those who also see a mystery about its nature. But, as I shall argue in Section III, we can indeed give a more informative account of qualitative character, and thus dispel such mystery.

II. Emergence and Experience

Physicalists often hold that physical reality gives rise to conscious experience by combining in various ways. No physical ultimates are themselves experiential, but the experiential emerges from combinations of physical components.

Strawson contests this wide-spread idea. Whenever emergence occurs, he urges, 'there must be something about the nature of the emerged-from (and nothing else) in virtue of which the emerger emerges as it does and as what it is' (p. 15). This condition is plainly satisfied, for example, in the case of molecular properties' giving rise to liquidity. But Strawson sees the experiential as crucially different; we have no idea, he insists, of anything about the physical ultimates that could give rise to conscious experience.

Strawson urges that we should not model the emergence of the experiential from the nonexperiential on the emergence of liquidity from the molecular. We should see it instead as more like the extended emerging from the unextended or the spatial from the nonspatial. How could the extended emerge from that which is in no way extended, to say nothing of the spatial from what is not spatial?

But the extended does arguably emerge from the unextended. Geometrical points, though unextended, nonetheless constitute lines, planes and solids; and that might be enough for us at least to keep a provisionally open mind about the emergence of the experiential from the nonexperiential.

Strawson insists that the factors blocking the emergence of the experiential from the nonexperiential are 'not epistemological' (p. 15). Still, his discussion does often appeal to considerations that seem epistemological. To sustain a claim that the experiential emerges from the physical, he urges, we need some 'imaginative grip' on a suitable analogy that can ground that emergence (p. 15), and the features of what emerges must 'trace intelligibly back to' the features they emerge from (p. 18).

But these things aside, there are reasons to be particularly cautious about concluding, in advance of empirical investigation, that the experiential cannot emerge from the neurophysiological. As Strawson notes, when one phenomenon uncontroversially emerges from another, the emergent phenomenon 'arise[s] simply from the routine workings of basic physical laws' (p. 13). Since we don't now know the laws that govern the occurrence of conscious experiences, we cannot see at present how the experiential might emerge from the neurophysiological. The reason that currently contemplating such emergence may well 'boggle the human mind' (p. 15) is that without suitable knowledge of the laws governing conscious experiences we cannot give substance to the way such emergence might work. When we do discover laws and mechanisms that govern the occurrence of conscious experiences, we will be in a position to see whether, and if so how, experiences arise from neurophysiological functioning. That case may then come to seem rather like the emergence of liquidity from molecular interactions.

Strawson takes note of the argument that the emergence of the experiential from the nonexperiential is no more problematic than the emergence of life from nonliving matter. And he writes that '[t]his very tired objection' fails because the emergence of life has itself seemed problematic only insofar as experience is taken as an essential aspect of life (p. 20).

But that is far from obvious. Henri Bergson argued, for example, that an *élan vital* is needed to explain evolution and the development of organisms, independent of conscious experience (Bergson, 1955). Rather, comparing the emergence of mind with the emergence of life is useful because we can now see how the living emerges from the nonliving only because, unlike those earlier vitalists, we now know the relevant laws and mechanisms. And it is just such knowledge that we do not yet have in the case of the experiential and the neurophysiological.

Can we discover laws and mechanisms that govern the way the experiential arises from neurophysiological functioning? Perhaps not if Strawson is right that we know about conscious experiences only by our first-person conscious access to them. Laws or mechanisms that govern the way conscious experiences arise from neurophysiological functioning would give us another, competing way for us to know about experiential states, which would undermine Strawson's claim. This doubtless explains his scepticism about such laws and mechanisms and the emergence they might sustain. But, as argued in Section I, there is compelling reason to doubt that we can know about conscious experiences only by way of our first-person access to them.

Because Strawson holds that the experiential cannot emerge from the physical ultimates, he sees physicalism as defensible only if those ultimates are themselves in some way 'intrinsically experiential' (p. 19). And he seeks to make room for that possibility by endorsing a view of physical reality that he traces to Bertrand Russell and Arthur Eddington. Russell held that our knowledge about physical reality is solely mathematical; Eddington goes farther, insisting that our knowledge of the physical is solely an operationalist matter of meter readings (Russell, 1948, p. 240; Eddington, 1929). On neither view does physics speak to the intrinsic nature of ultimate physical reality; we have only indirect, structural knowledge about that nature. And that, Strawson urges, leaves it open for us to ascribe experiential properties to the physical.

But we know more about physical reality than these views claim. We learn about physical reality by formulating successful theories that ascribe properties, not all of them mathematical or purely structural, to the things those theories posit. Physics and the other theoretical sciences tell us more about physical reality than mere meter readings, and more than purely mathematical and structural properties.

Strawson quotes Russell's claim that '[p]hysics is mathematical... because we know so little [about the physical world]' (Russell, 1948, p. 240; quoted by Strawson at p. 10). But that's very likely not why physics proceeds mathematically. Physics characterizes things mainly in mathematical terms because it abstracts from properties such as colour and sound, which, because they are special to the particular ways we sense things, capture only superficial aspects of their objective nature. And abstracting from such properties leaves us with mainly mathematical descriptions of physical reality. Still, physics does posit nonmathematical properties as well, such as mass, spin, charge, and the like.

The ultimates that physics describes seem in any case poorly suited to be bearers of experiential features, however primitive or elemental. Indeed, it is arguably harder to see how such ultimates could have intrinsically experiential features than to see, even at our current stage of empirical and theoretical knowledge, how ordinary conscious experiences might emerge from neurophysiological functioning.

III. Experience and Physicalism

It is a striking phenomenon, to which Strawson usefully calls attention, that many recent advocates of physicalism have tacitly assumed an opposition between mental and physical. Indeed many saw the defence of physicalism as largely a matter of overcoming that opposition.

Some physicalists overcame the opposition simply by adopting the eliminativism that Strawson rightly rejects. Others, such as J.J.C. Smart and D.M. Armstrong, sought instead to overcome it by construing mental descriptions of experiences in topic-neutral terms. If mental descriptions are topic neutral, they would be neutral about whether experiences are physical. Mind-body correlations together with Ockham's razor would then show that experiences are physical.

But that strategy wildly overshoots, since if mental descriptions are topic neutral, they are also neutral even about whether experiences are mental. Topic-neutral construals thereby in effect eliminate the distinctively mental. One would seek to defend physicalism this way only if one held that being distinctively mental is actually incompatible with being physical.[5]

Strawson's own argument, however, also appeals to a kind of incompatibility between mental and physical. We cannot, he insists, know about conscious experiences except from our own first-person access, a restriction that holds of nothing that we describe as physical. To dissolve that incompatibility, Strawson urges that we see physics as fixing only the mathematical and structural properties of physical reality, which then allows us to ascribe experiential properties to the physical ultimates. Earlier physicalists sought to overcome the opposition they saw between mental and physical by construing experiential descriptions topic neutrally. Strawson seeks to overcome a somewhat different opposition between mental and physical by a topic-neutral construal of physics. And, whereas the topic-neutral construals of earlier physicalists undercut ordinary notions about mental phenomena, Strawson's purely structural construal of the physical arguably conflicts with established views about ultimate physical reality.

But we needn't adopt either the topic-neutral translations of earlier physicalists or Strawson's structural view of physical reality. As argued above, there is reason to resist Strawson's claim that we cannot describe conscious experiences objectively, which underlies the need to ascribe experiential properties to the physical ultimates.

Nor does the commonsense contrast of mental with physical that concerned earlier physicalists undermine physicalism. That contrast implies no incompatibility between mental and physical, and so is compatible with those occurrences' being a special case of the physical. The contrast between mental and physical is in this way similar to other contrasts with the physical, which also imply only that the contrasted phenomena are a special case of the physical, not that they fail themselves to be physical. Chemical processes and combinations contrast with physical processes and combinations, but are plainly special cases of the physical. And we contrast living things with what is merely physical and the virtual memory of computers with their physical memory, even though the things we contrast with the

---

[5] It is crucial that these authors did not argue that the relevant descriptions are neutral simply about whether experiences are physical, but that they are topic neutral in general.
For more on the way topic neutrality has these results, see Rosenthal (1976), Section II.

physical are again special cases of the physical. The term 'physical' simply applies in these cases to the more inclusive realm with which we contrast a range of special phenomena.

We have no reason to see the commonsense contrast of mental and physical as at all different. Nothing about mental phenomena is indisputably nonphysical, at least when we understand being physical in the broad way that includes macroscopic physical objects.[6] The mental is a special case of the physical, just as chemical and biological phenomena are, and just as cows are animals.[7] So understood, '[t]he physical ... includes', as Strawson urges, 'everything that concretely exists in the universe' (p. 8), and with no need to ascribe experiential properties to the physical ultimates.

In Section I, I urged that we can determine what qualitative character an experience has independently of the way one is conscious of that experience. I'll close by briefly expanding on this.

Experiences are states that enable us to perceive things in our environment and conditions of our own bodies. To do this, experiences must register relevant similarities and differences among the perceptible properties of the things we experience. Mental qualities are the properties experiences have in virtue of which they register those similarities and differences.

This holds of the mental qualities of both conscious experiences and nonconscious perceptual states. Qualitative states, whether conscious or not, function perceptually, and to do so the similarities and differences among their mental qualities must reflect the perceptible similarities and differences among the things we perceive. So the mental qualities of each sensory modality must occupy positions in a quality space that matches the quality space defined by the similarities and differences among the properties that the modality in question enables one to perceive.

The mental quality of any perceptual state is that property which enables an individual that is in that state to perceive the corresponding perceptible property. Each mental quality occupies a position in the quality space of its sensory modality that matches the position of the

---

[6] For more on this, see Rosenthal (1976), Section I.

[7] One might object that we do not speak of a contrast between being red and being coloured or between being a cow and being an animal as we speak of the contrast between mental and physical. But we do contrast being red with merely being coloured, and being a cow with merely being an animal. And we have no reason not to see the commonsense contrast between mental and physical as a contrast of the mental with the merely physical, i.e., the physical in general.

For more on this understanding of the mental-physical contrast, see Rosenthal (1980), Section II.

perceptible property in the space of properties to which that modality enables perceptual access. So we can determine the mental quality of any experiential state by fixing the position in the relevant quality space of the perceptible property to which that mental quality is a normal perceptual response.

We can, moreover, fix the space of objective perceptible properties to which each modality is responsive by seeing which properties an individual can discriminate (see, e.g., Clark, 1993, chs. 4 and 6; Goodman, 1966, chs. 11 and 12; and Goodman, 1972, pp. 423–36). And, since we can do this independently of the individual's first-person access to the relevant perceptual states, we can determine the mental qualities that figure in these discriminations independently of first-person access. A mental quality is conscious, then, when one comes to be conscious of oneself as being in a state that has that mental quality. What it's like for one to have a conscious experience is a matter of being conscious of oneself as having that experience.[8]

On this account, your mental quality of red and mine are automatically the same if our abilities to discern physical colour properties are the same. Indeed, this must be so. Mental qualities are fixed by their relative position in the quality space of the relevant sensory modality. So any inversion of qualities between your quality space and mine would be undetectable by third-person means only if that inversion preserved the relations of similarity and difference that the switched qualities bear to all the others. If those relations of similarity and difference changed, so would the perceptual function of the switched qualities, and others would then be able to detect that switch. Because qualities are the same if their relations of similarity and difference are the same, undetectable quality inversion is not possible.[9]

Why, then, has it seemed to so many that it is possible?[10] If one sees mental qualities solely in terms of the way one is conscious of them, the ties those qualities have with perceptible properties will seem to be merely contingent. One will then conclude that there can be no way to tell whether your mental qualities are the same as mine. The alleged

---

[8] For more on this, see Rosenthal (2005). The similarities and differences among perceptible properties that matter here are those which can be perceived, not similarities and differences that hold in respect of psychophysical descriptions of those properties.

[9] The asymmetry in human quality spaces that Shoemaker acknowledges makes quality inversion impossible for us (see n. 2, above) is no accident; it is required for mental qualities to play their distinctive perceptual roles.

[10] The hypothesis of undetectable inversion famously appears in Locke (1975), II, 32, 15. But it can be found even in Sextus Empiricus *Against the Logicians*, I, (*Adversus Dogmaticos*), 95–198, describing the views of the Cyrenaics; Sextus Empiricus (2006), pp. 40–41. I owe the reference to Sextus to Ben-Yami (unpublished).

possibility of undetectable quality inversion is of a piece with the view that mental qualities can be known only by the way one is conscious of them.[11]

Strawson's probing article seeks to reconcile our first-person access to qualitative character with our third-person understanding of objective reality. This is the fundamental challenge in understanding experiential phenomena. Strawson's strategy is to accommodate the claim that we can know about the qualitative only by first-person access by adjusting our view of the physical. But we can more successfully square our first- and third-person access to reality by rejecting that claim. In that way we can understand how our first-person access to conscious experience fits with the objective role that conscious qualitative character plays in perceiving.

References

Ben-Yami, Hanoch (unpublished), 'The origins of the inverted spectrum hypothesis'.
Bergson, Henri (1955), *An Introduction to Metaphysics*, tr. T.E. Hulme (Indianapolis: Bobbs-Merrill [original 1903]).
Breitmeyer, Bruno G., Tony Ro, and Neel S. Singhal (2004), 'Unconscious color priming occurs at stimulus — not percept-dependent levels of processing', *Psychological Science*, **15** (3), pp. 198–202.
Clark, Austen (1993), *Sensory Qualities* (Oxford: Clarendon Press).
Eddington, Arthur (1929), *The Nature of the Physical World* (New York: Macmillan).
Goodman, Nelson (1966), *The Structure of Appearance*, 2nd edn. (Indianapolis: Bobbs-Merrill).
Goodman, Nelson (1972), 'Order from indifference', in Nelson Goodman, *Problems and Projects* (Indianapolis: Bobbs-Merrill), pp. 423–436.
Kim, Chai-Youn, and Randolph Blake (2005), 'Psychophysical magic: Rendering the visible "invisible" ', *Trends in Cognitive Sciences*, **9** (8), pp. 381–388.
Locke, John (1975), *An Essay Concerning Human Understanding*, edited from the fourth (1700) edition by Peter H. Nidditch (Oxford: Clarendon Press).
Marcel, Anthony J. (1983), 'Conscious and unconscious perception: An approach to the relations between phenomenal experience and perceptual processes', *Cognitive Psychology*, **15**, pp. 238–300.
Rosenthal, David M. (1976), 'Mentality and neutrality', *The Journal of Philosophy*, **LXXIII** (13), pp. 386–415.
Rosenthal, David M. (1980), 'Keeping matter in mind', *Midwest Studies in Philosophy*, **V**, pp. 295–322.
Rosenthal, David M. (2005), 'Sensory qualities, consciousness, and perception', in David M. Rosenthal, *Consciousness and Mind* (Oxford: Clarendon Press) pp. 175-226.
Russell, Bertrand (1948), *Human Knowledge: Its Scope and Limits* (New York: Simon and Schuster).

---

[11] For more about the impossibility of undetectable quality inversion, see Rosenthal (2005), Section 7.

Sextus Empiricus (2006), *Against the Logicians*, translated by Richard Bett (Cambridge and New York: Cambridge University Press).

Shoemaker, Sydney (1982), 'The inverted spectrum', *The Journal of Philosophy*, **LXXIX** (7), pp. 357–381; reprinted in Shoemaker (1984), pp. 327–357.

Shoemaker, Sydney (1984), *Identity, Cause, and Mind: Philosophical Essays* (Cambridge: Cambridge University Press, [2nd edn.: Oxford: Clarendon Press, 2003]).

Shoemaker, Sydney (1996), 'Intrasubjective/intersubjective', in Sydney Shoemaker, *The First-Person Perspective and Other Essays* (Cambridge and New York: Cambridge University Press), pp. 141–154.

Strawson, Galen (1994), *Mental Reality* (Cambridge, MA: MIT Press/Bradford Books).

Strawson, Galen (2006), 'Realism monism: Why physicalism entails panpsychism', *Journal of Consciousness Studies*, **13** (10–11), pp. 3–31.

William Seager

The 'Intrinsic Nature' Argument for Panpsychism

1. Intrinsic Properties and Panpsychism

Strawson's (2006) case in favour of panpsychism is at heart an updated version of a venerable form of argument I'll call the 'intrinsic nature' argument. It is an extremely interesting argument which deploys all sorts of high calibre metaphysical weaponry (despite the 'down home' appeals to common sense which Strawson frequently makes). The argument is also subtle and intricate. So let's spend some time trying to articulate its general form.

Strawson characterizes his version of panpsychism, or 'real physicalism', as the view that 'everything that concretely exists is intrinsically experience-involving'. He approvingly quotes several of Russell's remarks, the general upshot of which is that 'we know nothing about the intrinsic quality of physical events except when these are mental events that we directly experience' which sentiment is echoed by various pronouncements of Eddington, such as 'science has nothing to say as to the intrinsic nature of the atom'. In whatever way the argument is going to proceed, it evidently depends upon some conception of the intrinsic nature of things. What is this supposed to be?

The philosophical literature on the distinction between intrinsic and extrinsic properties (or relational properties) is vexed and very far from settled (see Humberstone, 1996, for an extensive review and discussion). The core intuition would seem to be the idea that the intrinsic properties of X are the properties that all duplicates of X would have. Thus, for example, any duplicate of me would have the same mass as I do (so mass looks like an intrinsic property) but would differ from me in not being an uncle (so uncle-hood looks — as it should — to be a non-intrinsic or extrinsic property). But there does not seem to be any way to define duplication in the relevant sense without circular reference back to intrinsic properties. Another way to get at the idea is to characterize the intrinsic properties as those which X would persist in exemplifying were it absolutely alone in the universe. That is, the intrinsics are the properties X has 'all by itself' or 'of its own nature'. For example, the clearly extrinsic (or relational, I will not attempt to forge a distinction between these notions here) property of 'being an uncle' is not a property one can have if one is absolutely alone in the universe. This suggests that a simple characterization of the notion of intrinsic property would be something like 'F is an intrinsic property of x just in case Fx does not imply the existence of anything distinct from x'. Unfortunately, this won't quite do. In the first place, necessary existents are entailed by anything having any property. Kim (1982) amended the condition to require that Fx not entail the existence of a distinct, contingent thing. But as Lewis (1983) pointed out the property of *loneliness* (being absolutely alone in the world[1]) is obviously extrinsic and yet its possession does not entail the existence of any distinct contingent beings. Langton and Lewis (1998) suggest that the intrinsics are the properties which are logically independent of both loneliness and accompaniment, that is, F is an intrinsic property of x just in case Fx is compatible with loneliness and accompaniment and ~Fx is similarly compatible with both loneliness and accompaniment. On the face of it however, this criterion appears to make the relational property of '. . . loves a' (where a is some object) into an intrinsic property, for a can love itself or not whether or not it is accompanied. Be that as it may, the concept of the intrinsic properties of an object seems intuitively intelligible, despite the difficulties philosophers have in spelling it out precisely in non question-begging terms. It may well be that the concept of intrinsicness is a 'primitive' notion. The point here is simply that a difficult metaphysical question lurks within the issue of intrinsicness itself, even prior to putting the concept to any argumentative use.

This example of the relational property of '. . . loves a' raises another issue. If there were extrinsic properties which supervened upon an object's intrinsic properties, then they would show up as intrinsics themselves according to our test criteria. While it may seem like an idle worry, since it appears obvious that extrinsic properties such as 'being an uncle' do not supervene on their subject's intrinsic properties, the history of philosophy reveals a powerful current of

---

[1] Or, taking into account again the problem of necessary existents, loneliness should be defined as being unaccompanied in the world by any distinct contingent beings.

thought which endorses exactly this kind of supervenience which is also important for the intrinsic nature argument for panpsychism.

The most famous and extreme proponent of this view was Leibniz, who held that the entire world could be reconstructed given the information contained within, or determined by, the intrinsic properties of any individual (see e.g. Leibniz, 1714/1989). For Leibniz, it would be possible for God, at least, to tell whether or not I am an uncle merely by examining me and me alone. It's worth pointing out a slight subtlety in the claim that the intrinsic properties determine the relational properties. The rather hackneyed example is the relational property belonging to Socrates of being shorter than Plato. It is plausible (assuming that height is an intrinsic property) that this property is indeed determined by the intrinsic features of the subjects taken together — once we know the heights of Plato and Socrates, we can know that Socrates is shorter than Plato.[2] The kind of determination here is that of the relational properties of things depending on the intrinsic properties of the relevant relata. This is a relatively weak view. Leibniz maintained the stronger view that the distribution of relational properties throughout the world is determined entirely by any individual's intrinsic properties. The mechanism behind this miracle is, of course, that the intrinsic properties of anything are determined by the intrinsic properties of anything else!

Leibniz's panpsychism stems from his view that only mental features have the right characteristics to perform the reduction of the relational to the intrinsic. Essentially, it is only via mental *representation* that an entire world can be wrapped up inside a single individual so that all the relations can be 'read off' the intrinsic properties of that individual.

But leaving aside the mechanism, let us codify the key feature of Leibniz's position at issue in a principle, the Principle of the Reducibility of Relations, PRR:

> PRR: All extrinsic properties are determined by intrinsic properties.

PRR can then be divided into strong and weak versions, whenever it matters, as discussed above.

The sheer audacity of PRR is most apparent if we think of the most purely extrinsic relations we are familiar with: spatial relations. The

---

[2] Thus 'x is shorter than y' is what Bradley (1893) called an internal relation. Notoriously he went on to argue that these were the only kinds of relations there could be. So, crudely speaking, Bradley too joins the philosophers who believe that all relational properties supervene on intrinsic properties. For a classic discussion of this issue, which much relevance to this paper as well, see Moore (1919).

idea that my intrinsic properties somehow determine that I am fifty miles from a burning barn seems ludicrous. Even on the weak thesis, spatial distance and arrangement seem precisely to be relations that can vary independently of the nature of the objects that enter into them. Leibniz is of course famous for the doctrine that space (and time) is nothing more than a set of relations. But in his view these relations themselves are determined by the intrinsic properties of things. Once again we can see the attractiveness of panpsychism, for it does not seem altogether hopeless to define space and time in terms of perceptual contents, given that there are sufficient perceiving subjects to 'nail down' the infinite and continuously varying spatial and temporal relations that structure our world.[3]

As pointed out by Moore (1919) PRR implies that intrinsically identical things are numerically the same thing (this is the Leibnizian principle of the identity of indiscernibles). For if the principle of the identity of indiscernibles is false then the relational property of 'being identical to a' will be an extrinsic property (or involve an external relation) which is not determined by intrinsic properties. The principle of the identity of indiscernibles is thought to be implausible by many, but it does follow of course from the Leibnizian idea that the entire structure of the world is encoded in the intrinsic properties of each individual thing.[4]

Strawson does not endorse the Leibnizian metaphysics, but the intrinsic nature argument he advances has interesting affinities with it. One question is to what extent Strawson has to endorse PRR in the extreme form stated above. It may be that a weaker or circumscribed version will suffice for his argument. For example, the limited claim that *causal relations* are determined by intrinsic properties might suffice. However, once it is allowed that there are some relational properties that are not grounded in intrinsic properties, it may be hard to dispute the cogency of views that assert that the power of matter to generate

---

[3] One can attempt to avoid the postulation of this continuous infinity of minds by appeal to certain modal facts about space and time, along the lines of 'position x,y,z,t is defined in terms of what a possible perceiver would perceive under certain circumstances...'. But this is merely a relational characterization of spatial and temporal structure. On the kind of views we are considering, this would require some intrinsic ground, which seems to reinstate the need for something actual corresponding to each possible position in space and time.

[4] A theological argument helps to make this clear: if the principle of the identity of indiscernibles was false and there were two intrinsically identical things, God would have no reason to place one at X and the other at Y, or *vice versa* and thus his creation of the world — governed completely by reason — would be frustrated. Leibniz illustrated this point with his lovely tale of the nobleman at Herrenhausen unsuccessfully searching for two leaves of identical appearance.

consciousness is one case of such an ungrounded extrinsic. But not to get ahead of ourselves, let us turn to the argument. It is possible to regard Strawson's argument as a kind of supplement — a crucial one — to the argument for panpsychism advanced by Nagel (1974). The form of Nagel's argument is a destructive dilemma.

(1) Either consciousness emerges from non-conscious features of the world or else consciousness is a fundamental and ubiquitous feature.
(2) Emergence is impossible.

Therefore, consciousness is a fundamental and ubiquitous feature (= panpsychism).

The argument is obviously valid but both premises are problematic. The claim of the second premise, that emergence is impossible, will seem implausible to many, especially in light of how fashionable it has become to throw around the term 'emergence'. Nagel says surprisingly little about this but the bottom line is 'there are no truly emergent properties of complex systems. All properties of complex systems that are not relations between it and something else derive from the properties of its constituents and their effects on each other when so combined' (Nagel, 1974, p. 182).

The problem with the first premise is that while the fundamentality of the mental seems to follow from the failure of emergence, the ubiquity of the mental is not so easily established. Why couldn't a fundamental feature appear here and there throughout the world rather than everywhere (perhaps in the way that electric charge or mass are features of some fundamental particles but not others).

Strawson addresses both problems. He argues first that a concept of emergence powerful enough to undercut the argument for panpsychism is incoherent, and then proceeds (perhaps rather cursorily) to address the issue of the ubiquity of the mental. It is in the attack on emergence that the connection with intrinsic properties appears.

First, Strawson thinks that the kind of emergence needed to undercut panpsychism is *brute* emergence. Why is that? One argument is that if consciousness emerged from physical processes in the normal manner, in the same sort of way that liquidity emerges from the details of inter-molecular relationships or that tornadoes emerge from the dynamics of heated atmospheres, then there would be no explanatory gap between the physical and consciousness. Since the gap is undeniable, ordinary emergence cannot be the story of consciousness. This invites the reply, which has been made by several philosophers, that the gap is merely a feature of our cognitive limitations, either temporary, awaiting further science to build the bridge between consciousness and the brain (see e.g. Hardcastle, 1996) or permanent, reflecting an unfortunately inbuilt weakness of the human mind (see e.g. McGinn, 1989). As for the first sort of reply, as Yogi Berra said, it is hard to make predictions, especially about the future. The thing is, though, that while we are ignorant about all sorts of cases of ordinary emergence, it is not that hard to see in a rough way how the explanations might go. We don't understand, for example, how high temperature superconductivity emerges, but we have some idea about how this might go and lots of detailed ideas about the low level interactions that underlie this sort of emergence. The case of consciousness really does seem to be uniquely different. While it is not hard to see how neural activity could possibly underlie all sorts of complex behaviour, we have no clue how it could be that certain patterns of neural activity could constitute phenomenal consciousness. One of the nice features of panpsychism is how it evades this problem by being able to assert that the patterns of neural activity have consciousness already built in to them. Still, as we shall see, the width and depth of the explanatory gap depends upon how intrinsic natures are deployed in the anti-emergence argument.

The second sort of reply — that we are constitutionally unable to understand how matter generates consciousness — is little more than an expression of faith in a physicalism which endorses the metaphysical principle Strawson labels NE: physical stuff is, in itself, in its fundamental nature, something wholly and utterly non-experiential. While it is true that almost everyone — at the level of the proverbial 'man in the street' — accepts that there are things that completely lack mental properties (stones, cars, planets, etc.) hardly any of these people have heard of the explanatory gap or have ever thought about the problem of consciousness. I expect that most in this blissfully pre-reflective state would be initially happy to endorse the 'wait for science to explain emergence' line of thought, until the sheer size and unique nature of the explanatory gap is revealed. The subsequent retreat to 'cognitive closure' then really does seem based on stubborn faith in NE plus physicalism.

I think it's a good idea to try to elucidate the argument against faith-based physicalism from Strawson's article. The core premises of this argument are, first, that physics reveals to us only the relational properties of matter (or 'the physical') and, second, our old friend: relational properties (at least a relevant set of them) are determined by intrinsic properties. Strawson cites the august authority of Eddington and Russell in support of the first premise, but the view is very widely held and very plausible. If someone asks what an electron is, all we can say is that is a 'particle' with a certain mass ($9.10938188 \times 10^{-31}$ kilogram), electric charge -1, spin ½, etc. Each of these attributes can only be defined relationally and all we know about them is what these relations provide. A mass of m is just that property such that something with it will obey the relation that $m = F/a$ for a force F and acceleration a, and so on. Another way to put this, in line with the way Russell views things, is that all that science provides, or can provide, is structural or purely mathematical information about the world. To add a quote to those Strawson assembles, but one that might seem to cut against the move towards panpsychism, Russell says: 'the only legitimate attitude about the physical world seems to be one of complete agnosticism as regards all but its mathematical properties' (Russell, 1927, p. 270).

If we couple the idea that physics provides us insight only into relational properties of matter with the (appropriate form of the) reducibility principle, we are forced to postulate an intrinsic ground for the relational, structural or mathematical properties. One central and vitally significant element of the structural properties of matter as revealed to us by physics is the causal relationships which matter can form. Indeed, these structural features may *exhaust* what science can tell us about the physical world. These fall naturally into fundamental and emergent 'levels'. The force experienced by an electron in an electric field is fundamental, the capacity of hurricanes to wreak destruction is derived and appears a very long way from the fundamental forces at work. These derived causal relations are the province of 'ordinary', non-brute emergence — they are determined by, and in ways that are, in principle, intelligible to us, the fundamental causal relationships. A lot of science is the investigation of these mechanisms of emergence. But we eventually reach the fundamental causal relations which are, according to the reducibility principle, in need of an intrinsic ground.

However, we have, it seems, absolutely no knowledge of the intrinsic properties of matter which underwrite their causal relations. But surely, all this rationally licenses is, as Russell says, agnosticism about these intrinsic properties rather than the radical endorsement of an experiential interpretation of them. There are some suggestive hints and significant constraints here though. First, since it is evident that certain configurations of matter generate or constitute conscious states, the intrinsic properties of matter must encompass this power. Second, our own introspective awareness of consciousness reveals it to be an intrinsic property, and given that consciousness is materially realized, consciousness is then an intrinsic property of matter — at least of certain organized material systems.

The realization that states of consciousness are intrinsic properties is of great significance. Although he did not use this language, I think Descartes was the first person to argue for this thesis. His sceptical worry that it was impossible to tell whether or not he was alone in the universe on the basis of the contents of his consciousness clearly suggests that my duplicates — even if the only thing in the world — would share all my states of consciousness. The philosophical problem of the external world and the coherence of solipsism entail that consciousness is an intrinsic property of things. We do not have to embrace Descartes's dualism to share this insight. We must also recognize that our states of consciousness, though intrinsic, are not at all simple. This might evoke the worry that our states of consciousness are relational structures themselves, whose identity depends upon nothing more than a certain inter-related set of conscious 'parts'. While in a certain sense it is quite true to say that states of consciousness have parts, the phenomenon of the unity of consciousness shows that the nature of a state of consciousness is more than the mere inter-relatedness of these parts. William James pointed this out in a famous passage: 'Take a sentence of a dozen words, and take twelve men and tell to each one word. Then stand the men in a row or jam them in a bunch, and let each think of his word as intently as he will; nowhere will there be a consciousness of the whole sentence' (James, 1890/1950, p. 160). We can read this as an argument against the idea that conscious states are 'structural' in any ordinary sense. Contrast the situation of conscious states with the obvious fallacy in this analogue of James's remark: take a hundred men and give them each a girder. Let them jam the girders together however they will, nowhere will there be a bridge made out of the girders. The emergence of bridges is ordinary fare but the emergence of consciousness seems altogether different.

According to the reducibility principle, matter must have an intrinsic nature to ground the relational or structural features revealed to us by physical science. We are aware of but one intrinsic property of things, and that is consciousness. It is plausible to assert physicalism — we are physical beings and our consciousness is a feature of certain physical structures.[5] Therefore, consciousness is an intrinsic property of matter. To show panpsychism we need to show that

---

[5] This is a merely empirical premise, but it is well supported by common sense and scientific evidence of the vital link between brains and consciousness. It is a virtue of panpsychism

consciousness is both fundamental and ubiquitous. The fundamental features of matter are the intrinsic properties that are exemplified by the most basic constituents. The non-fundamental features are determined by them. If consciousness was not fundamental it would have to be determined by the fundamental intrinsic properties of matter.

It may be possible to postulate that matter possesses fundamental intrinsic properties which are entirely non-experiential and which nonetheless permit matter to subvene states of consciousness and of which we are entirely ignorant. I think this calls out for rather too much faith in physicalism. The faith-based physicalism which endorses NE retained some plausibility when it maintained that our cognitive powers were too weak to see how matter as science reveals it could generate consciousness. But given the reducibility thesis, matter as science reveals it is simply not the kind of thing that could generate anything, except via the inapplicable and irrelevant mechanisms of ordinary emergence. We have to turn to the intrinsic properties for that. We have one ready to hand in consciousness itself. Nothing justifies the brute posit of additional intrinsic properties with this power, except the verbal demand that it be 'non-mental'. It is not clear to me that there is anything more than a merely verbal issue at this point, whether to *call* the intrinsic property at issue phenomenal or proto-phenomenal or absolutely and definitely non-mental (though we know nothing about it whatsoever) yet capable of producing conscious states.

And notice that the production of consciousness cannot be accomplished in such a way as to make conscious states merely structural or relational. For we know they are intrinsic features of things.

Ubiquity remains unproven. As Strawson notes, considerations of parsimony and elegance discourage us from doling out fundamental mentality to electrons but not positrons. After all, it seems we could build a conscious being out of anti-matter no less than matter. One might still worry that to the extent that there are physical entities which play no role whatsoever in the constitution and operation of the brain — neutrinos as it may be — there is ground to deny any experiential aspect to at least these elements of the material world. But again, that forces us, by the reducibility principle, to invent new intrinsic natures that are absolutely unknowable and uninvestigatable. On the other hand, if this line of argument is acceptable, then it seems to suggest that *all* of matter's relational properties should be traced

---

that it permits us to be physical beings in the face of the difficulties of conceiving of consciousness as a physical phenomenon.

back to experiential intrinsic properties. This is to get on the road to Leibniz's position, I think.

2. Radical Relationalism

I hope the argument I've sketched expresses and elucidates something of Strawson's argument for panpsychism. If the way I'm looking at it is right, then the argument depends upon the heavy duty metaphysical premise of reducibility (although presumably not such an extreme version as PRR). There are two core issues that arise here. The first, explored in Part 1 is about how the intrinsic nature of matter should be understood. The second is: why should matter have any intrinsic properties at all? An alternative view is what may be called 'relationalism' which asserts that *all* there is to matter is the set of inter-relationships which science reveals. Relationalism is undeniably plausible in certain domains. In graph theory — a mathematical discipline exemplified by Euler's treatment of the Königsberg bridge problem — for example, there is nothing more to a graph than the set of relationships between the nodes; and the nodes themselves are defined by their place within the overall set of relationships. Within the realm of the mathematics, relationalism seems to be the generally correct ontological account. Does the number 2 have intrinsic properties apart from its relational place within the system of numbers? Relationalism within the realm of the concrete is much more controversial, and it has not been much discussed among philosophers with the exception of the so-called structural realists in the philosophy of science (see e.g. Worrall, 1989; Ladyman, 1998; see also Dipert, 1997, for a distinctive treatment of the issue of relationalism).

Structural realism can be quickly, if superficially, characterized in terms of the Ramsey method of eliminating theoretical terms from a syntactical specification of a theory. In this well known procedure, Ramsey replaces each referring term in the theory with an existentially bound variable; the Ramsified version of a theory says only that there exist certain entities that are inter-related thus-and-so. An ontological reading of structural realism then goes on to assert that there is nothing more to the entities (which are asserted to exist) than their place within this relational system. A theory becomes rather like a mathematical 'graph' and the nature of the entities involved reduces to their place within this structural description. The reducibility principle is simply rejected.

How does the argument for panpsychism look if we deny that relational properties (or some relevant subset of them) need to be determined by intrinsic properties? Quite different. If all there is to matter is its relational or structural properties then the impetus to seek an intrinsic 'background' to underpin them completely evaporates. Does this really affect the overall argument however? The explanatory gap remains, since we lack any glimmering of an explanation of how matter as science reveals it to be (the relational structure) could generate or constitute states of consciousness. But now it is *much* more plausible to claim that this is a merely temporary failure which awaits advances in neuroscience. This idea is hopeless from the point of view of the intrinsic nature argument, for the only thing that science will ever reveal are more of the relational properties of matter. But once we deny the reducibility principle, then the kind of structure which science is so good at discovering is all there is to discover. In particular, according to relationalism, there is nothing more to consciousness than its place in a system of relations linking it to events in the material world as well as other mental states. Presumably, nothing stands in the way of neuroscience discovering this system of relations in the brain.

Relationalism thus construed sounds like functionalism, but there is a big difference. Standard functionalism buys into the line of thought that leads to the intrinsic nature argument. Functionalism requires that the system of relations it specifies be implemented or realized by some appropriately organized system of entities whose own properties permit them to 'mimic' the functional specification of the system. The affinity with the reducibility principle is clear. But relationalism dispenses with the realization requirement: the system of relations is enough all by itself to underpin the reality of the entities at issue. The electron, according to relationalism, is just, as it were, a node in a system of relations whose identity and complete nature is fixed simply by its place in that system. There is no need for a 'qualitative something' to realize the electron. The same would be true of consciousness, and its proprietary system of relations might well be a neural structure (relationalism need not deny that there is a hierarchy of relational structures within nature, but it is 'relations all the way down').

So, on the relationalist view, the explanatory gap could easily be nothing but an artifact of the immaturity of theoretical neuroscience. It is also possible for the relationalist to endorse the stronger claim that the explanatory gap is unbridgeable because of innate human cognitive limitations. Here too, relationalism lends extra plausibility to this account of the explanatory gap, for the limitations at issue will now be of the rather straightforward kind of intractable complexity of relational structure rather than some mysterious failure to grasp the 'intrinsic connection' between the purely spatial and causal features of matter and phenomenal experience.

Relationalism undercuts the claim that the ontological investigations of science are essentially deficient and thereby seriously weakens the intrinsic nature argument for panpsychism. No intrinsic experientiality need be accorded to matter, any more than intrinsic anything else. Physicalists might therefore wish to embrace it as a doctrine which enshrines science as the final ontological arbiter and dissolves the explanatory gap. Perhaps, too, physicalists would hope that relationalism would relegate the intrinsic nature argument to the discredited realm of transcendent metaphysics, thus explaining why panpsychism exudes, as Nagel put it, 'the faintly sickening odor of something put together in the metaphysical laboratory'.

However, relationalism faces a number of difficulties and I would like to conclude this paper by outlining some of them.[6]

Strawson advances some considerations that might provide the starting point of an attack on relationalism where he says 'one needs to abandon the idea that there is any sharp or categorial distinction between an object and its propertiedness. One needs to grasp fully the point that "property dualism", applied to intrinsic, non-relational properties, is strictly incoherent insofar as it purports to be genuinely distinct from substance dualism, because there is nothing more to a thing's being than its intrinsic, non-relational propertiedness' (Strawson, 2006, p. 28). The argument that this suggests to me is this. Relations require relata. There is no real distinction between having intrinsic properties and being an individual. The relata of any relation are individuals. Therefore they possess intrinsic properties. Unfortunately, I think the relationalist is likely to complain that the second premise here begs the question. Why could not the relata be ontologically constituted out of the relations they stand in (as are the nodes in a graph conceived of as an abstract mathematical object). Nonetheless, the identification of individuality and possession of intrinsic properties is intuitively correct. In a discussion of Leibniz, where Kant comes close to making the relationalist claim about objects as we experience them, he goes on to say: 'besides external presence, i.e. relational properties

---

[6] One point worth mentioning is that there are two senses in which relational properties require an intrinsic ground (see Moore, 1919). One is, as we have been exploring, the intrinsics determine the relational properties. The other is a weaker view: relational properties require relata with intrinsic properties but these properties do not determine the relational properties. The weaker view is more plausible, but will not serve to motivate the intrinsic nature argument (as I see it), except in the sense that we already know that consciousness is intrinsic and (presumably) cannot be determined by merely relational properties (see below).

of the substance, there are other intrinsic properties, without which the relational properties would not exist, because there would be no subject in which they inhered' (Kant, 1756/1986, p. 123).[7] Kant merely states this — there is no argument.

Another argument against relationalism appeals to some fairly conventional philosophical wisdom that itself possesses close ties to the reducibility principle: that dispositions require a categorical base. A good number, perhaps most or conceivably even all of the relational properties which science discovers about matter are causal dispositions. If dispositions require (metaphysically) a base of intrinsic properties which determines their powers then we have an argument from the relational structures revealed by science to the need for some intrinsic nature which subvenes these powers. Science itself provides innumerable examples of how dispositions are based upon lower level *structures*. But it does not directly follow that the hierarchy of causal relationships has to bottom out in intrinsic properties rather than some fundamental system of relational properties (see Lycan's 1987 discussion of 'levelism'). Strawson perhaps advances a version of this argument where he asserts against brute emergentism that causal powers require an intrinsic base. Against this, the relationalist is likely to assert that there really is nothing more to causation than certain patterns of relations amongst events. It may be yet another argument against relationalism that philosophers have so dismally failed to produce any such structural analysis of causation despite many years of efforts.

One of the core intuitions about intrinsic properties is that they are the properties that things have 'in themselves', the properties that something would retain even if it was the only thing in the universe. If we add the premise that things can exist as the sole denizen of a world (in some appropriately weak modal form — that fact that I need oxygen to survive will not prevent me from having intrinsic properties) we have an argument against relationalism. Individuals can, on this view, be 'pulled out' of the relations they may find themselves in and exist entirely apart from them. This is possible only if they possess intrinsic properties. So relationalism must be false. Here, the relationalist would have to deny the modal premise despite its intuitive plausibility. If relationalism is true then no entity can exist by itself — all entities metaphysically imply the existence of other things, just as it is

---

[7] For this passage and many ideas that influenced this paper, see Langton's remarkable *Kantian Humility* (1998). Although rather controversial as an interpretation of Kant (see Falkenstein, 2001) the book is a wellspring of ideas on the issue of intrinsicality and related areas.

impossible for a node of a particular graph to exist apart from the rest of the graph. However, the evident difference between concrete individuals and the merely abstractly specified structures of graph theory (and other mathematical constructs) tells against relationalism here. What is concreteness, if not the ability of concrete things to exist apart from other things?

One can further argue against relationalism by an indeterminacy argument (this argument has clear affinities with the first given above and the immediately preceding argument). The linchpin premise here is that structure is insufficient to nail down the particularity of the concrete entities which enter into any system of relations. Contrast this with the evident success of relationalism in the mathematical realm. It seems ludicrous to suppose that the mathematical concept of a particular graph is somehow deficient because the nodes are not specified in any way beyond their place in the system of connections which describe the graph. The current argument re-emphasizes that this is precisely a key difference between the concrete and the abstract.

A famous — at least within structural realist circles — argument by the mathematician Max Newman (1928), originally directed at Russell's claim that science provides only structural information about the world, can be deployed to put some real meat on the bones assembled above. The conclusion of Newman's argument, when interpreted to bear on relationalism, is that the existence of a system of relations is trivially true of a set of objects, so unless there is something, as Newman says, 'qualitative' (I read this as involving intrinsic properties) about the relata, relationalism says exactly nothing about the world, beyond an assertion of cardinality.[8] This is because, assuming there are enough entities, it follows from pure logic that any system of relations over those entities is instantiated. How can that be? Because, conceived apart from considerations of the intrinsic properties of the relata, relations are simply sets of ordered sequences of entities (e.g. a two-place relation is a set of ordered pairs) and, given the entities, those sets and sequences will automatically exist. Newman puts it

---

[8] The idea that the number of things is independent of the relations into which they fall perhaps by itself comes close to providing an argument against relationalism, suggesting that there is some definite 'thingness' which anchors the metaphysical possibility of counting individual things. In this light, it is interesting that particle number is *not* in general an eigenvalue of states in quantum field theory. A defender of the reducibility principle might well take this as a sign that the intrinsic properties are quite different than the kinds of things studied by physical science. The example of Leibniz once more comes to mind. On the other side, and frequently voiced, is the claim that quantum mechanics is trying to tell us that relationalism of some kind is true (see Teller, 1986). I find it somewhat strange however that we should use physical science, which is limited to discovering the relational structure of things, in an argument that all there is, is relational structure.

thus: 'any collection of things can be organized so as to have the structure W, provided there are the right number of them' (Newman, 1928, p. 144).

It is very satisfying to see that the intrinsic nature argument is exactly what is required to avoid Newman's problem, and one would want it to be the case that both Russell and Eddington's deployment of consciousness as an intrinsic nature was explicitly directed at this issue. Alas, things are not nearly so clear cut (see Demopoulos and Friedman, 1985; Russell, 1968 [in which a letter to Newman appears on p. 176]; Braithwaite, 1940; Eddington, 1941; and Solomon, 1989). Nonetheless, this seems a very powerful argument against relationalism.[9]

Finally, we might appeal to the phenomenology of consciousness itself as evidence that some things (such as *me*) have intrinsic properties. As discussed above, my current state of consciousness seems to be something that could exist even if I was the only thing in the universe.[10] Its causal conditions are no doubt richly connected to a host of other things tracing back to the big bang, but in itself it appears serenely independent of everything else. Therefore, there are at least some beings which are not 'purely relational' and therefore pure relationalism is false. But, comes the objection, relationalism could be true of matter. Not if I am a material thing, as I appear to be.[11] Consciousness itself provides perhaps the best argument that there are

---

[9] It's worth pointing out that some philosophers (e.g. Maxwell, 1971; perhaps Chalmers, 2005) have taken *causation* to be among the qualitative properties needed to evade Newman's problem. To the extent that causation itself is seen to be purely relational this obviously won't do, so this manoeuvre appears to require that causal powers be determined by intrinsic features. It is also worth noting that Kant explicitly denied this (see Langton, 1998) and maintained that causal powers could vary across intrinsic identicals. There is some intuitive support for this too. Couldn't God have made G slightly different and thence have altered the causal powers of things without changing anything intrinsic about them? In fact, could God not set G to zero, and thus remove the causal power to gravitate without changing the intrinsic properties of matter? Other anti-Newman ploys depend upon distinctions between 'real' versus 'merely logical' or 'Cambridge' relations or other more or less dubious metaphysical manoeuvres.

[10] Perhaps it would be safer, given worries about content externalism, to assert only that a state of consciousness introspectively indistinguishable from my own is the sort of thing that could exist even if its subject was the sole inhabitant of the universe. Perceptual states provide more examples of extrinsic properties for which it is hard to see how they could be determined by intrinsic properties. Any state with external content will likewise provide an example of a seemingly pure extrinsic property.

[11] This raises the spectre of strange metaphysical view, the inverse of right-thinking philosophical common sense (as I see it), in which the intrinsic properties of things are determined by their relational properties. This would make it possible for consciousness to be an intrinsic property of material objects without requiring that the fundamental physical things have intrinsic properties of their own. One argument against this is the possibility that there are metaphysical *simples* which have no spatial parts but which do have

intrinsic properties, and this is exactly why consciousness appears as so troubling or even alien to the scientific picture of the world which deals exclusively with relational or structural features, leaving aside any attempt to grapple with intrinsic natures. This is a positive feature so long as science is dealing with, as Eddington puts it, 'meter readings' or the relations amongst observable events. But it prevents science from being able to even begin to address the problem of consciousness. Maybe this is the ultimate source of the explanatory gap, which shows the limits of the methodology of empirical science. This is perhaps the most important lesson we should draw from Strawson's article.

References

Bradley, F. (1893), *Appearance and Reality* (London: Swan Sonnenschein).

Braithwaite, R. (1940), 'Critical notice: *the philosophy of physical science*', *Mind*, **49**, pp. 455–66.

Chalmers, D. (2005), 'Russell, Newman and the mind-body problem', at *Fragments of Consciousness* (weblog), URL: http://fragments.consc.net/djc/2005/10/russell_newman_.html.

Demopoulos, W. and Friedman, M. (1985), 'Critical notice: *Bertrand Russell's The Analysis of Matter: its historical context and contemporary interest*', *Philosophy of Science*, **52**, pp. 621–39.

Dipert, R. (1997), 'The mathematical structure of the world: the world as graph', *Journal of Philosophy*, **94** (7), pp. 329–58.

Eddington, A. (1928), *The Nature of the Physical World* (New York: MacMillan).

Eddington, A. (1941), 'Discussion: group structure in physical science', *Mind*, **50**, pp. 268–79.

Falkenstein, L. (2001), 'Langton on things in themselves: a critique of *Kantian Humility*', *Kantian Review*, **5**, pp. 49–64.

Hardcastle (1996)

Humberstone, I. (1996), 'Intrinsic/extrinsic', *Synthese*, **108**, pp. 205–67.

James, W. (1890/1950), *The Principles of Psychology*, Vol. 1 (New York: Dover).

Kant, I. (1756/1986), 'Physical monadology', in *Kant's Latin Writings: Translations, Commentaries and Notes*, ed. and trans. L. Beck *et. al.* (New York: Lang).

Kim, J. (1982), 'Psychophysical supervenience', *Philosophical Studies*, **41**, pp. 51–70.

Ladyman, J. (1998), 'What is structural realism?', *Studies in the History and Philosophy of Science*, **29** (3), pp. 409–24.

Langton, R. (1998), *Kantian Humility: Our Ignorance of Things in Themselves* (Oxford: Oxford University Press).

Langton, R. and Lewis, D. (1998), 'Defining "intrinsic" ', *Philosophy and Phenomenological Research*, **58**.

---

properties of phenomenal consciousness. While I am not such a metaphysically simple entity, a state of consciousness indistinguishable from my own could be possessed by such a thing. Therefore it can't be the case that all intrinsic properties are dependent upon relational properties.

Leibniz, G. (1714/1989), *Monadology*, in *G.W. Leibniz: Philosophical Essays*, ed. and trans. R. Ariew and D. Garber (Indianapolis: Hackett).

Lewis, D. (1983), 'Extrinsic properties', *Synthese*, **44**, pp. 197–200.

Lycan, W. (1987), *Consciousness* (Cambridge, MA: MIT Press [Bradford Books]).

Maxwell, G. (1971), 'Structural realism and the meaning of theoretical terms', in *Analyses of Theories and Methods of Physics and Psychology*, ed. M. Radner and S. Winokur, Minnesota Studies in the Philosophy of Science, Vol. 8 (Minneapolis: University of Minnesota Press).

McGinn, C. (1989), 'Can we solve the mind–body problem?', *Mind*, **98**, pp. 349–66.

Moore, G. (1919), 'External and internal relations', *Proceedings of the Aristotelian Society*, **20**, pp. 40–62.

Nagel, T. (1974), 'Panpsychism', *Philosophical Review*, **53**. Reprinted in Nagel's *Mortal Questions* (Cambridge: Cambridge University Press, 1979, page references to the reprint).

Newman, M. (1928), 'Mr. Russell's causal theory of perception', *Mind*, **37**, pp. 137–48.

Russell, B. (1927), *The Analysis of Matter* (London: Routledge).

Russell, B. (1968), *The Autobiography of Bertrand Russell*, Vol. 2 (London: Allen and Unwin).

Solomon, G. (1989), 'An addendum to Demopoulos and Friedman (1985)', *Philosophy of Science*, **56** (3), pp. 497–501.

Strawson, G. (2006), 'Realistic monism: Why physicalism entails panpsychism', *Journal of Consciousness Studies*, **13** (10–11), pp. 3–31.

Teller, P. (1986), 'Relational holism and quantum mechanics', *British Journal for the Philosophy of Science*, **37**, pp. 71–81.

Worrall, J. (1989), 'Structural realism: the best of both worlds?', *Dialectica*, **43**, pp. 99–124. Reprinted in *The Philosophy of Science*, ed. D. Papineau (Oxford: Oxford University Press, 1996).

Peter Simons

The Seeds of Experience

Stripped of detail and rhetoric, here is an attempt to capture the gist of Galen Strawson's argument for panpsychism (Strawson, 2006).

(1) We cannot deny the existence of experience.
(2) Experience appears to emerge from physical phenomena that are not themselves experiential.
(3) Wholly non-experiential phenomena are not by their physical nature capable of giving rise to experience.
(4) Therefore either experience emerges magically from wholly non-experiential phenomena or the physical phenomena from which experience emerge are in some way themselves experiential.
(5) Magical or brute emergence is absurd.
(6) Therefore the physical phenomena from which experience emerge are in some way themselves experiential (Micropsychism).
(7) It is implausible to suppose that nature is so fragmentarily constituted that some physical phenomena are experiential while others are not.
(8) Therefore all physical phenomena are in some way experiential.
(9) But all phenomena are physical (Physicalism).
(10) Therefore all phenomena are in some way experiential (Panpsychism).

Put in this rather skeletal but I think helpful and reassuringly rational way, its conclusion may seem a little less shocking. But only a little. Panpsychism, at least in caricature, is one of the most immediately counterintuitive and off-putting of metaphysical positions. The idea of electrons making decisions about how to spin, nuclei harbouring intentions to split, or photons with existential Angst, makes idealism seem positively sane. That great philosophers such as Leibniz or Whitehead have been panpsychists is insufficient recommendation: everyone makes mistakes. That panpsychism is argued by Strawson to follow soberly from physicalism is guaranteed to outrage many who happily and proudly call themselves physicalists.

But let us leave affect and rhetoric aside and look at the argument. It is not deductively valid because there are certain steps, such as 5 or 7, which invoke plausibility considerations. There may be ways to tighten this up, but I shall not chase them. It is enough if the argument is relatively persuasive in form, and I think it is good enough to work with. That leaves the premisses, which are, I take it, 1, 2, 3 and 9. I shall not dispute 2 or 9. I agree with Strawson's rejection of what he calls physicSalism, the idea that the vocabulary and theory of physics suffices to describe everything, and for that reason I personally prefer to avoid the term 'physicalism' and prefer 'naturalism', but I will acquiesce in Strawson's terminology here.

Which leaves the exposed premisses 1 and 3. Premiss 1 accepts the existence of experience and appears to be pretty difficult to deny, though it has been done. Premiss 3 formulates the key idea that something (here: experience) whose nature is wholly alien to the nature of something else (here: physical phenomena) cannot emerge by virtue of the nature of the basis of the latter, that there must be something about the basis which is metaphysically apt to give rise naturally to the emergent. In metaphorical words, the basis must contain the seeds of the emergent phenomenon, for otherwise the emergence would be absurd, magical, miraculous, supernatural, contrary to nature: pick your insult.

For the moment I shall leave premiss 1 in place and assume there is such a thing as experience. Certainly anyone who has pondered the inner recesses of Husserlian phenomenology or felt the frustration at the belittlement of conscious experience by the behaviourists will be inclined to accept it. That leaves the question of emergence. Irrespective of the shockingness of Strawson's conclusion, the attention he directs on the concept of emergence is timely and valuable. Emergence is one of the most eel-like of metaphysical concepts. Not only is it invoked *in extremis*, as precisely when a non-dualist is trying to account for the existence of experience in a physical world, but there has to my knowledge never been anything approaching a decent analysis of what it means. One try at a definition is that a property is emergent which an object has that it neither inherits it directly from its parts nor in a straightforward summative way from the properties of its parts. This rough account may work for certain phenomena such as liquidity but it is frustratingly inept, because as Strawson points out, it is clearly of the nature of certain molecules that in certain circumstances when in proximity they slide past one another and that is what liquidity *is*. In other words, the parts may not have the property but they contain the seeds of the property. Whether we can predict the emergent property or not is neither here nor there: it is not epistemology but metaphysics that matters here. Liquidity is emergent, but intelligibly and illuminatingly so.

By contrast, the gulf between the physics of the brain and conscious experience appears to be so wide and their natures so radically heterogeneous that natural emergence appears to be excluded. That leads to philosophers adopting desperate measures: accepting dualism, or denying experience. If emergence *is* at work in the transition from fundamental particles to the experience of listening to Mozart's *Requiem* then it is an emergence which is so radical, so immense, as to be wholly mysterious and beyond our fathoming. Rather than admit defeat and impotence, Strawson turns the argument around and infers that the seeds of the Mozart experience must be in the molecules already. I hope it is clear that adopting panpsychism is of the same order of desperation as denying experience or accepting dualism, because to all appearances there is nothing like experience down among the quarks and leptons.

I think at this point the most sane and sober conclusion is that we simply do not know enough to see how experience emerges from the non-experiential, if it does. Admitting ignorance is unsatisfying and perhaps professionally embarrassing for philosophers, but there are times when it is folly or hubris to do more. Let us mention some sobering facts. We have been investigating the human nervous system seriously for a mere one hundred and fifty years or so. The human brain is the most complex object in the known universe and, professional pride of some neuroscientists notwithstanding, we have probably only begun to scratch the surface of what there is to know. I see no reason why a decisive breakthrough is likely to be round the corner in five or ten or fifty years. Perhaps it will be several hundred years before we have even a fair grasp of the detailed facts about how the brain works. Perhaps we will never know because its scale and complexity will defeat our collective intellectual capacity. Perhaps it will require us to come up with concepts of an audacity that rivals that of quantum theory or a complexity which requires us to rely on computers to do almost all of the work before we can adequately capture the facts. In these circumstances, it would I think be presumptuous to suppose that because we are currently unable to see how the emergence might work, that there can be no natural emergence. The emergence of consciousness is arguably *the* most untransparent transition in the history of the universe: it is uniquely difficult, so to suppose we can from our present state of rather doleful ignorance pronounce the natural emergence of conscious experience from the non-experiential as conclusively impossible is in its own way as presumptuous as those who think we'll have the answer in the next five years.

This deep and extensive lack of knowledge is very unexciting and disappointing, but in this case I think it is better to admit we don't know than to try and close the gap. If premiss 3 is not rejected but placed in sober *epoché*, Strawson's argument does not go through.

Let me return to premiss 1. We all know, from the inside, what it is like to be conscious. It does not automatically follow that there is such a thing as experience. For a start, experience is a very fleeting, intangible phenomenon, as numerous philosophers, such as Hume, Brentano, Husserl and William James have pointed out. When we try to nail down what it is, we find it elusive, even though we have such immediate lived experience. This raises the conceptual possibility that what we mean by 'experience' is not a qualitatively distinct phenomenon at all, that while there are acts of perception, judgment, desire, hate and so on, there is no common feature they all have which renders it apt to call them all 'experience'. I happen to think this scepticism is probably false, but it suggests that it is not necessarily pure absurdity or bloody-mindedness to deny the existence of experience, but could be metaphysical fastidiousness. Again, I do not think it is right to issue a blunt denial of experience, but I think a little more care is needed. Brentano's characterization of consciousness as being presentations or being based on presentations still strikes me as a reasonable shot. If that is right, we begin to see how one form of consciousness might be based or built on another. Allowing a central role to presentation would slightly reduce the explanatory gradient we need to climb, because it is highly plausible that presentation is a widespread animal phenomenon, and is almost certainly not attended everywhere by the limpid self-awareness that accompanies, or we fancy accompanies, our own distinguished version.

From our privileged intellectual and imaginative vantage point we can *perhaps* gain an inkling of what it is like, from the inside, to be a dog or a fish or an ant: almost certainly they can have no inkling of what it's like to be a human. The inkling takes the form of imagining being aware of something, visually, aurally, or by some other sensory mode. Now, presentation is, at least in large part, a causal phenomenon, whereby events outside the organism give rise to events inside it, and — and here we have the explanatory gap — as a result of *being* the organism with those events going on inside it, the events outside are *discerned*. I know these are all names which are just as problematic as 'experience' but it pays to ring the changes and reduce the impression that experience has to be some well-distinguished phenomenon. It can be as dull as the headache you know you have even while you are asleep. The key point is that there is no substitute for *being* the thing in question. Our limited empathy with other animals draws a blank long before we get to stones and rivers. While there are stones and rivers, they are not of a fashion for it to be possible for them to have presentations: we think they do not, which is why when we try to imagine what it's like to be a stone we draw a blank. Since they in fact do not, panpsychism is false. Not absurd: there *might* be a very dull what-it's-like-to-be-a-stone, as Leibniz imagined, well below the threshold of what we can imagine, but I think the physics tells pretty conclusively against it.

So, while I am still not inclined to deny premiss 1, I do not think it is foolish to question it. The concrete thing is the whole living presenting organism, and only that *lives through* the experience, as Husserl put it, but the experience is at best a rather thin trope-like property or process. So despite the gulf between quarks and listening to Mozart, I do think we can make a good guess at what might be significant milestones along the way: causal processes, specifically which serve to echo, mirror and finally present external events in such a way that the bearer of the resulting processes is, by being that bearer, thereby aware of different events and things. I know the slippery transition comes along here and there is no minimizing its importance: I am not trying to pull any wool over any eyes. We may wait for hundreds of years to find out how natural emergence works. We may never find out and yet it still could exist. Or Strawson could be right and it's experience all the way down. I might be wrong, but my money is on him being wrong.

Reference

Strawson, G. (2006), 'Realistic monism: Why physicalism entails panpsychism', *Journal of Consciousness Studies*, **13** (10–11), pp. 3–31.

David Skrbina

Realistic Panpsychism
Commentary on Strawson

Galen Strawson's 'Realistic monism' (2006) is a crisp and compelling paper, one that should put to rest many of those lingering concerns about how 'unreasonable' a view panpsychism is. Strawson succeeds in giving a most succinct and illuminating argument that radical emergence is untenable, and, therefore, that the experiential quality of mind is inherent in the material structure of the universe. This has profound and far-reaching implications, both within and outside conventional philosophy.

For many, this will be a shocking conclusion. The old intuitions about the uniqueness of humanity are deeply embedded in all of us, both through our cultural heritage of Christianity and via the secular, mechanistic worldview that predominates in official circles today. The Christian view grants mind and soul only to human beings, who, alone among the things of creation, are formed in the image of God. Apart from certain radical theologians, most notably St Francis, Christian orthodoxy allows no role for any psychic qualities in non-human nature. The mechanistic worldview is equally hostile to any form of panpsychism; matter is inherently lifeless and insentient, merely pushed about by the laws of physics. The emergence of mind and consciousness is an unexplained, and perhaps unexplainable, mystery.

It is a form of hubris to hold that only human beings, or (in slightly more enlightened thinking) only the 'higher animals', are capable of experiencing the world. There is certainly no evidence for this. Where it is held, whether explicitly or implicitly, it is pure presumption. Those who insist that panpsychism cannot be true are heirs to Descartes and his Christianized ontology. These neo-Cartesians see humans (or humans and higher animals, etc.) as miracles of existence, alone in possessing subjectivity, alone in holding a qualitative perspective on the world. Perhaps they get some satisfaction from this. Perhaps it is comforting to believe that consciousness is a rare thing — an *extremely* rare thing — in this universe, and that we are among the blessed few. But one should at least expect a modicum of evidence before accepting such a view. As Strawson points out, 'there is *absolutely no evidence whatsoever*' for an utterly non-mental reality.

Evidence in favour of panpsychism must of necessity be indirect and circumstantial, since we can have no direct access, nor even useful indirect access, to the internal world of non-humans. Arguments can be roughly grouped into positive and negative forms. The former attempt to show why panpsychism is likely to be true; the latter why non-panpsychism — i.e. neo-Cartesianism — is likely to be false. In the pages of this journal (Skrbina, 2003) and in my recent book *Panpsychism in the West* (Skrbina, 2005) I have argued that, historically, there have been many such arguments; in my book I identify twelve distinct ones. This is so because many of our greatest thinkers and philosophers have held to some version of panpsychism. Among the more notable luminaries we can include: virtually all the pre-Socratics, Plato (arguably), Aristotle, the Stoics, Bruno, Campanella, Spinoza, Leibniz, Diderot, Schopenhauer, Nietzsche, James, Peirce, Bergson (arguably), Whitehead and Hartshorne — not to mention a host of British scientist-philosophers: Priestley, Eddington, Jeans, Huxley, Haldane, Sherrington, Waddington, Bohm and Bateson.

Strawson introduces a new take on one of these arguments, that of Non-Emergence. This argument claims that mind or consciousness cannot plausibly emerge from non-conscious matter, and thus that the only alternative is to see all matter as in some sense enminded. This view originated long ago, in the work of Epicurus (ca. 300 BCE), who argued that the spontaneous action of the human will can exist only because will is exhibited by atoms themselves. Several other important thinkers accepted some form of this argument; among them, Telesio, Patrizi, Gilbert, Campanella, Fechner, Paulsen, Clifford and Strong. Even Thomas Nagel (1979) put forth a tentative version — though without endorsing it. But these non-emergence arguments were, for the most part, never really developed. Rather, they are typically stated in passing, as if the author finds the view almost self-evident — or at least as not requiring much elaboration.

To Strawson's credit, he sees that there is much more to be said; and much more at stake. If I may hazard a simplified reconstruction of his main points:

- There is one ultimate reality to the universe, which encompasses all real and concrete phenomena (monism, or 'realistic monism', or 'real physicalism', as he prefers).
- 'Mental' (experiential) phenomena are a part of this monistic reality, and hence are 'physical' (as distinct from 'physicSal', i.e. reality as described by modern physics).
- 'Radical kind', or brute, emergence is impossible, i.e. mental phenomena cannot arise from any purely non-mental stuff (which exhibits only shape-size-mass-charge-etc. phenomena).
- Therefore, the one reality is inherently experiential.

Many common forms of emergence, Strawson argues, *are* comprehensible; liquidity reduces to shape-size-mass ('P') phenomena, life does the same, spatial things reduce to spatiality, and so on. The emergent quality is 'conceptually homogenous' with the underlying phenomena. But experientiality is not like this; it is conceptually *heterogeneous* with any supposedly non-experiential physical reality. Unless we allow for miracles, mind cannot emerge from a non-mental substrate. So we either deny our own consciousness (absurd, in Strawson's view), or we accept that experientiality is a fundamental part of reality.

Though he doesn't use the term, Strawson's view is essentially that of *dual-aspect monism*. This is an approach that dates back at least to Spinoza: There is one ultimate substance of the world, but it exhibits two faces — 'physical' from the 'outside', and 'mental' from the 'inside'. Since the one reality is simultaneously physical and mental, the problem of emergence does not arise. Furthermore, dual-aspect monism strongly urges one toward panpsychism, since it is very hard to see how such a world could exclude mind from any part of reality. In fact, many dual-aspectists have reached a panpsychist conclusion; among these, Schopenhauer, Peirce, Royce, as well as Spinoza.

Strawson's citations from Eddington are instructive. They demonstrate the insightful metaphysical sensitivities of that great British astrophysicist. The passage from Eddington's *Nature of the Physical World* (1928) is especially interesting (which I partially repeat from Strawson's piece for purposes of comparison):

> The recognition that our knowledge of the objects treated in physics consists solely of readings of pointers and other indicators transforms our view of the status of physical knowledge in a fundamental way ... Take the living human brain ... I know that I think, with a certainty which I cannot attribute to any of my physical knowledge of the world. ... How can this collection of ordinary atoms be a thinking machine? ...

> [N]ow we realize that science has nothing to say as to the intrinsic nature of the atom . . . Why not then attach it to something of a spiritual nature of which a prominent characteristic is *thought* . . . We have dismissed all preconception as to the background of our pointer readings, and for the most part we can discover nothing as to its nature. But in one case — namely, my own brain — I have an insight which is not limited to the evidence of the pointer readings. That insight shows that they are attached to a background of consciousness. [Furthermore] I may expect that the background of other pointer readings in physics is of a nature continuous with that revealed to me in this particular case . . .

Such words recall the thinking of Schopenhauer, one of the most prominent panpsychists in Western thought. Eddington's 'foreground' Schopenhauer interprets as a collection of sensory impressions, as per classical idealism. Eddington's 'background' is, for Schopenhauer, that which we perceive within ourselves, namely, *will*. Compare his words:

> The double knowledge [of the world as both will and idea] given in two completely different ways . . . has now been clearly brought out. We shall accordingly make further use of it as a key to the character of every phenomena in nature, and shall judge all objects which are not our own bodies . . . according to the analogy of our own bodies, and hence shall assume that as on one hand they are idea, just like our bodies, . . . so on the other hand, what remains of objects when we set aside their existence as idea of the subject, must in its inner nature be the same as that in us which we call *will*. For what other kind of existence or reality should we attribute to the rest of the material world? Where should we obtain the elements from which to put together such a world? Besides will and idea nothing is known to us, or is even conceivable. (Schopenhauer, 1819/1995, pp. 37–8)

His 1928 work was not the only time Eddington ventured into panpsychist territory. His earlier book, *Space, Time, and Gravitation* (1920) included this observation:

> In regard to the nature of things, this knowledge [of relativity] is only an empty shell — a form of symbols. It is knowledge of structural form, and not knowledge of content. All through the physical world runs that unknown content, which must surely be the stuff of our consciousness. Here is a hint of aspects deep within the world of physics, and yet unattainable by the methods of physics. (p. 200)

Later, in 1939, his *Philosophy of Physical Science* noted the following:

> The recognition that physical knowledge is structural knowledge abolishes all dualism of consciousness and matter . . . When we take a structure of sensations in a particular consciousness and describe it in physical terms as part of the structure of an external world, it is still a

structure of sensations. It would be entirely pointless to invent something else for it to be a structure of. Or, to put it another way, there is no point in inventing non-physical replicas of certain portions of the structure of the external world and transferring to the replicas the non-structural qualities of which we are aware in sensation. The portions of the external universe of which we have additional knowledge by direct awareness amount to a very small fraction of the whole; of the rest we know only the structure, and not what it is a structure of.

Although the statement that the universe is of the nature of 'a thought or sensation in a universal Mind' is open to criticism, it does at least avoid [a] logical confusion. It is, I think, true in the sense that it is a logical consequence of the form of thought which formulates our knowledge as a description of a universe. (pp. 150–1)

This latter passage seems to incline toward absolute idealism, but is certainly consistent with a panpsychist ontology.

Strawson grants that problems remain with a panpsychist real physicalism. He accepts that 'whatever problems are raised by this [panpsychism] are problems a real physicalist must face'. He spells out three noteworthy concerns. First is the 'combination problem', the question of how many, small atomic experiencers can combine to form, for example, our singular sense of consciousness. He cites James (1890) as elucidating this issue, but by the time of his 1907 Hibbert lectures James seems to have reconciled himself with it.

James's 1890 work was critical of panpsychism because he could not, at that time, envision how lower-level bits of 'mind-dust' or 'mental atoms' could combine into higher-order minds. As he recounts:

> I found myself obliged, in discussing the mind-dust theory, to urge [the view that higher-level states are not combinations.] The so-called mental compounds are simple psychic reactions of a higher type. The form itself of them, I said, is something new . . . There is thus something new in the collective consciousness . . . The theory of combination, I was forced to conclude, is thus untenable, being both logically nonsensical and practically unnecessary . . . The higher thoughts, I insisted, are psychic units, not compounds. (James, 1909/1996, pp. 188–9)

He notes that 'for many years I held rigorously to this view', a view compelled by ordinary reasoning and logic. But by 1907 he was ready to give it up. If compounding of mental entities is not allowed, he said, then we are forced to accept a radically discontinuous picture of the world:

> [I]f we realize the whole philosophic situation thus produced [by non-compounding], we see that it is almost intolerable. Loyal to the logical kind of rationality, it is disloyal to every other kind. It makes the universe discontinuous . . . I was envious of Fechner and the other

pantheists [read: panpsychists] because I myself wanted the same freedom that I saw them unscrupulously enjoying, of letting mental fields compound themselves and so make the universe more continuous ... In my heart of hearts, however, I knew that my situation was absurd and could be only provisional. That secret of a continuous life which the universe knows by heart, and acts on every instant, cannot be a contradiction incarnate. If logic says it is one, so much the worse for logic. (James, 1909/1996, pp. 205–7)

James thus found himself 'compelled to give up the logic, fairly, squarely, and irrevocably' that demanded a non-compounding of mind. 'Reality, life, experience, concreteness, immediacy, use what word you will, exceeds our logic, overflows and surrounds it' (p. 212). Ultimately we find ourselves swept into 'the great empirical movement towards a pluralistic, panpsychic view of the universe' (p. 313). All this suggests that ordinary analytical logic cannot adequately grasp the notion of panpsychism. Perhaps this is why so many philosophers today have such a hard time accepting it.

Strawson's second problem is that we need to better understand the vast range of qualia experienced by objects ranging from atoms to humans to (perhaps) higher-order entities. Just as we see a spectacular range of physical complexity in things, so we should expect, and perhaps be able to describe, a correspondingly spectacular range of subjectivity in things.

Third, he suggests that we need to develop a more refined metaphysics, one that can engender a more 'vivid sense' of this panpsychist nature of the universe. He refers to the process nature of things, suggesting that perhaps a process philosophy view, such as that of panpsychists Whitehead, Russell (arguably), Hartshorne, or David Ray Griffin, might be fruitful.

These last two issues are not so much problems as tasks, areas needing further development. The combination problem is significant but not insurmountable; certainly it is less daunting than articulating a comprehensible theory of radical emergence of mind from utterly mindless matter.

I would add that, in addition, we might encourage a greater understanding and respect for the panpsychist insights of those leading thinkers of the past. Panpsychism has a long and noble legacy in Western thought, but this legacy is almost completely unknown — to the point that even as eminent a philosopher as John Searle could, in 1997, call panpsychism 'breathtakingly implausible', and state that 'there is not the slightest reason to adopt panpsychism'.

Clearly we have far to go. But work is proceeding on several fronts. Strawson's analytical approach is one of at least six distinct lines of thought pursuing panpsychist ontologies. Others include:

- The *quantum physics* approach, as articulated in various forms by Haldane, Bohm, Seager and Hameroff.
- The *information theory* approach, as developed by Bateson, Bohm and the early Chalmers.
- The *process philosophy* approach, originated by Bergson and James, and further elaborated by Whitehead, Hartshorne, Griffin, DeQuincey, Manzotti (2006a, 2006b) and many others.
- The *part-whole 'holarchic'* approach, as conceived by Cardano in the sixteenth century and given new life by Koestler and Wilber.
- The *nonlinear dynamics* approach, envisioned by Peirce (1892) and further developed in various forms by myself, Deiss (2006), and others.

To be sure, a fully articulated metaphysic of panpsychism remains far off. But Strawson's (2006) groundbreaking paper goes a long way toward this end.

References

Deiss, S. (2006), 'What it all means: in search of the UNCC', http://appliedneuro.com/WIAM_Submission_Draft.pdf
Eddington, A. (1920), *Space, Time, and Gravitation* (Cambridge: Cambridge University Press).
Eddington, A. (1928), *The Nature of the Physical World* (New York: Macmillan).
Eddington, A. (1939), *The Philosophy of Physical Science* (Cambridge: Cambridge University Press).
James, W. (1890), *Principles of Psychology* (New York: Dover).
James, W. (1909/1996), *A Pluralistic Universe* (Lincoln, NE: University of Nebraska Press).
Manzotti, R. (2006a), 'Consciousness and existence as a process', in *Mind and Matter*, **4** (1), pp. 7–43.
Manzotti, R. (2006b), 'Outline of an alternative view of conscious perception and representation', *Journal of Consciousness Studies*, **13** (6), p. 7–41.
Nagel, T. (1979), *Mortal Questions* (Cambridge: Cambridge University Press).
Peirce, C. (1892), 'Man's glassy essence', *Monist*, **3** (1) (reprinted in *The Essential Peirce*, Vol. 1 (1992), ed. N. House and C. Kloesel [Bloomington: Indiana University Press]).
Schopenhauer, A. (1819/1995), *The World as Will and Idea* (London: J.M. Dent).
Searle, J. (1997), 'Consciousness and the philosophers', *New York Review of Books*, **44** (4), pp. 43–50.
Skrbina, D. (2003), 'Panpsychism as an underlying theme in Western philosophy', *Journal of Consciousness Studies*, **10** (3), pp. 4–46.
Skrbina, D. (2005), *Panpsychism in the West* (Cambridge, MA: MIT Press).
Strawson, G. (2006), 'Realistic monism: Why physicalism entails panpsychism', *Journal of Consciousness Studies*, **13** (10–11), pp. 3–31.

J.J.C. Smart

Ockhamist Comments on Strawson

Despite admiration for his brilliant rhetoric and his ontological seriousness, I am very much opposed to Galen Strawson's panpsychism or even to what he calls 'micropsychism'. He calls himself a real physicalist but from my point of view he is not really a physicalist. He holds that 'experiences' should count as a primitive term of the vocabulary of physics. This, to me but not to him, smells a bit like what Russell called the advantage of theft over honest toil (Russell, 1919, p. 71). Still, I do not deny that experiences exist. I believe that experiences are brain processes and since brain processes exist so must the relevant experiences. What I deny is that experiences have non-physical properties (qualia). Let us say that the experience of seeing a red tomato involves having a red sense datum. The experience is not red. It is just the sort of experience that I have when a red tomato is before my eyes, illuminated in bright sunlight. What I deny is that experiences have (in my sense) non-physical qualities. (There may be a sensible sense of 'quale' used by cognitive scientists whereby a quale is a point on a similarity space or something like that but I shall ignore this because it is obviously harmless to my sort of physicalist.)

Strawson has his use of 'physicalism' because he simply adds on 'experience' to the terms of physicalism. His use of the word 'experience' does not fit into physics in the way that 'brain process' does. There are physical and chemical theories of nerve transmission and of how neurons behave like switching devices and so on. Nothing like that with qualia.

It seems that in philosophy there are rarely knock-down arguments (Smart, 1993). So though Strawson's speculations seem bizarre to me I do not aspire to convince him, though if I do so much the better. The reason why there are hardly ever completely knock-down arguments, except between very like minded philosophers, is that philosophers, unlike chemists and geologists, are licensed to question everything, including methodology. There are even philosophers who question the law of non-contradiction, and bizarre though this may seem to most of us, one can admire the way that they develop a dialethic logic which prevents the denial of the law of non-contradiction from trivializing proof. Of course, in philosophy knock-down argument may be possible when there is already a fair amount of agreement between the proponents. Failing this we may end by trading off plausibilities. Even this may not work and we end up with a restrained (no cheap debating tricks allowed) rhetoric and there is no doubt that Strawson is a fine rhetorician.

One mistake that I see in Strawson is that he is too empiricist. In claiming knowledge of the properties of experience he can get into disastrous metaphysics. One must remember Mach and the logical positivists who tried to reduce this great universe of ours to actual and possible sense data. I do not mean to say that Strawson is reductionist in the way Mill (who said that matter is a permanent possibility of sensation) or the logical positivists were. Indeed quite the reverse. But about the nature of experience he is like them in being too empiricist. In some respects F.H. Bradley had a better epistemology of science than had Mill. We need elements of a coherence theory of knowledge (but emphatically *not* of truth).

For lack of space I cannot give a full defence of my position, nor would it be appropriate here to do so, but I will indicate why I think differently and why others should think more or less in my way and differently from Strawson. In particular we should be very suspicious of intuition and of phenomenology. I use this last word in the sensible way that Strawson might (see the second paragraph of his 2006 article) and not as the name of a school of unintelligible German philosophers.

In physics we test our theories by experiment and observation. However it would be wrong to say that we test them by our experiences. We test them by experiment and observation. Let us adopt David Armstrong's account of perception as coming to believe by means of the senses (Armstrong, 1993). Unimportantly for present purposes he came to prefer to talk of information rather than belief. We come to believe about the blue and white bird on the gatepost, not about our experience. On the higher order theory of consciousness developed by Armstrong, Rosenthal, Lycan and others we can perceive without consciousness, that is, we can be on 'automatic pilot' and aware of events, for example a car approaching us on the wrong side of the road, but not be aware of our awareness We can think of consciousness as awareness of awareness. We can think of the second order awareness as the coming to believe by one part of our brain about a process in another part of our brain. Since the brain is part of our body, Armstrong has compared inner sense with kinesthesis. The awareness of the awareness is not awareness of a quale. The bird on the gatepost may be blue and white, but not our experience of it. In this sort of way I would deny the existence of any (for me mysterious) qualia. The understanding of experience will come from neuroscience, not from inner sense.

Nearly fifty years ago I used the words 'topic neutral' to denote neutrality between materialism and dualism. I borrowed the words from Gilbert Ryle, who had used them in a more general sense. Ryle used them to denote logical words such as 'if', 'and', 'not', 'all', 'because'. If you heard only such words in a conversation you would have no idea of the topic of discourse. In my restricted sense the neutrality is between dualism and materialism or physicalism (in my sense of the latter word, not Strawson's). The topic neutral idea is absolutely vital to the identity theory of mind (Smart, 1999). Since otherwise we would be landed with property dualism. I believe that Strawson with his extended sense of 'physical' conceals this. Some people think that the identity theory has been supplanted by functionalism, but there is little ontological difference between them since most functionalists would be happy to say that the categorical bases of the functional states and processes are purely physical brain states and processes.

What needs to be done now is to discuss Strawson's appeal to phenomenology. His qualities of experience seem to me to be the same as properties of what used to be called 'sense data' or of mental images. Having a mental image is the brain putting itself through similar motions that occur in having a sense datum: it is a sort of pretence seeing. Now a brain process cannot be red, white and blue as is a union jack. One might talk of a red, white and blue sense datum, but I contend that there are no such thing as sense data and mental images. There is only havings of them. The having of a red, white and blue sense datum is the functionally described process that typically occurs when a union jack is before the eyes of a normal human percipient in clear light. 'Normal human percipient' can be defined without circularity. Colours themselves are very disjunctive and idiosyncratic physical properties of the surfaces of objects. Some easily made epicycles have to be made to deal with such things as radiant light and the colour of the sun at sunset. So anthropocentric and disjunctive, probably of no interest to denizens of other planets elsewhere in the universe. Maybe not quite as disjunctive as I thought since David R.

Hilbert has argued for at least approximately identifying the colours of surfaces with reflectances. Reflectances are well defined physical properties (Hilbert, 1987). The above considerations enable me to avoid phenomenology and provide what I contend is a much more plausible alternative to Strawson's speculations. In some ways Strawson's panpsychism reminds me of Whitehead's *Process and Reality*, for neither of them are the experiential properties 'emergent': on the contrary they are universal. Like Strawson I distinguish a harmless sense of 'emergence' (which Strawson illustrates well with the existence of liquidity) from a highly dubious sense of this word. The ability of a radio to tune in the BBC is emergent in this harmless sense (Smart, 1981). The functioning of the receiver can be explained by physics plus wiring diagrams and the physics of the various components. In the same way I regard the biochemical core of biology to be physics and chemistry plus generalizations (not laws) of natural history, even though these generalizations of natural history may be known only with sophisticated apparatus. C.D. Broad in his *The Mind and its Place in Nature*, mentioned by Strawson, espoused the dubious sense when he said that no amount of knowledge of the properties of sodium and chloride would explain the properties of sodium chloride, seemingly unaware that scientists were busy working on the theory of the chemical bond. Strawson's *qualia* have presumably always existed and so in a temporal sense they cannot have emerged (though I wonder about the milliseconds after the big bang) but in an atemporal sense they might be thought to have emerged in the dubious sense of this word.

A root of my disagreement with Strawson about the value of phenomenology comes from my assertion of the topic neutrality of ordinary talk about experience. In any case the notion of experience is itself elusive (Farrell, 1950). We can certainly say that we have experiences and I say that they are physical in my narrower sense of 'physical' with no need for Strawson's extended sense of this word. There can be metaphysical illusions (Smart, 2006; Armstrong, 1998). I hold that for there to be a metaphysical illusion there must be in addition a mistake in logic, but there must be a strong psychological pressure to make it.

I apologise for so many references to my own ideas but I think that the only way to criticize Strawson's bold speculation is to indicate, if only briefly, an alternative that may appeal to some readers as more believable. I am not clear that I have sufficiently understood Strawson and if I have not I expect that he will put me right in his Response.

References

Armstrong, D.M. (1968), 'The headless woman illusion and the defence of materialism', *Analysis*, **29**, pp. 48–9.
Armstrong, D.M. (1993), *A Materialist Theory of the Mind* (London: Routledge).
Farrell, B.A. (1950), 'Experience', *Mind*, **59**, pp. 170–98.
Hilbert, D.R. (1987), *Color and Color Perception* (Stanford, CA: CSLI).
Russell, B. (1919), *Introduction to Mathematical Philosophy* (London: Allen and Unwin).
Smart, J.J.C. (1981), 'Physicalism and emergence', *Neuroscience*, **6**, pp. 109–13.
Smart, J.J.C. (1993), 'Why philosophers disagree', in *Reconstructing Philosophy: New Essays on Meta Philosophy*, ed. Jocylyne Couture and Kai Nielsen (Calgary, Alberta: University of Calgary Press), pp. 67–82.
Smart, J.J.C. (2000), 'The identity theory of mind', in *The Stanford Encyclopedia of Philosophy* (on line).
Smart, J.J.C. (2004), 'Consciousness and awareness', *Journal of Consciousness Studies*, **11** (2), pp. 41–50.
Smart, J.J.C. (2006), 'Metaphysical illusions', *Australasian Journal of Philosophy*, **84**, pp. 167–75.
Strawson, G. (2006), 'Realistic monism: Why physicalism entails panpsychism', *Journal of Consciousness Studies*, **13** (10–11), pp. 3–31.

H.P. Stapp

Commentary on Strawson's Target Article

Strawson's primary claim is that 'physicalism entails panpsychism' (Strawson, 2006).[1] This claim would be surprising if it meant what it seems to mean. But it does not.

According to Strawson's words, 'physicalism' is the doctrine that every real temporally located existent *is physical*; and 'panpsychism' is the assertion that *the existence of every real temporally located existent involves experiential being*. Here, and throughout, I have, in accordance with the meanings specified at the beginning of Strawson's article, replaced 'concrete phenomena' and 'concrete thing' by 'temporally located existent' (p. 3).

According to Strawson, the key phrase 'is physical' in the definition of physicalism *normally* means '*can in principle be fully captured in the terms of physics*' (p. 4). However, Strawson emphasizes that experiences are temporally located existents, and claims that this fact forces him to distinguish the usual meaning of physicalism, which he dubs 'physicSalism', from 'real physicalism'. Thus he writes:

> It follows that real physicalism can have nothing to do with *physicSalism*, the view — the faith — that the nature or essence of all real temporally located existents can in principle be fully captured in the terms of *physics* ... It is unfortunate that 'physicalism' is today standardly used to mean physicSalism because it obliges me to speak of 'real physicalism' when really I only mean 'physicalism' — realistic physicalism.(pp. 3–4)

This *twisting* of the meaning of Strawson's primary claim (produced by relabelling the usual concept by the awkward term 'physicSalism', and then using the word 'physicalism' to denote a differing concept of 'real physicalism') is furthered by his use of the loose term 'involved'

---

[1] Unreferenced page numbers refer to this target article.

in his characterization of panpsychism, which he defines as 'the view that the existence of every real temporally located existent *involves* experiential being'. But the existence of any real thing in the universe may *involve* every other real thing, in some general sense, for one cannot simply pluck one part of reality out from the rest. Reality may exist only as a whole, in which case each reality may *involve* experiential being in some way, which in many cases could be quite indirect. Consequently, Strawson's claim of 'panpsychism' is very weak compared to the *normal* claim, characterized (Honderich, 1995) as 'the doctrine that each spatio-temporal thing has a mental or "inner" aspect'. The assertion that each spatio-temporal thing *has an experiential inner aspect* is a more stringent condition than the assertion that each such thing is merely *involved* — perhaps from afar — with experiential being.

In view of these shifts in the specified meanings of the key words, it must be recognized that Strawson's claim that '*physicalism* entails *panpsychism*' does not mean what it would mean if more normal meanings of the two nouns were used.

In commenting on Strawson's article, it is therefore important to determine: (1) *what*, in normal terms, is Strawson actually claiming to prove; (2) *how does* he claim to prove it; (3) *is* his proof valid; and (4) *what* does contemporary science have to say about the matter.

Strawson does not explicitly define what *real physicalism* means. But conditions on its meaning are imposed by how it is used in his proof of his primary claim.

Strawson's proof goes as follows. First, he assumes that physicalism is true: he assumes that every real temporally located existent *is physical*. He then emphasizes that experiences are such existents, and thus concludes that *experiences are physical*. He then argues:

> It follows that real physicalism can have nothing to do with *physicSalism*, the view — the faith — that the nature or essence of all concrete reality can in principle be fully captured in the terms of *physics*. Real physicalism cannot have anything to do with physicSalism unless it is supposed — obviously falsely — that the terms of physics can fully capture the nature or essence of experience. (p. 3)

Thus Strawson assumes physicalism to be true, claims physicSalism to be 'obviously' false, and concludes that physicalism cannot be normal physicSalism. This means that the assumed-to-be-true physicalism must be something else, which he calls also *real physicalism*. This argument rests squarely on Strawson's claim that it is 'obviously' false that '*the terms of physics* can fully capture the nature or essence of experience'.

But is this really 'obvious'?

The truth of this claim depends upon what the terms of physics actually are, and what they represent. It is, of course, certainly true that the nineteenth-century classical-physics extension of the seventeenth-century physics of Galileo and Newton does not include among its terms any representations of experiential realities. But *contemporary* physical theory differs fundamentally from that earlier classical physics, basically because it incorporates explicitly into the dynamics both *conscious choices* made by human experimenters about how they will act, and also the *experiential increments of knowledge* that constitute the experiential feedbacks from these consciously chosen actions. Orthodox contemporary physics is in fact built upon these experiential realities, and it has terms that represent these realities, as they actually exist in our streams of consciousness. So if by 'physics' is meant orthodox contemporary valid physics, rather than the known-to-be-fundamentally-false classical physics, then Strawson's argument fails. Certain terms of orthodox contemporary physics do denote, precisely, various kind experiences — namely our experienced choices and experienced feedbacks — that actually appear in our streams of consciousness.

Thus if by 'physics' one means valid contemporary physics, rather than invalid classical physics then *ordinary* physicalism — i.e. physicSalism — is valid. But the validity of ordinary physicalism does not entail the validity of ordinary panpsychism. It does not entail that each spatio-temporal thing has a mental or 'inner' aspect. For in orthodox quantum physics the explicitly experiential aspects that enter the theory are mental or 'inner' aspects *of human beings*: actual experiences are not associated indiscriminately with every spatio-temporally located existent. Hence, if the normal meanings of the terms are used, and 'physics' means orthodox contemporary physics, then physicalism does *not* entail panpsychism, and Strawson's primary claim fails.

In view of the importance of the difference between classical physics and quantum physics in analysis of Strawson's arguments, and, by extension, in the general arena of the study of the connection between mind and matter, it may be useful to elaborate upon this difference within the general philosophical context of Strawson's article.

I begin by giving a very brief description, from a philosophical standpoint, of the essential differences between classical physics and orthodox quantum theory. Classical physics represents the physical world by assigning a set of numbers to every point in space for every time t in some interval, and it specifies the evolution of the universe between any two times t_1 and t_2 in this interval by giving rules that specify how each number at each point x changes (with the passage of time) in terms of the numbers assigned to points in the immediate neighbourhood of x. These local rules suffice to determine, unambiguously, the values of all the numeric variables associated with any later time t from the values of all numeric variables associated with any earlier time. That is, the physical dynamics is deterministic: it manifests causal closure of the physical. All physically described properties are unambiguously determined by earlier physically described properties, where 'physically described' means described in terms of the definite numbers attached to the various points of space-time.

Full reality includes, however, our human experiences. These experiences are *not* represented in the terms used in classical physics, and their developments in time are not specified by the rules of classical physics. The causal closure of the physical manifested by classical physics renders these unconstrained experiential aspects redundant, or epiphenomenal: given the numeric variables at any one time (or some very short interval of time), it makes no difference in the physical future what the flows of our conscious experiences have been, are, or will be.

Orthodox quantum theory is different in a fundamental way. The dynamics is described in terms of an *evolving quantum state*, and a sequence of *abrupt quantum events*. The evolving quantum *state* represents 'our (evolving) knowledge' and also 'objective tendencies for future quantum events to occur'. Each quantum *event* represents 'an increment in knowledge' coupled with a reduction of the quantum state to a form compatible with this new knowledge. This reduction changes the objective tendencies associated with future experiential events.

Von Neumann calls the abrupt changes in the quantum state by the name 'Process 1'. Between these sudden events the quantum state evolves in accordance with a quantum generalization of the physically deterministic local laws of classical mechanics. Von Neumann calls this evolution by the name 'Process 2'. But there is a dynamical gap in the theory. It pertains to the form and timing of the Process 1 events. There are *no rules in the theory, statistical or otherwise*, that fix either *which* of all the mathematically possible Process 1 actions actually occurs, or *when* it occurs. In actual practice the Process 1 action constitutes a *probing action* performed upon a probed quantum system by a probing conscious experimenter, and the choice of the occurring

Process 1 is specified by the consciousness of the probing person/agent, on the basis of his agenda or preferences. In the von Neumann formulation, in which the entire physical universe, including the bodies and brains of the experimenter/observers, are taken to be the probed quantum system, the conscious choices correspond directly to Process 1 events *acting upon* the brain/bodies of the experimenter/observers. These actions upon human brains appear to come from the mental realm, and they are not determined by any yet-known rules or laws.

This description of the essential conceptual structure of (ontologically construed) orthodox von Neumann quantum theory is a compact summary of more detailed descriptions given elsewhere (Stapp, 2003; 2005; Schwartz *et al.*, 2005). But this brief account will be sufficient for the present, primarily philosophical, purposes.

The main consequence of identifying 'physics' with 'contemporary orthodox quantum theory' has already been spelled out. Orthodox ontologically construed quantum theory injects our conscious experiences into both the dynamics and the basic ontology of contemporary physical theory; and these experiences, as they actually occur — or might occur — in our streams of consciousness, are represented in the terminology employed by the theory. Thus quantum theory meets Strawson's demand that 'full recognition of the reality of experience ... is the obligatory starting point for any remotely realistic version of physicalism'.

On the other hand, due to the explicit incorporation of our experiences into physics achieved by quantum theory, the entire contentious issue pertaining to 'physicalism' effectively evaporates. That issue was created by the fact that the terms of classical physics left out an important aspect of reality, namely our experiences, and this omission raised the problem of how to reconcile the existence of these definite realities with our science-based understanding of nature. That problem has been resolved, not by abstruse philosophical analysis, but by a profound advance in physics that recognized and exploited philosophical understandings that were prevalent already in the early twentieth century.

Quantum theory can thus be viewed as a brand of physicalism in which the physics encompasses human experience. The concept of 'physicalism' is thereby enlarged, and rescued. On the other hand, since quantum theory is built upon experiences, one might also be justified in calling it idealistic. Indeed, the actual events of quantum theory are experienced increments in knowledge, and hence are idea-like, and the evolving quantum state represents *a state of knowledge*, which is also an idea-like reality. The quantum state represents also a set of tendencies for new experiences to occur, and this is ontologically like an imagined state of potentialities for future experiences. Hence quantum theory could quite fairly be characterized as a dualistic idealism, with its two idea-like components being experiencial increments in knowledge and the evolving mathematically described 'potentialities' for such experiences to occur.

These considerations mean that a thinker with one viewpoint could call quantum theory a brand of *physicalism*, whereas a thinker with another viewpoint could call this very same theory a brand of *idealism*. The two labels merely emphasize two different aspects of one logically coherent contemporary understanding of nature, within which, in Strawson's words, 'no problem of incompatibility arises'.

If one adopts the idealistic viewpoint, in which the evolving quantum state of the universe is essentially idea-like — like an imaginary idea for what future experiences might be — then it becomes reasonable to say that the theory is 'panpsychic', in the weak sense introduced by Strawson.

What all this means is that the terms 'physicalism', 'idealism' and 'panpsychism' carry the huge baggage loaded upon them by centuries of essentially unquestioned acceptance of the basic concepts of classical physics; and that the way out of the dilemmas generated by using the intuitions arising from the earlier false physics is to use in philosophical studies of the nature of reality only concepts compatible with empirically adequate contemporary physics.

Within the quantum context Strawson's observations about Descartes seem justified and pertinent. Descartes proposed that the aspects of nature characterized in idea-like (psychologically described) terms and those characterized in terms of mathematical properties localized in space-time *interact within the brains of human beings*. This is exactly what happens in orthodox ontologically construed von Neumann quantum theory.

Strawson is also justified in pointing to the irrationality of 'the most fervent revilers of the great Descartes ... who have made the mistake with most intensity' of themselves believing what they (incorrectly) ascribe to Descartes, and revile him for. They accept that 'the experiential and physical are [so] utterly and irreconcilably different, that they [the revilers] are prepared to deny the existence of experience' (p. 5). But quantum mechanics reveals how these two aspects of nature can co-exist and interact within human brains in essentially the way proposed by Descartes.

Strawson interprets the word 'physical' broadly, so that it encompasses also the mental/experiential aspects of reality. Then everything we know about can be called physical. He can then say, with other physicalists, that 'experience is "really just neurons firing" '. But he goes on to insist that this does mean that 'all characteristics of what is going on, in the case of experience, can be described by physics and neurophysiology or any non-revolutionary extension of them'. He claims that 'there is a lot more to neurons than physics and neurophysiology records (or can record)'.

Strawson's solution is close to being just a word game: 'the physical' is asserted to encompasses our experiences *but physics and neurophysiology do not*. So Strawson hangs on to 'physicalism' by allowing what he calls 'the physical' to go — by virtue of a contravention of both the traditional and natural meaning of this word — beyond physics and neurophysiology. But the resolution of these problems provided by quantum mechanics is not just a shuffling of the meanings of words. It is an explicit conceptual structure that combines aspects that are described in physical terms (i.e. by assigning mathematical properties to points in space-time) with aspects of reality that are described in psychological terms, in such a way as to produce very accurate and useful predictions about future experiences from knowledge derived from past experiences.

References

Honderich, T. (1995), *Oxford Companion to Philosophy* (Oxford: Oxford University Press), p. 641.

Schwartz, J., Stapp, H. and Beauregard, M. (2005), 'Quantum physics in neuroscience and psychology: a neurophysical model of the mind/brain interaction', *Phil. Trans. Royal Society*: B 360 (1458) (1309–27).

Stapp, H.P. (2003), *Mind, Matter, and Quantum Mechanics* (Berlin, Heidelberg, New York: Springer), Chs. 6 and 12.

Stapp, H.P. (2005), 'Quantum interactive dualism: an alternative to materialism', *Journal of Consciousness Studies*, **12** (11), pp. 43–58.

Strawson, G. (2006), 'Realistic monism: why physicalism entails panpsychism', *Journal of Consciousness Studies*, **13** (10–11), pp. 3–31.

Daniel Stoljar

Comments on Galen Strawson
*'Realistic Monism:
Why Physicalism Entails Panpsychism'*[1]

There is at least one element in Strawson's extremely rich paper that seems to me to be correct and important, and Strawson is absolutely right to bring it out. This is the point that people in philosophy of mind go around assuming that they know what the physical facts are, if not in detail then in outline: '… they think they know a lot about the nature of the physical' (Strawson, 2006, p. 4). This assumption is false, or at any rate implausible, or at any rate un-argued for. To make the assumption, Strawson says, is a 'very large mistake. It is perhaps Descartes's, or perhaps rather "Descartes's", greatest mistake, and it is funny that in the past fifty years it has been the most fervent revilers of the great Descartes, the true father of modern materialism, who have made the mistake with most intensity' (p. 5; footnote omitted).

Strawson says that the mistake is not only large: it is fatal. Here too I agree, though I think I would express the fatality somewhat differently from him. In my view, the mistake is fatal because, on the assumption that we are ignorant of some of the crucial facts, the central pieces of reasoning in philosophy of mind collapse. For example, consider the zombie argument against materialism, or, as Strawson would say for reasons slightly opaque to me, the *Australian* (p. 22, fn. 37) zombie argument. Its first premise is that it is conceivable that I have a zombie duplicate; that is, there is someone identical to me in respect of every physical fact, but different from me in respect of some experiential fact. Its second premise is that if this is conceivable it is possible. Its conclusion is that physicalism is false, for physicalism (setting aside some technicalia) entails that zombies so described are impossible. This argument is unpersuasive if we take seriously the hypothesis that

---

[1] Thanks to Fiona Macpherson and David Wall for written comments on a previous draft.

we are ignorant of some of the physical facts. For suppose that the hypothesis is true, and there are some physical facts of which we are ignorant but which are relevant to the nature of experience. We therefore face two (and only two) options: either we include these facts within the scope of the quantifier 'every physical fact' which occurs implicitly or explicitly in both premises of the argument, or we don't. Suppose they are included; then the claim of conceivability loses all credibility. If there are relevant physical facts unknown to me, I will simply not be imaginatively acquainted with the putative possibility of zombies in the way that is required of me if the argument is to be plausible. Of course, this is not to deny that there may be *a* sense of 'conceiving' according to which I can truly conceive of zombies even if I am ignorant; it is only to insist that this sense is not the one that counts. On the other hand, suppose they are not included; then the physicalist can fairly claim that not all of the relevant facts have been taken into account. It is not at issue that experiential facts bear a contingent relation to *some* physical facts; the issue is whether they bear a contingent relation to all. Either way, the argument fails. (See Stoljar, 2006, for my own development of these issues.)

So on this point — on the point of out-Descartes-ing Descartes (fn. 21, p. 11) — I fall into line with Strawson, as he in fact falls into line with others such as Russell and Chomsky. But unfortunately, this element of Strawson's position sits very uncomfortably with other, and perhaps (for him) more important, elements of his overall view. We may illustrate this if we concentrate on what I take to be a central argument — if not *the* central argument — in Strawson's paper: the argument against the hypothesis that the world is in some sense fundamentally utterly non-experiential.

As I understand it, this argument proceeds as follows. First, it is asserted that if any experiential fact is wholly dependent on a non-experiential fact, then it must be the case that the non-experiential fact in question is, as Strawson says, 'intrinsically suitable' — that is, it must be the case that the non-experiential fact, whatever it is, has internal to it a potentiality to, as we might say, *wholly yield* the experiential fact. Next, it is asserted that no non-experiential fact has this potential. In short, the argument is this:

(1) If an experiential fact *e* is wholly dependent on a non-experiential fact *n*, then *n* must be intrinsically suitable (i.e. be intrinsically such to wholly yield an experiential fact).
(2) There is no non-experiential fact *n* such that it is intrinsically suitable.

(3) Ergo, no experiential fact e is wholly dependent on any non-experiential fact n.

There is clearly nothing wrong with this from a logical point of view. So what is up for discussion is the truth of (1) and (2).

Now, when he discusses this argument, Strawson is mainly concerned to discredit two kinds of opponent. The first kind — the emergentist — denies (1). The emergentist wants to say that e is wholly dependent on n, and yet insist also that this tells us nothing about the intrinsic or essential nature of n. Strawson says this idea is 'incoherent' (p. 12). I am not sure that it is incoherent, since it is not clear to me what the theory of coherence is that classifies it as incoherent. But I do agree that we are here in the realm of something that — for what it is worth — large numbers of people find decidedly odd. To give the phenomenon a name Strawson does not, we are here in the realm of 'necessary connections between metaphysically distinct existences'. The emergentist thinks that there are necessary connections between metaphysically distinct existences. Strawson is saying that if you think this you think something mistaken.

The emergentist is one kind of opponent discussed by Strawson; the other is the eliminativist, i.e. the person who denies that experiences exist at all. (The eliminativist may be construed as denying a presupposition of (1–3) above as standardly presented; that is, that there are experiential facts.) Strawson is pretty severe on eliminativism. He says (pp. 5–6) it is 'the strangest thing to ever happen in the whole history of human thought … the greatest woo-woo of the human mind'. Once again I am not sure about this, because I don't know what the theory of strangeness is according to which this one is the strangest. Moreover, I am not at all sure that Strawson *himself* is entitled to the strangeness claim, given what he says elsewhere. In an earlier paper, Strawson discusses Chomsky's methodological naturalism, and comments that he can feel it creeping over him (see Strawson, 2003, p. 63). Someone over whom Chomsky's methodological naturalism is creeping is so far as I can see not entitled to say that eliminativism is so strange, though in fairness I suppose this depends on how much it — i.e. methodological naturalism — *has* crept over you. As I understand it, Chomsky's methodological naturalism not only leaves it open that there might be no experiences: it positively encourages the thought. On Chomsky's view, a central feature of science is its divorce from commonsense; since there is nothing more solidly commonsense than experiences, Chomsky seems to be getting you ready for the thought that there are none. (Though of course he doesn't say this, nor does what he says entail it.)

Methodological naturalism aside however, I do agree with Strawson that both eliminativism and emergentism are things to be avoided if possible. However, in concentrating so much on emergentism and eliminativism, Strawson spends less time than he might have on a more resourceful opponent of the argument we just set out. This opponent denies (2) with something like the accompanying commentary: 'Look, I agree there are experiential facts; I am not an eliminativist. I agree too that it is impossible for two facts to participate in the relation of *being wholly dependent on* without being intrinsically suitable for doing so; I am not an emergentist. However, I think nevertheless that experiential facts *are* wholly dependent on non-experiential facts and in consequence that at least some of those facts are intrinsically such as to yield the experiential facts. What I don't understand is (2), the assertion that no non-experiential fact is intrinsically suitable. Where does *that* claim come from?'

Strawson seems to me to have something to say to this, but not much. Most of what he says comes up in the context of a discussion of liquidity. Suppose we advance an argument about liquidity that is analogous to Strawson's about experience. The argument would go like this:

(4) If a liquidity fact l is wholly dependent on a non-liquidity fact m, then m must be intrinsically suitable (i.e. be intrinsically such to wholly yield a liquidity fact).
(5) There is no non-liquidity fact m such that it is intrinsically suitable.
(6) Ergo, no liquidity fact l is wholly dependent on any non-liquidity fact m.

This argument is clearly unsound because its conclusion is false: the facts about something's being a liquid — for example the facts about water's being a liquid — do indeed depend on facts not about liquid, for example facts about the nature of various chemical elements and their properties. Moreover, the *first* premise of this argument, (4), has whatever force (1) has and for the same reason. After all, what got us to believe (1) had nothing to do with experience or liquidity *as such*. What got us to believe (1) was a perfectly general piece of metaphysical reasoning, viz. that two facts may not stand in the sort of relation required by dependency without being of such and such a sort. *That* piece of reasoning, if it were intact, would remain so whether the facts under discussion concerned experiences or liquids. In sum, if the liquidity argument is unsound, and (4) is true, the culprit must be (5). It is this that opens the door for the position I stated in the previous paragraph: according to this position, just as the second premise of the liquidity argument is false or without foundation, so too is the second premise of the experience argument.

Now, Strawson agrees that (5) is false, but insists that (2) is true. Why? Well one thing he says is that liquidity is no help if you are an emergentist. That is true, but irrelevant. The position I am imagining *agrees* with Strawson about emergentism, and so agrees with both (1) and (4). The issue concerns, not (1), but (2), and about this emergentism has nothing to say. Another thing he says is that it is quite transparent in the liquid case how liquidity facts might be necessitated by facts not about liquidity. But again: true but irrelevant. The point is not that it is *easy* to see how non-experiential facts are intrinsically suitable to yield experiential facts, nor that we can get an imaginative grip on how this might happen. The point was to nevertheless insist that it does happen, or at any rate to insist that we have been given no reason that it can't. Why does 'we have no imaginative grip on p' entitle us to say 'it is not the case that p'?

Of course Strawson might just *insist* that (2) is true. Is there something that can be done to counter this insistence? Well, one thing that can be done is to remind him of the large numbers of people who go around thinking they know about the nature of the physical and what *he himself* says about such people. Strawson says, as we have seen, that such people are making a large and fatal mistake. But the mistake is presumably not that of thinking that they know in detail about every single physical fact. That position is absurd and nobody holds it. The mistake that such people are making — we agreed earlier that it *was* a mistake — is to think that they 'know enough to know' (as it is often put; see also fn. 21, p. 11) that no physical fact is such as to yield *any* experiential fact, or at any rate that it is very controversial whether it is. More generally, the mistake is to think that you know what *kind* of fact a physical fact is, and moreover that you know that no fact of *that kind* is such as to yield anything experiential. But isn't Strawson's claim about non-experiential facts directly analogous to *this* claim about physical facts? Isn't *he* simply insisting that he knows enough to know about non-experiential facts that they are not intrinsically suitable? Why then isn't his position on non-experiential facts directly analogous to the mistaken position about physical facts that he himself so correctly identifies and criticizes?

For Strawson to insist on the truth of (2), therefore, seems to me to be dialectically weak. But — if I understand matters correctly — his overall position is in some ways worse than this. For let us reconsider what Strawson means by 'physical'. What are the physical facts according to him? Well, as he explains in detail, Strawson has a slightly idiosyncratic (though none the worse for that) way of introducing the term 'physical'. For him a physical fact is a kind of fact that includes as sub-kinds both experiential facts and non-experiential facts. This account of what it is to be a physical fact permits Strawson to officially distinguish himself from those philosophers who out-Descartes Descartes. Those philosophers think we *know* that no physical fact is such as to yield any experiential fact. Strawson denies this. For him 'physical fact' includes experiential facts and experiential facts *are* such as to yield experiential facts. So, in Strawson's terminology, it is not the case that we know that no physical fact is such as to yield any experiential fact.

However, while Strawson may be distinguished from his opponents in this way, we need also to keep our wits about us. For Strawson's way of introducing the term 'physical' is, as I have said, idiosyncratic. His opponents don't mean what he means by 'physical'. What do they mean? I think they mean, near enough, 'non-experiential'. Factoring this into the issues we have been considering, it turns out that 'the' mistake of out-Descartes-ing Descartes might be one of two quite separate mistakes. On the one hand, it might be the mistake of supposing that you know enough to know about *both* the experiential *and* the non-experiential that no fact of that kind, i.e. either experiential or non-experiential, is such as to yield any experiential fact. On the other hand, it might be the mistake of supposing that you know enough to know about the non-experiential *alone* that no fact of that kind, i.e. the non-experiential, is such as to yield any experiential fact.

Now, which mistake does Strawson intend to attribute to the philosophy of mind establishment when he talks so compellingly in (e.g.) the passage I quoted at the beginning, of 'the most fervent revilers of the great Descartes'? I think it very unlikely that anyone, even the pillars of that establishment, make the first mistake, i.e. the one about both the experiential and the non-experiential. For surely everybody agrees that experiential facts *themselves* are sufficient to yield experiential facts. So the only mistake that could be at issue is the second mistake, i.e. the one about the non-experiential alone. And as I have said, I agree with Strawson that this mistake is large and fatal, though again I would express the fatality differently from him. On the other hand, while this is a large and fatal mistake, so far as I can make out one philosopher who makes it is, unfortunately, Strawson himself. For, as we have seen, Strawson insists on (2), and (2) is the claim that no non-experiential fact is intrinsically such as to yield an experiential fact. When we ask what grounds this insistence, however, all we seem to find is that we know enough to know.

I argued a moment ago that Strawson's confidence that (2) is true is dialectically weak because it is analogous to a confidence that he rightly sees as misplaced. Once we adjust for the terminological difference between him and his opponents, however, this charge of dialectical weakness may be upgraded to one of contradiction: part of Strawson's overall account entails he is ignorant of non-experiential facts; another part entails he is not. More generally — and here I am summarizing my overall reactions to Strawson's paper — there appear to be two rather different views struggling for dominance here. The radical view is that we really are ignorant of the nature of the physical or non-experiential, and moreover that this ignorance has a significant impact on philosophy of mind. This is the view that we find hinted at but not developed in Russell and Chomsky, a view that I believe has the potential to completely transform philosophy of mind. The conventional view is a kind of dualism about the experiential and the non-experiential, a dualism not mitigated by the idea that there might be a super-kind of fact that includes both, and that deserves the name 'physical'. I have been suggesting that these two views are inconsistent. If that is right, Strawson needs to give up something; I hope it is obvious what I at least think should go.

References

Stoljar, D. (2006), *Ignorance and Imagination: The Epistemic Origin of the Problem of Consciousness* (New York: Oxford University Press).

Strawson, G. (2003), 'Real materialism', in *Chomsky and His Critics*, ed. L. Antony and N. Hornstein (Oxford: Blackwell), pp. 49–88.

Strawson, G. (2006), 'Realistic monism: Why physicalism entails panpsychism', *Journal of Consciousness Studies*, **13** (10–11), pp. 3–31.

Catherine Wilson

Commentary on Galen Strawson

I

The ancient philosopher Anaxagoras (500–428 BCE) wondered how hair could be made from what was not hair, and flesh from what was not flesh. He concluded that the 'seeds' of everything were in everything, and his defiantly nonreductionistic theory of mind was ridiculed by the atomist Lucretius as follows:

> If we cannot explain the ability of each animate being to experience sensation without attributing sensation to its constituent elements, what are we to say of the special atoms that compose the human race? Doubtless they shake and tremble with uncontrollable laughter and sprinkle their faces with dewy tears; doubtless they are qualified to discourse at length on the structure of compounds, and even investigate the nature of their own constituent atoms. (Lucretius, 2001, p. 59)

Lucretius in turn found emergentism unproblematic:

> [S]ince we perceive that the eggs of birds change into live chicks, and that worms come seething out of the earth when untimely rains have caused putrefaction, it is evident that the sensible can be produced from the insensible. (Lucretius, 2001, p. 58)

Galen Strawson sides with Lucretius on the question of hair and flesh, but with Anaxagoras on the question of mind. He asserts that the real physicalism to which he subscribes entails that the basic constituents of the physical world, though not entities like tables and chairs, have experiences. According to Strawson, it is not necessary to have a brain to have experiences; some physical things that exist outside of and are not identical with living creatures, must have experiences too. I take his argument to comprise a thesis and two conditionals. The thesis is that explanans and explanandum need to share properties and characteristics for there to be explanation at all. The first conditional is that if there are experiences, they must be explicable. The second conditional is that if we can't in principle explain how the brain generates experiences, the brain can't be a necessary condition of experience. It follows, according to Strawson, that there are sub-brain experiencing entities that permit or necessitate the emergence of experiences. I am not sure why the sub-brain experiencing entities are identified with ultimate particles; there is no separate argument for this that I found in Strawson's text, but I leave this objection aside.

In this comment, I will attempt to bring some historical perspective to this issue to show that Strawson's Anaxagorean micropsychism, while not impossible, is unlikely and is not entailed by real physicalism. It conflicts with some of the most exotic, as well as some of the most popular views about consciousness on record, including the views of Descartes, Kant and also Spinoza and Leibniz. Historically, defenders of 'thinking matter' have been concerned to show that, even if atoms could not think, suitably organized congeries of them could do so.

The possibility of thinking matter was much discussed after the publication of Descartes's *Meditations*, with its claim to have demonstrated the existence of incorporeal thinking substance. Long before Locke ventured the seemingly timid, but actually rather inflammatory suggestion in the second edition of the Fourth Book of his *Essay* that we could not know that matter could not think, critics of Descartes urged an open-minded stance, for example, the authors of *Objections VI*. '[S]ince we do not know,' they said, 'what can be done by bodies and their motions, and since you confess that without a divine revelation no one can know everything which God has imparted... how can you possibly have known that God has not implanted in certain bodies a power or property enabling them to doubt, think, etc.?' (Descartes, 1985–1993, VII, p. 421). The author of a notorious anti-Cartesian broadsheet circulating in Holland proposed that thought and extension were modes of a single substance. This single-substance position was explicitly developed by Spinoza. Other critics thought Descartes's restriction of experiences to humans was arbitrary and implausible. Leibniz in turn developed that line of thought. Spinoza and Leibniz might at first appear to be natural allies for Strawson. However, neither Spinoza nor Leibniz sanctions the view that physical particles are or even might be conscious.

Spinoza insisted against Descartes that we cannot conceive of disembodied minds: 'An idea that excludes the existence of our Body cannot be in our Mind, but is contrary to it.' (Spinoza, 1985, I, p. 500) He did not treat the question of the distribution of consciousness in the universe directly, but he seems to have ascribed thoughts of an intellectual, conceptual type only to God (identical with the totality of extended substance) and to humans, and to have ascribed experiences and feelings only to humans and animals. He took a definite anti-Cartesian stand in referring to the lusts and appetites of insects, fish and birds (Spinoza, 1985, I, p. 528). Though Spinoza failed to say why one had to be a certain kind of 'finite mode' to possess consciousness, or even what kind of finite mode one had to be, he implied that one had to be a fairly complex entity of a certain degree of maturity. Some human bodies, he thought, possess very little by way of consciousness:

> He who, like an infant or child, has a Body capable of very few things, and very heavily dependent on external causes, has a Mind which considered solely in itself is conscious of almost nothing of itself, or of God, or of things. On the other hand, he who has a Body capable of a great many things, has a Mind which considered only in itself is very much conscious of itself, and of God, and of things. (Spinoza, 1985, I, p. 614)

Spinoza is accordingly best read as an ordinary physicalist who thinks that human and animal minds depend on their brains, and as an unusual 'macropsychist' who considers the entire corporeal world as a divine thinking subject. It is hard to see him as an Anaxagorean micropsychist.

With Leibniz, the case is more difficult. Leibniz's famous 'mill' argument, though not identical to Strawson's, appeals similarly to the inexplicability of experience by reference to the combined effects of non-experiencing entities in interaction:

> We must confess that the perception, and what depends on it, is inexplicable in terms of mechanical reasons, that is, through shapes and motions. If we imagine that there is a machine whose structure makes it think, sense, and have perceptions, we could conceive it enlarged, keeping the same proportions, so that we could enter into it, as one enters into a mill. Assuming that, when inspecting its interior, we will only find parts that push one another, and we will never find anything to explain a perception. And so, we should seek perception in the simple substance and not in the composite or in the machine. (Leibniz, 1989, p. 215)

So Leibniz claimed that, because there are experiences and because experiences cannot be generated by complex machines, perception must pertain to the simple. There must be monads, unextended, simple, indivisible entities, with perception and appetition. Leibniz was a panpsychist in the sense of believing that everything that is a real, individual thing, and not an entity by convention, has experiences.

(For Leibniz, a table or chair is no more a single, real, individual thing than a salt-and-pepper set, or, for that matter, a knife and a fork.)

Monads are not, however, like Strawson's ultimate physical constituents. First, they are not physical — everything the natural scientist encounters in the physical world, said Leibniz, is complex and divisible. Second, they are supposed to give rise to the phenomenal world of extended bodies, and Strawson says that the emergence of extension from nonextension is incoherent. The monads are *metaphysical* and are the basis of an incoherently articulated emergentist account. Therefore, in insisting on them, Leibniz disqualified himself as a Strawsonian real physicalist.

II

Leibniz however held another version of panpsychism that has given commentators headaches because it does not seem to fit with the monadological theory. This version is well worth considering, because it suggests that the ultimate physical particles do not *as a matter of fact* have experiences, though, for all we know, they *could*.

Leibniz' other version of panpsychism adopts Spinoza's anti-Cartesian dictum that there are no experiences that do not belong to living, organic bodies. Additionally, there is no lower limit to the size of organisms. Everything that exists is either an animated organism, and itself a federation of smaller animated organisms, or an entity by convention composed of such organisms.

Though Leibniz's theory has been judged unacceptable for centuries, it is in some ways less problematic than Strawson's, insofar as Leibniz supposed that a power of autonomous change — 'appetition' always accompanied the power of 'perception'. Here is another passage from the *Monadology* that has special relevance to that point:

> I . . . take for granted that every created being, and consequently the created monad as well, is subject to change, and even that this change is continual in each thing . . . It follows from what we have just said that the monad's natural changes come from an internal principle, since no external cause can influence it internally . . . One can call all simple substances or created monads entelechies, for they have in themselves a certain perfection . . . ; they have a sufficiency . . . that makes them the sources of their internal actions, and, so to speak, incorporeal automata. (Leibniz, 1989, p. 215)

To a biologically-minded philosopher, this dual ascription of perception and agency makes a good deal of sense. Aristotle pointed out that there is no use having senses if you cannot move, for, if you could perceive and feel but not move, you would be powerless either to escape painful stimuli or to approach objects that appeared beneficial. Your senses would have no function, and there would have been no reason for frugal Nature to have endowed you with them. If ultimate particles do not approach, avoid, interact with and defend against other objects, they do not need senses. If they are simply knocked about by other particles, or undergo whatever the subatomic equivalent of being knocked about involves, they do not need senses. They don't have to react and respond to a life-world.

To this point, someone sympathetic to Strawson might reply that ultimate particles are not objects of natural selection and do not undergo embryogenesis. Accordingly, giving them useless experiences might not be expensive in terms of basic energetics — they might have them just 'by chance'. Ultimate particles in that case might passively experience some features of the world, while the changes in their micro-experience if any would be due to whatever forces operate at that level, not to their own micro-exertions or micro-appetitions. But what would their experiences be like? If the ultimate particles are not alive and self-moving, their experiences are most easily conceived as resembling those of disembodied Cartesian selves — presumably restricted to pure mathematics or pure conceptual thought without visual imagery.

Alternatively, someone sympathetic to Strawson might insist that animation, the capacity to initiate changes in the universe by moving oneself — is, like consciousness, an emergent phenomenon that must exist in ultimate particles. How else could agency emerge from some aggregation of physical stuff lacking agency? According to this line of reasoning, we should conclude not only panpsychism, but full pananimism.

I can imagine Strawson resisting this elaboration of his theory. Animation, he might respond, is a phenomenon like life; it can be explained as emerging from chemical and physiological processes. Animal activity can be understood to emerge from something similar — namely, chemical activity; and, as he points out, many seventeenth-century philosophers, including Descartes, thought it possible to explain life, though not consciousness 'mechanically'.

Life is also relatively easily conceived as emergent from a nonliving substratum, because we are now thoroughly familiar (as our ancestors were not) with many entities that 'aren't quite there yet' where life is concerned, like viruses, prions and antibodies, and also with certain kinds of automatic machines. Similarly, an entity can be not quite there yet when it comes to agency. Many animals, for example, seem to have some modest ability to control themselves and to decide what to do, but human 'autonomy' can achieve spectacular results in this direction. Consciousness, however, seems resistant to explanation because it seems impossible to have something on the way to an experience that isn't quite there yet.

So, we may not need to ascribe animation to micropsychic particles; at the same time, a real physicalist shouldn't rule it out. The fact that we can *explain* animation as an emergent phenomenon of complex chemical structures can't by itself entail that ultimate particles don't have animation; my ability to explain P by Q doesn't imply that P couldn't come about elsewhere in some other way without Q being involved. And, since animation has always been found, as far as our experience reaches, together with consciousness, there is good reason to think, on inductive grounds, that if ultimate particles are conscious, they are animated too.

III

To conclude, Strawson's version of panpsychism is historically unprecedented, though his arguments for it bear some resemblance to Leibniz'. Though Strawson maintains that his argument is not epistemological and does not concern the limits of our current understanding but the limits of what is truly possible in nature, it is hard to see why it is *impossible* that what one needs for there to be experiences in the universe is a brain made of insentient molecules put together in a certain way — and that's all. Perhaps we have no models in natural science for the emergence of experiencing creatures from the states and activities of unexperiencing physical entities, but it is hard to see why the nonexistence of available models entails that this cannot happen. The claim that real physicalism not only permits but entails micropsychism invites us to ask, in a Humean spirit: which is more likely? That some things happen in a way that seems impossible but is in fact only incomprehensible to me? Or that subatomic particles have experiences somewhat like mine, and are probably animated to boot?

Another version of real physicalism — more modest and more tenable by Humean empiricist criteria — would depart from the observation that we do not know, as the anti-Cartesians of the seventeenth century maintained, what physical entities by their own powers, without the addition of souls or mental substance, can bring forth. At the same time, we have good reason to think that experiences only occur in living creatures with nervous systems, that are products of natural selection and that are able to move and to interact with the world. This is not the conclusion of a deductive argument, and it possesses only high probability, not certainty; but, as Kant repeatedly pointed out, one cannot expect certainty about ultimate entities from a priori reasoning. It is meanwhile suggested by the kinds of deficits that occur in people whose brains are damaged, deficits that do in fact help us to understand what it is like to be 'not quite there yet' as a conscious entity. This other version of real physicalism also falls out of a more general scientific conception of the world that is the outcome of the tendency to de-ascribe consciousness, purpose, intention and mentality to entities that are not, in the vernacular sense of the word, animals.

References

Descartes (1985–1993), *Philosophical Writings*, ed. and trans. J. Cottingham, R. Stoothoff, D. Murdoch and A. Kenny (3 vols., Cambridge: Cambridge University Press).
Leibniz (1989), *Leibniz: Philosophical Essays*, trans. Roger Ariew and Daniel Garber (Indianapolis: Hackett).
Lucretius (2001), *De rerum natura*, trans. Martin Ferguson Smith (Indianapolis: Hackett).
Spinoza (1985), *The Collected Works of Spinoza*, trans. E. Curley 2 vols. (Princeton: Princeton University Press).

Galen Strawson

Panpsychism?
Reply to Commentators
with a Celebration of Descartes

1. Introduction

I'm very grateful to all those who have written replies to my paper 'Realistic monism: why physicalism entails panpsychism' ('RMP'), and to the editors of the *Journal of Consciousness Studies* who have provided a forum for the debate. I enjoyed all the papers, and the good humour that characterized most of them, and I know that it is a great privilege to have one's views scrutinized in this way in an age in which there is so much good work in philosophy and in which almost all of us feel that our work is neglected.[1] The sense of neglect is often justified; too much is being written. Things haven't improved since 1642, when Descartes — the hero of this piece — observed that 'it is impossible for each individual to examine the vast numbers of new books that are published every day'.[2]

My experience since I first lectured on the 'mind-body problem' in the late 1980s has been one of finding, piece by piece, through half-haphazard reading, that almost everything worthwhile that I have thought of has been thought of before, in some manner, by great philosophers in previous centuries (I am sure further reading would remove the 'almost'). It is very moving to discover agreement across the centuries, and I quote these philosophers freely, and take their agreement to be a powerful source of support. Almost everything worthwhile in philosophy has been thought of before, but this isn't in any way a depressing fact (see p. 200 below), and the local originality that consists in having an idea oneself and later finding that it has

---

[1] This can happen to anyone; my father, for example, had a particular affection for a book — *Subject and Predicate in Logic and Grammar* (1974) — that received very little attention.
[2] 1642, p. 386; 'books' had a wide reference.

already been had by someone else is extremely common in philosophy, and crucial to philosophical understanding.[3]

This is a long paper; my hope is that it is easier and quicker to read than a shorter one. The main reason for its length is not Pascal's — it isn't 'longer simply because I have not had time to make it shorter' (1656–7, letter 14), although I have had little time; it is rather Abbé Terrasson's, who remarked — as Kant pointed out when seeking to justify the length of his *Critique of Pure Reason* — 'that if the size of a volume be measured not by the number of its pages but by the time required for mastering it, it can be said of many a book, that it would be much shorter if it were not so short' (Kant 1781, Axviii). I have tried to write out the problem in a new way as it came to me, in particular in §§10–14, and it has stretched in the telling. I have made free use of footnotes to help the flow of the main text. There are those who look down on footnotes, but I think they are one of the great pleasures of life.

My aim is not to try to convince anyone of anything, but to record the truth of the matter as far as I can. I am aware, down to the details, I think, of this paper's vulnerability to unsympathetic reading or constructive misunderstanding, but it would take too long to try to block it all. The only lesson of science that I apply is the general lesson that we are profoundly in the dark about the nature of things, and in particular the nature of the non-experiential. This is a very old lesson, and one that Locke knew well, having learnt in particular from Descartes's difficulties,[4] but the magnificent science of the last hundred years has found overwhelming and bewildering new ways to drive it home. And yet some still largely ignore it.

Anthony Freeman and I agreed on the title of this book — a simple modification of the title of C.D. Broad's book *The Mind and its Place in Nature* (1925) — at the end of 2005. In April 2006, examining Sam Coleman's PhD thesis, I discovered that David Chalmers had in 2003 published a piece with the same title. It was too late to change, but there seemed to be nothing but good in further homage to Broad.

---

[3] It is not particularly saintly not to be disappointed that one has been anticipated (although it helps if the anticipator is a little in the past); that kind of disappointment is knocked out of anyone who survives as a philosopher after writing a doctoral thesis on free will — a process that invariably involves living through the problem in such a way that one feels that it is peculiarly one's own. A little inconsistency, furthermore, allows one to derive considerable gratification both from observing that one has powerful allies among the heroic shades of philosophy and from noting, in a deflationary way, that the views of one's living colleagues were put forward long ago by others.

[4] As had many others, as Wilson observes on p. 178 (all simple page references are to the present volume).

2. Conditional Physicalism (*ad hominem* physicalism)

I have a number of preparatory tasks. The first is to note the published ancestors of RMP, because although RMP is self-standing it cites its predecessors on matters that are relevant to the objections but couldn't be set out in sufficient detail in the space provided. They are, in reverse order, 'What is the relation between an experience, the subject of the experience, and the content of the experience?' (M2003b — the 'M' marks the piece's ancestral status), 'Real Materialism' (M2003a), 'Realistic Materialist Monism' (M1999a), and 'Agnostic Materialism', chapters 3 and 4 of *Mental Reality* (M1994), a book for which I feel affection, although it fell more or less dead-born from the press. In *Mental Reality* I was already arguing for the panpsychist (or at least '(micro)psychist') view that there must be experientiality at the bottom of things (see e.g. pp. 60–2, 68–9), in such a way that Chalmers correctly classified me as holding a version of the position he now calls 'Type-F monism',[5] but in those days one felt considerably more abashed about doing such a thing. Chalmers talked in the same spirit of the 'threat' of panpsychism, in spite of his own respect for it (1997, p. 29).

One thing the four 'M' works have in common with each other and with RMP, and that needs to be mentioned now, given some of the replies to RMP in this book, is that they are all conditional in their overall form: they are dialectically *ad hominem*, in the non-aggressive sense of the term. That is, they are directed to someone who is assumed to hold a certain position, and their principal arguments are designed to have whatever force they have because the person to whom they are addressed holds that position. The position in question is materialism or physicalism, as defined on p. 3, and the general form of argument in each case is '*If* you accept physicalism, if, that is, you are a serious and realistic materialist, then you also have to accept this' (as before I take 'materialist' and 'physicalist' to be synonymous).

In order to run the conditional, *ad hominem* argument smoothly[6] I assume in each of these works that physicalism is true, while setting no great store by the word 'physicalism'. I point out the sense in which my use of the term risks rendering it descriptively vacuous (at least so far as non-experiential phenomena are concerned) and suggest that this seeming calamity is just what is needed at the current stage of the

---

[5] See Chalmers (2003). Chalmers read the typescript of M1994 before it was published, although he subsequently made no mention of it, and judged then that we were broadly in agreement, although we differed about how to use the word 'materialist'.

[6] As things have turned out, it has led to misunderstanding.

debate.[7] To those who baulk at such a use of 'physicalism' I offer, each time, 'experiential-and-non-experiential monism', or even 'experiential-and-non-experiential ?-ism', as more cautious names for the position I am assuming to be true: this is the position of someone who

> [a] fully acknowledges the evident fact that there is experiential being in reality, [b] *takes it* that there is also non-experiential being in reality, and [c] is attached to the [stuff-] 'monist' idea that there is, in some fundamental sense, only one kind of stuff in the universe (p. 7, emphasis added)

where 'stuff' is a completely general term for concrete being.[8]

3. Equal-status Monism

In *Mental Reality* I set out a version of this position called *equal-status* monism, according to which

> reality is irreducibly both experiential and non-experiential, while being substantially single in some way W that we do not fully understand, although we take it that W is a way of being substantially single that does not involve any sort of *asymmetry* between the status of claims that reality has non-experiential aspects and claims that reality has experiential aspects. [On this view] it is not correct to say (a) that the experiential is based in or realized by or otherwise dependent on the non-experiential, or (b) vice versa. The truth is rather (c) that the experiential and non-experiential coexist in such a way that neither can be said to be based in or realized by or in any way asymmetrically dependent on the other; or if there is any sense in which one can reasonably be said to be dependent on the other, then this sense applies equally both ways ....
>
> To get an explicitly materialist form of equal-status monism one

---

[7] M1994, pp. 99, 105; M2003a, p. 73; RMP, p. 8. Compare in particular Crane and Mellor 1990. McGinn objects to it, as does Macpherson, at least in part; but I embrace it, for I am only trying to set out what one has to say if one is a (realistic) physicalist.

[8] In M1994 I characterized my materialism as follows: 'I believe that experience is not all there is to reality. I believe that there is a physical world that involves the existence of space and of space-occupying entities that have nonexperiential properties. I believe that the theory of evolution is true, that once there was no experience like ours on this planet, whether panpsychism is true or false, and that there came to be experience like ours as a result of processes that at no point involved anything not wholly physical or material in nature. Accordingly, I believe that however experiential properties are described, there is no good reason to think that they are emergent, relative to other physical properties, in such a way that they can correctly be said to be nonphysical properties. Finally, with Nagel (1986, p. 28), I believe that one could in principle create a normally experiencing human being out of a piano. All one would have to do would be to arrange a sufficient number of the piano's constituent electrons, protons, and neutrons in the way in which they are ordinarily arranged in a normal living human being. Experience is as much a physical phenomenon as electric charge' (p. 105). Since then I have given up the view that materialism requires belief in any non-experiential intrinsic properties or stuff.

simply has to add in the words 'properties of the physical' in (c) to get 'The truth is that the experiential and non-experiential properties of the physical coexist in such a way that neither can be said to be based in, or realized by, or in any way asymmetrically dependent on, the other, etc.'....

I will restate ... this ... although it is clear enough. [i] All reality is physical (the basic materialist premise). [ii] There are experiential and non-experiential phenomena (unavoidable realism about the experiential, plus the assumption (!) that there is more to physical reality than experiential reality). [iii] Among physical phenomena, experiential physical phenomena do not depend on non-experiential physical phenomena ..., or do not depend on them in any way in which non-experiential phenomena do not also depend on experiential phenomena.[9]

My reason for mentioning this straight away is that I take it that real (realistic) physicalists must be equal-status monists, given the argument in RMP that the experiential cannot possibly emerge from the wholly and utterly non-experiential. If one is a realist about the experiential, a *real* realist about the experiential (see p. 3), one faces the fact that any asymmetry or one-way dependence or reducibility must be to the detriment of the non-experiential.

I am going to continue to assume for purposes of argument that monism is true, in spite of the difficulties in the notion:[10] both insofar as I continue to assume for *ad hominem* purposes that physicalism is true (for whatever physicalism is it is a monist position), and on my own account — at least until the term 'physical' falls apart (see p. 228 below). No monism can be 'neutral', however, given that there is no sense in which experience considered just as such can be mere appearance, in the sense of not being really real at all,[11] and given that this is so — given that neutral monism is out — , it looks as though the only monism that really makes sense, given the certain existence of experience, is experiential or panpsychist monism. If so, continued use of the word 'physicalist' will sound ever more oddly, with an increasingly *reductio ad absurdum* ring, and equal-status monism will turn out to be a pipe-dream — unless, that is, Spinoza can save it.

---

[9] Strawson M1994, pp. 73-4; the assumption marked with an exclamation mark is indeed that, an assumption; unlike the acknowledgement of experiential phenomena.

[10] See e.g. M1994, pp. 43-4, M2003, pp. 74–5

[11] See e.g. M1994, pp. 50-1, M1999a, pp. 24-5, M2003a, p. 54. The point is not only that we can know that experience considered just as such is real; it is also that we can and do know the intrinsic nature of the experiential, at least in certain respects, simply in having it. This will become important later on.

4. 'Panpsychism'

My characterization of panpsychism in RMP was intentionally imprecise: 'all physical stuff is ... an experience-involving phenomenon'.[12] This is too loose as it stands, as Stapp observes (p. 164), and for the moment I am happy to solidify it by fusing it with Nagel's characterization — 'by panpsychism I mean the view that the basic physical constituents of the universe have mental [and in particular experiential] properties'.[13] Let me stress that I make — find — no distinction between panpsychism and panexperientialism, because the word 'panpsychism' doesn't have any implications that the word 'panexperientialism' doesn't also have. 'Psyche' was a mass term before it was a count noun, and the word 'panpsychism' doesn't in itself imply that there are subjects of experience in addition to experiential reality, or indeed that everything that exists involves the existence of a subject of experience in addition to the existence of experiential reality.[14]

5. Subjects of Experience

In fact, though, it wouldn't matter if the word 'panpsychism' did carry this implication, because it is 'an obvious conceptual truth that an experiencing is necessarily an experiencing by a subject of experience, and involves that subject as intimately as a branch-bending involves a branch'.[15] There cannot be experience without a subject of experience simply because experience is necessarily experience *for* — for someone-or-something. Experience necessarily involves experiential 'what-it-is-likeness', and experiential what-it-is-likeness is necessarily what-it-is-likeness *for* someone-or-something. Whatever the correct account of the substantial *nature* of this experiencing something, its *existence* cannot be denied. 'An experience is impossi-

---

[12] p. 25; compare M1994, pp. 76-7, where I distinguish between 'experience-involving' and 'experience-realizing' versions of panpsychism.

[13] 1979, p. 181. That said, I am going to avoid talking about properties as far as possible, for reasons that will emerge; and I certainly don't want to rule out the version of panpsychism according to which there is no non-experiential being at all.

[14] I make the same provision in M1994, p. 76. Here I disagree terminologically with Coleman (pp. 48–50). Skrbina (2005, p. 21) traces the term 'panexperientialism' to Griffin 1977.

[15] Shoemaker 1986, p. 10. The point is made briefly in RMP p. 26. Here, it seems, I may have a substantive disagreement with Coleman (pp. 48–50), but I still like to think it is really only terminological.

ble without an experiencer' (Frege 1918, p. 27). To understand this claim in the sense in which it is intended is to see that it is true.[16]

This is not to commit oneself to any view about the ontological status of the necessarily existing subject. It is certainly not to commit oneself to the idea that it must be a substance in any conventional sense of that word, i.e. any sense in which a substance is understood to be something stands in fundamental ontological contrast with a property. One can be certain that an experience is impossible without an experiencer while knowing nothing more than Descartes knows in his *Second Meditation* when he says 'I know that I exist; the question is, what is this "I" that I know?'. Descartes makes it as explicit as he can that he is at this stage entirely uncommitted on the question of the ontological nature of what gets referred to when he says 'I'.[17] Kant does the same in his Paralogisms, using for this purpose the terms of the conventional substance/property distinction. Certainly one knows that one exists, but it is 'quite impossible' for one, he says, given one's self-conscious experience of oneself as a mental phenomenon, 'to determine *the manner* in which [one] exist[s], whether it be as substance or as accident' (1781/7, B420; my emphasis). Certainly 'the I who thinks or is conscious must *in such thought or consciousness* always be *considered* as a *subject*, and as something that does not merely attach to thought or consciousness like a predicate', but — this is Kant's central point — nothing follows from this about how things actually are, metaphysically speaking.[18] What we have to do, then, is acknowledge the certainty of the existence of the subject, the experiencing 'someone-or-something', in a way that is wholly metaphysically neutral as to ontological category, even while the essentially discursive, subject-predicate tenor of our thought and language revolts against the attempt at such neutrality.

It may be thought to be misleading to make the experience-entails-an-experiencer point while using individual-substance-suggesting noun phrases like 'experiencer', 'subject of experience', or

---

[16] It is in fact analytic, if not obviously so. To understand what experience is is to understand that it is essentially experience-*for*, in the intended sense. Note that I take 'experience' to cover not just sensory episodes but all conscious mental goings on, including the most abstract conscious thoughts (see e.g. Strawson M1994, ch. 1).

[17] 1641, p. 18; for all he knows at this point in his meditations, he observes, he may be nothing more than his body.

[18] 1787, B407; first two emphases mine. Kemp Smith (1933) and Guyer & Wood (1999) mistranslate this sentence. They have 'can' instead of 'must' (the mistake results indirectly from a mistaken reading of 'gelten' as 'to be valid' rather than 'to consider as' or 'to count as'). Pluhar (1996) gets this right, and the correctness of his translation has been confirmed to me by many native speakers, including professional translators, and — among philosophers — Fred Beiser, Han-Jo Glock and Michael Rosen.

'someone-or-something', and I think it is worth stressing that nothing in Buddhism conflicts with the point when it is understood as it is here (certainly the notion of a subject carries no implication of long-term persistence). One can put the point paradoxically by saying that if *per impossibile* there could be experience without any experiencer, if there could, say, be pain-experience — massive, appalling, avoidable, wholly useless pain — without any subject of experience, there would be no point in stopping it, because no one, no someone-or-something, would be suffering.

If someone is prepared to grant that there is necessarily *subjectivity*, when there is experience, but not that there is necessarily *a subject of experience*, we have a merely terminological disagreement, for I understand the word 'subject' in a maximally metaphysically neutral way (with Kant and the Descartes of the *Second Meditation*) given which the existence of subjectivity entails the existence of a subject.[19]

6. Thin Subjects

It is plain to most philosophers that there cannot possibly be experience — experiencing, experiential reality, experiential being (I will use these terms interchangeably) — without a subject of experience.[20] What is less plain, or less remarked on, is that there is an important use of the term 'subject of experience' given which the converse is also true. I will record it here because it is helpful in many contexts, and helps to diminish resistance to the necessary truth recorded in the last section.

There are two common conceptions of what a subject of experience is. First,

> [a] the *thick* conception according to which it is only human beings and other animals *considered as a whole* that are properly said to be subjects of experience.

Second,

---

[19] See further M2003b, pp. 293–4. This is part of the explanation of why Lichtenberg's famous objection to Descartes is no good.

[20] Let no one think that Hume thought otherwise. His target in 'Of personal identity', section 1.4.6 of the *Treatise*, is not this view, which is after all a necessary truth. It is the view, standard in his time, that the self or subject is something that has 'perfect identity and simplicity' and that 'continue[s] invariably the same, through the whole course of our lives'; together with the view that this is something that we can know to be the case. See Strawson (2011a).

> [b] the *traditional* conception of the subject, the traditional *inner* conception according to which the subject *properly or strictly speaking* is some sort of persisting, inner, mentally propertied entity or presence.

I take it that [a] and [b] both build in the assumption that a subject may and standardly does continue to exist even when it is not having any experience (for whether you think that human subjects are whole human beings or whether you think they are inner loci of consciousness, you are likely to allow that they can continue to exist during periods of complete experiencelessness — in periods of dreamless sleep, say), and it is this that creates the need for the third, relatively unfamiliar conception of the subject

> [c] the *thin* conception according to which a subject of experience, a true and actual subject of experience, does not and cannot exist without experience also existing, experience which it is having itself.

The thin conception stands opposed to both [a] and [b] precisely because they both contain the natural assumption that a subject of experience can be said to exist in the absence of any experience. It doesn't, though, offer any support to the idea that thin subjects are short-lived or transient entities. I believe that they are short-lived or transient, momentary, in the human case, as a matter of empirical fact, but Cartesian subjects also qualify as thin subjects by the present definition, and they are long-lived, possibly immortal.[21]

There is a problem of exposition here, because most are so accustomed to [a] and/or [b], and to the idea that they exhaust the options, that they cannot take [c] seriously. And yet [c] simply makes a place for a natural use of the term 'subject' according to which it is a necessary truth, no less, that

> there cannot be an actual subject *of experience*, at any given time, unless some *experience* exists for it to be a subject *of*, at that time.

On this view, there can no more be a subject of experience without an experience than there can be a dent without a surface.

Most think that to talk of the subject of an experience is necessarily to talk of something ontologically distinct from the experience or

---

[21] See further §8. Other thinkers whose subjects are 'thin' in this sense include Leibniz, whose subjects are like Descartes's long-lived. On the short-lived side we find Hume and William James, Buddhists, who are often supposed to deny the existence of subjects of experience altogether, and, arguably, Fichte.

experiencing itself (in this book see e.g. Coleman pp. 48-9). Others think that to talk of the subject of experience is necessarily to talk of something that can be said to perceive, or to be in intentional states. I reject both of these views (if, that is, intentional states are externalistically construed). So far as the first is concerned I am inclined to agree with Descartes that to talk of a subject of experience in the fundamental thin sense is not only not *necessarily* to talk of something ontologically distinct from the experiencing ('thinking'), but is also not *in fact* to talk of something ontologically distinct from the experiencing, and is indeed *necessarily not* to talk of something ontologically distinct from the experiencing.

This will be clarified in the next three sections, which will also constitute my main reply to Macpherson's very helpful paper. Here let me just state that I had thin subjects in mind when I noted in RMP that panpsychism, conjoined with the assumption that there are many 'ultimates' or fundamental constituents of reality, leaves us with 'a rather large number of subjects of experience on our hands' (p. 26). The semi-humorous 'rather large' acknowledged that this claim would be thought by many people to constitute a special extra problem for panpsychism, but then as now I am quite sure that it does not, however many ultimates there are (each with its own feeling-hum of existence and/or representation of its environment). I think, in fact, that it constitutes exactly as much as a problem for panpsychism as the claim that there are n ultimates does for ordinary physicalism — whatever the value of n.

7. Objects and their Properties

I need now to explain how I conceive the relation between objects and their properties — something mentioned only in passing in RMP (p. 28). This will allow me to explain why I say on p. 28 that 'property dualism' is incoherent insofar as it claims to be distinct from substance dualism — unless it is nothing more than the claim that there are two very different sorts of properties. It will also allow me to explain why I have to turn down Macpherson's invitation (on p. 81) to accept that I am really a property dualist. After that (in §8) I will explain why I think we need to return to Descartes, the thrilling Descartes, if we want to go anywhere with the mind-body problem.[22]

Objects have properties, we say. Our habit of thinking in terms of the object/property distinction is for everyday purposes ineluctable.

---

[22] I have revised this section slightly for this second edition. It is adapted from M2003b, in which I used Descartes's terms without realizing that I was also expressing his view about the object/property relation.

And it is perfectly correct, in its everyday way. But ordinary language is not a good guide to metaphysical truth, and as soon as we repeat the observation — *objects* — *have* — *properties* — in philosophy we risk a great error, the error of thinking that there is a fundamental categorial distinction between objects and their properties (we compound the error if we think such a distinction is fundamental to ordinary thought). The truth about the relation between objects and their properties eludes sharp formulation — millennia of vehement philosophical disagreement testify to this fact — but I think it's possible to express it in a philosophically respectable fashion. The key is not to say too much (I think my profound ignorance of the traditional debate gives me a head start).

In setting this out I will consider only concrete phenomena, although the idea has general application. So my concern with properties will be only with concretely existing properties, concrete propertiedness, and only with intrinsic, natural (non-conventional) properties of objects.[23] I'll use the word 'property' — this is a definition — as it is used with no knowledge of philosophy, although I think this crucial use may be inaccessible to many philosophers. I could try to convey the point about being concerned only with concretely existing properties by saying that I'll be concerned only with 'property-instantiations' or 'property-concretions', but these terms are already problematic inasmuch as they imply a contrast with properties considered as universals considered as abstract objects.

What is at issue, then, is the relation between a particular concrete object and its properties, i.e. its whole actual qualitative being,[24] and the proposal is that one has already gone fatally wrong if one thinks that there is any sort of ontologically weighty distinction to be drawn according to which there is the object, on the one (concrete ontological) hand, and the properties of the object, on the other (concrete ontological) hand: according to which one can distinguish between the existence or being of the object, at any given time, and its nature, at that time; between the thatness of the object and the whatness or howness of the object, at any given time. (One of the deep agents of confusion is counterfactual thinking, which I will come to in due course.)

Plainly objects without properties are impossible. There can no more be objects without properties than there can be closed plane

---

[23] I am taking the general propriety of such notions for granted. For some recent discussion, see Lewis & Langton 1996 and the ensuing debate in *Philosophy and Phenomenological Research* 2001, pp. 347–403. See also note 162 below.

[24] 'Qualitative' has nothing in particular to do with experience — my teapot has many qualities.

rectilinear figures that have three angles without having three sides. 'Bare particulars' — objects thought of as things that do of course *have* properties but are *in themselves* entirely independent of properties — are incoherent. To be is necessarily to be somehow or other, i.e. to be some way or other, to have some nature or other, i.e. to have (actual, concrete) properties.

Rebounding from the obvious incoherence of bare particulars, one may think that the only other option is to conceive of objects as nothing but collections or 'bundles' of properties — property-concretions. And this option may seem no better. Mere bundles of properties seem as bad as bare particulars. Why accept properties without objects after having rejected objects without properties? But this is not what we are asked to do. The claim is not that there can be concrete instantiations of properties without concrete objects. It is, rather, and to repeat, that concrete objects are nothing but concrete instantiations of properties.

This still sounds intolerably peculiar, though, and it is not in the end a helpful thing to say.[25] What is helpful, I think, is to compare the point (1) that there can no more be concrete instantiations of properties without concrete objects than there can be concrete objects without concrete instantiations of properties with the point (2) that there can no more be dispositional properties without categorical properties than there can be categorical properties without dispositional properties.[26] To these one can add the stronger point that there is at bottom no 'real distinction', in Descartes's sense — see the next section — between a thing's categorical properties and its dispositional properties: nothing can have the categorical being that it has and not have the dispositional being that it has, and nothing can have the dispositional being that it has and not have the categorical being that it has.[27] Transferring this point back to the case of objects and properties, we can say with Descartes that there is no real distinction between an object and

---

[25] I said it in M2003b and in the first edition of this book.

[26] The two points are deeply related.

[27] Quick thoughts about the 'multiple realizability' of certain functional properties may spark the idea that two things can be dispositionally identical without being categorically identical, and this may lead to the idea that a thing's categorical properties could be changed without its dispositional properties being changed; but a moment's more thought reveals that this cannot be so. (Many philosophical thought-experiments assume that a thing can be thought to retain its identity across different nomic environments. It is not clear that this makes sense, but even if it does it is superficial to think that its fundamental dispositions will change; for these fundamental dispositions will include the disposition to behave in way A in nomic environment 1, the disposition to behave in way B in nomic environment 2, and so on.)

its properties, although there is no doubt a useful and workable conceptual distinction between them. (The counterfactuals are coming.)

When Kant says that

> in their relation to substance, accidents [or properties] are not really subordinated to it, but are the mode of existing of the substance itself

I think he gets the matter exactly right.[28] Nothing more needs to be said (language will make a mess if you try). Consider an actual object in front of you. There is no ontological subordination of properties to object, no existential inequality or priority of any sort, no dependence of either on the other, no independence of either from the other (the counterfactuals are coming). There is, in other terms, no ontological subordination of the total *qualitative* being of the object to the object *an Sich*, 'in itself', no ontological subordination of its nature to its existence. We can as Armstrong says '*distinguish* the particularity of a particular from its properties', but

> the two 'factors' are too intimately together to speak of a *relation* between them. The thisness and the nature are incapable of existing apart from each other. Bare particulars are vicious abstractions...from what may be called states of affairs: this-of-a-certain-nature.[29]

Nagarjuna talks in the same vein of the complete codependence of things and their attributes,[30] Nietzsche is admirably brief — 'A thing = its qualities' (1885–8, p. 73; see also pp. 88, 110, 104–5) —, and P.F. Strawson's use of the suggestive phrase 'non-relational tie' can profitably be extended from a logico-linguistic application (to grammatical subject-terms and predicate-terms) to a straightforwardly metaphysical application (to objects and their properties).[31]

I believe it should be. One should — must — accept the 'non-relational' conception of the relation (!) between an object and its intrinsic properties, if one is going to retain words like 'object' and 'property' in one's metaphysics at all. This is entirely compatible with claiming that an object's properties — including its intrinsic or non-relational

---

[28] Kant 1781/7, A414/B441. It's important that 'mode of existing' cannot just mean 'the particular way a substance is', where the substance is thought to be somehow independently existent relative to its mode of existing; for that would be to take accidents or properties to be somehow 'subordinate' after all. (I'm assuming that here 'accident' means effectively the same as 'property-instance'.)

[29] 1980, pp. 109-110. Armstrong puts things this way for well known dialectical reasons to do with stopping 'Bradley's regress' (see Loux 2002, p. 39–40), but I take it that there are completely independent metaphysical reasons for saying it.

[30] c150 ce, chapter V (1995, pp. 14–15, see commentary pp. 149–152).

[31] Strawson 1959, pp. 167–178. 'Tie', though, is not a very good word for this non-relational mutual metaphysical involvement.

properties — may and do change through time, while it remains the same object.

'But we also want to be able to say that an object would have been the very object it is, at time t, even if its properties had been different, at t. We think that the (actual) object could have existed apart from some at least of its (actual) properties.'

Nothing here forbids this way of talking about the non-actual. To see this, all one needs to do is to lose any tendency to slip, even in one's underthought, from the evident fact

(i) that there are contexts in which it is entirely natural to take it that (some at least of) an object's properties might have been different from what they are while it remained the same object

to the entirely mistaken idea

(ii) that an object has — must have — *some* form or mode of being independently of its having the properties it does have.

'But we also want to be able to say that an object would still be the object it is even if (some at least of) its properties were other than they are in fact.'

True, but present-tense counterfactual talk is no more problematic than past-tense. In itself it's innocent, because it doesn't in any way license a shift from (i) to (ii). To think that it does is to build a whole metaphysics of object and property into counterfactual thought, a metaphysics that it does not contain or license as it stands, and that is simply incorrect, on the present view (although currently dominant).

That said, I have no special wish to defend ordinary thought. If ordinary counterfactual thought does after all harbour a deep error, when it comes to the object/property relation, so be it. The adequacy of ordinary thought and talk to represent reality is already in the dock, and already stands condemned on many counts. Appeal to it can't by itself ground any argument that the current Cartesian proposal is incorrect. Those who wish to reject the current proposal will have to produce independent (non-linguistic) metaphysical arguments in support of their view.

So I don't think that our ordinary understanding of counterfactuals is a source of metaphysical error. But a importantly related claim that does seem to be true: that when human beings *philosophize* about the object/property relation, certain features of language naturally lead them to think that (ii) is true.[32] I don't think Ramsey exaggerates when he says that 'the whole theory of universals is due to mistaking ... a characteristic of language ... for a fundamental characteristic of reality' (Ramsey 1925, p. 60). In this he agrees with Nietzsche, who writes that

> language is built in terms of the most naïve prejudices ... we read disharmonies and problems into things because we *think only* in the form of language — thus believing in the 'eternal truth' of 'reason' (e.g. subject, predicate, etc.)
>
> That we have a right to *distinguish* between subject and predicate — ... that is our strongest belief; in fact, at bottom, even the belief in cause and effect itself, in *conditio* and *conditionatum*, is merely an individual case of the first and general belief, our primeval belief in subject and predicate.... Might not this belief in the concept of subject and predicate be a great stupidity? (1885–8, pp. 110, 104–5)

These are powerful and dramatic ways to put the point (it may be that Nietzsche is claiming that metaphysical error is built into ordinary thought, but perhaps he is only pointing to the innocent grounds in ordinary thought of the distinctively philosophical error). But the best thing to do, I think, is simply to keep Kant's phrase in mind: 'in their relation to the object, the properties are not in fact subordinated to it, but are the mode of existing of the object itself'.[33] This is another of those points at which philosophy requires a form of contemplation, something more than theoretical assent: cultivation of a shift in intuitions, acquisition of the ability to sustain a different *continuo* in place in the background of thought, at least for a time. The object/property/state/process/event cluster of distinctions is unexceptionable in everyday life, but wholly superficial from the point of view of science and metaphysics.

Some think that conflict with ordinary ways of thinking is always an objection to a philosophical theory, but this is certainly untrue if it is anything more than a recommendation to keep in touch with common-sense conceptions. Philosophy, like science, aims to say how things are in reality, and conflict with ordinary thought is no more an automatic objection to a philosophical theory than it is to a scientific one. There are as remarked many areas in which we can see clearly that our ordinary concepts and ways of thinking can't be fully adequate to the reality they purport to represent. In the present case we can say that ordinary thought is inadequate to reality to precisely the

---

[32] It may also lead them to think its converse is true, the idea (iii) that properties have — must have — *some* form or mode of being that is independent of the being of the objects that have them.

[33] I have substituted 'object' and 'property' for 'substance' and 'accident' respectively.

(disputable) extent to which our easy everyday use of the object/property/state/process/event distinctions does indeed constitute a commitment on our part to the view that these distinctions mark fundamental categorial differences in reality.

One problem is that a commitment of this sort can exist in many parts of philosophy without causing any particular problems (the same is true of the use of Newtonian mechanics in physics). There are, however, areas in metaphysics where its inadequacy to reality is part of the problem at issue, explicitly or not, and then its uncritical use, its use in any robust form, wreaks absolute havoc, havoc aggravated by the ease and success of its employment in other areas, which understandably misleads many into thinking that it must be quite generally viable.

There is, in any case, no serious problem of universals and particulars — a realization which can be uncomfortable at first, if one has been exposed to the philosophical debate, but which settles out and matures powerfully. One looks at any ordinary object and it is deeply mysterious how there can be thought to be a problem. In discussing the mind-body problem, in particular, I think it best to talk in categorially more neutral terms of a thing's *being* or *reality*, rather than of its *properties*, and I will very often do so. Uncritical use of the object/property distinction has caused a truly huge number of unnecessary problems in the debate, and has wasted a vast amount of time.

8. Real Descartes [1][34]

Descartes, it appears — the magnificent, contumacious Descartes —, is the most distinguished holder of the view about the relation between objects and their propertiedness that I have just endorsed, at least in the Western tradition. This by itself is a good reason to start from his views, when entering into the mind-body debate, but it is not the only one. He still oversees the debate. He is still constantly referred to and is still its deepest thinker. Having recently read (and re-read) some of his writings, along with those of his contemporary critics and correspondents, and later commentators, I think that everyone engaged in the current discussion of the mind-body problem in philosophy should hold everything and read Descartes, especially his correspondence and his replies to objections — if, that is, they want to have their perspective on the problem sharpened, reinvigorated, simplified and deepened. (It is deepened by being simplified.)

---

[34] This section was inspired by reading Clarke 2003 and 2006 and fortified by Yablo 1990 (it may perhaps go a little further than either of them would approve). There is a huge and in many respects wonderful scholarly literature on Descartes of which I am almost entirely ignorant and to which I surely have nothing to add, except, perhaps, a slightly new arrangement.

To do so is to realize that there is really nothing radically new in the existing debate — nothing both new and true — ,[35] but this is a moment of illumination, not defeat. The fundamental positions in the mind-body debate have been marked out for a long time, and the quality of the present-day debate is embarrassingly lower than it was in the seventeenth century.[36]

It does not follow that there is nothing difficult and important left to do; nothing could be further from the truth. When Pascal imagined someone charging him with lack of originality, he replied:

> Let no one say that I have said nothing new: the organization of the subject matter is new. When we play tennis, we both play with the same ball, but one of us places it better.[37]

The point is of great importance and holds for all the discursive arts and sciences, even if it has special force in philosophy. The object of philosophy is not just to state the truth in a domain where matters are often so very difficult, but to make it shine out. To think that Pascal's dictum reflects badly on philosophy is comparable to thinking that the best science never produces new results; or like thinking that once someone had painted a picture of the Madonna and Child, or the Montagne Saint Victoire, there was no point in anyone else doing so.

This is why we should go back to the older debate before trying to go anywhere else. There is a great deal of time to be saved. Everything that matters can be put far more simply and more clearly than it is being put in the present debate, with its atrocious muddling of metaphysical issues with epistemological and semantic issues and its for the most part witheringly unhelpful, rococo, scholastic, multiply duplicative and multiply inconsistent terminologies. Descartes offers us one great anchor and framework, in spite of the terminological peculiarities of his time (far easier to master than those of the present day). If you want to get intimate with texts, get intimate with Descartes, Arnauld, and Princess Elizabeth of Bohemia. If you want to publish papers on the mind-body problem, there are fine papers to be written laying out and developing Descartes's treatment of currently fashionable issues. It's a tragedy for all those who coming into philosophy — a tragedy of waste, deprivation, stupidity and disrespect (it is

---

[35] Some might prefer to say: nothing both true and new and important.

[36] 'Things have clearly gone downhill in the last three hundred years' (M1994, p. 102, with reference to Locke).

[37] c1640-1662, §575. 'One might as well say that I've used old words', he continued. For 'just as the same words constitute different thoughts by being differently arranged, so too the same thoughts constitute a different body of work by being differently arranged'.

also a farce) — that there is so often a huge, sometimes grotesque, gap between the popular understanding of a great philosopher's treatment of a topic, by which I mean the popular philosophical understanding of the philosopher's treatment of the topic, and the understanding possessed by the clear-headed historians of philosophy. Descartes on substance, Locke on personal identity, Hume on cause — these are some salient examples. I found this out from twenty years of teaching history of philosophy (for which I had, otherwise, no special inclination) in tutorials at Oxford, and I can't see how it is ever going to get better, given the vast breeding population of misguided introductory texts and *idées fixes*. What I say here about Descartes will, in any case, be brief, and will take a basic grounding in his ideas for granted.

We start, then, with Descartes — but with the real Descartes, not the 'Descartes' of present-day non-historical philosophy.[38] And the first things to note about him, I think, given that we (the generality of philosophers) refer to him so much, and so freely, and so inaccurately, and in so many contexts, are that he is

[1] a *direct realist*

about perception, in the sense in which it is correct to be a direct realist,[39]

[2] an *outright externalist*

about the content of experience, in the sense in which it is correct to be an externalist,[40] and

[3] a *representationalist*

about sensations, in the new and current sense of the term, in holding that all sensations are representational, i.e. ideas or representations of, and intentional with respect to, objective physical properties.[41]

---

[38] I am not guiltless, when it comes to misrepresenting him.

[39] According to direct realism what you now see is a book, and not in any sense a mental intermediary, a representation of a book. There is of course a mechanism, on Descartes's view as on any sensible view, and there is also a certain sort of intermediary. But this is a wholly non-mental item for Descartes (it is a brain-pattern), and one's registration of that brain-pattern is one's seeing the book (today we might replace 'registration' by something like 'instantiation'). There is a beautifully clear exposition of Cartesian externalism, and of direct realism, in Arnauld 1683, especially pp 54–73 and the paragraph running from pp. 76–7; for a good discussion of this see Pyle 2003, ch. 4.

[40] The book itself is what my belief is about, part of the content of my belief.

[41] They are 'confused' ideas or representations rather than clear and distinct ones, on Descartes's view. (The new use of 'representationalism' is pretty disastrous, because it is often used to mean the opposite of what it used to mean. See Strawson 2005.)

He takes it, of course, that experiential states have, essentially, 'internal', private, subjective, qualitative-experiential, 'what-it's-likeness' (etc.) characteristics, over and above anything they are about or represent. He is, in other words,

[4] a *real realist about experience*

or consciousness ('thinking', in his terminology) as everybody agrees.[42] But there has never been any conflict between real realism about experience and any correct version of direct realism, externalism, or representationalism; contrary to what some have supposed.[43]

[1]–[3] are good to have in mind, but [4] is much more important for present purposes, and the next thing to note is that Descartes is

[5] *not a substance dualist*

in any conventional understanding of this term, for he does not think that the notion of substance has any meaning or intelligible reference or explanatory force insofar as a substance is supposed to be something whose existence is supposed to be in any way distinct from the existence of its properties (see e.g. Clarke 2003, chapters 1, 8, 9). In this sense the notion of substance is for him a dummy term, an empty placeholder.[44]

As is well known, Descartes thinks that there are only two kinds of substance: experiencing ('thinking') substance and extended substance.[45] Both have the maximally general 'attributes' or fundamental essential properties of duration and existence that any substance must have. They differ from each other only in their only remaining fundamental and essential property: the attribute of experience, in the one case, and the attribute of extension in the other. Many (including myself) find the term 'attribute' and its companion 'mode' off-putting and dusty-sounding, but the distinction between them is in fact very

---

[42] For 'real' realism as opposed to 'looking-glass' realism about experience see p. 5 n. 6.

[43] Nor is there any sort of conflict between real realism about the experiential and the thesis of the 'transparency' or 'diaphanousness' of perception (see e.g. Reid 1785, Essay II Chapter 16, Montague 2007, §8).

[44] Hume is gravely mistaken if he thinks he is ahead of Descartes in this matter.

[45] Descartes's identification of physical existence with extendedness is very far from foolish, given his overall understanding of extendedness, and seem profoundly in accord with leading present-day scientific conceptions of the physical. So too his view that the universe is a plenum, or in other words that there is no such thing as a vacuum ('plenum' is the opposite of 'vacuum') — inasmuch as the so called 'vacuum' of present-day physics is seething with activity (particle-pair creation and annihilation) and is defined simply as the lowest energy-state of the field system, which has no 'holes' or true vacua at all.

clear, simple and useful.[46] 'Mode' is used for particular types of experiencing, or extendedness, 'attribute' is a 'broader term' used when one is 'simply thinking in a more general way of what is in a substance' (Descartes 1645/6, p. 280, 1644, p. 211). Plainly one can think quite unspecifically about experience in general, or extension in general. Equally plainly, these attributes cannot actually be possessed at any given time without being possessed in some particular manner or mode at that time. If something is extended it must be either triangular, or pyramidal, or cubic, or giraffe-shaped, etc., and it must also have a certain specific size. If something is experiencing, it must either be thinking, doubting, fearing, hoping, imagining, willing, feeling pain, sensing, etc. If it is hoping, it must be hoping that p, or that q or that r, and so on; if it is sensing, it must be sensing visually or olfactorily, and so on; if it is sensing visually, it must be sensing redly or bluely, and so on. Modes, then, can be more or less precisely specified.

How do the two attribute terms relate to the term 'substance'? Discussing experience and extendedness in the *Principles* Descartes says that 'the distinction between these notions and the notion of substance itself is a merely conceptual distinction'.[47] By this he means that although it is a distinction that can indeed be made in thought, it is not a 'real' distinction, where to say that there is a real distinction between two things is simply to say that each can exist in reality without the other existing. They are in that sense ontologically unentangled. Each can exist 'separately' in complete ontological separation from the other.

There is nothing mysterious or difficult about the expression 'real distinction'; 'real' simply means 'in reality', 'in concrete reality'. A real distinction is a distinction or separation that can exist in concrete reality or as Descartes puts it 'outside our thought'. It is a distinction or separation that can exist independently of our thought as opposed to a distinction or separation that can be made only 'in our thought' (1645/6, p. 280). The 'can' means that a real distinction is not a matter of what things actually do exist separately, at any given time, but a matter of what things can concretely exist separately, a matter of what is possible as a matter of real, objective, mind-independent fact.[48]

---

[46] It is just a restriction of the equally old but currently more favoured distinction between 'determinable' properties like colour and their 'determinate' values like red or phthalo green or elephant's-breath grey (an example used by Bernard Williams, with a humorous reference to paint manufacturers, in his introductory lectures in Cambridge in 1972–3).

[47] *Principles* 1.63 (1644, p. 215). Cottingham et al. have 'conceptual distinction' in place of 'distinction of reason'.

[48] Note that the fact that an attribute cannot exist without existing in a certain mode means that there is no more a real distinction between a substance, considered at a given time, and

Consider triangularity and trilaterality (in a closed plane rectilinear figure). We can certainly make a genuine *conceptual* distinction between triangularity and trilaterality, but there is no *real* distinction between them, for neither of them can actually concretely exist without the other also existing. And the ground of this inseparability in reality is in fact a matter of identity, concrete identity, identity in the concrete, as it were. Any actually existing concrete case of triangularity is literally *identical* to the concrete case of trilaterality that it cannot exist without, for that in which the real existence of the one consists is the very same thing as that in which the real existence of the other consists.

The same goes, according to Descartes, for experiencing substance and the attribute of experiencing. Neither can exist without the other, any more than a thing can exist without itself. When he states, as he famously does, that the mind or soul or subject of experience does not and cannot exist in the absence of actually occurring conscious experience — when he holds that experience is an essential property of mind in this sense, a property that it can never lack — this is not some sort of odd and implausible stipulation on his part.[49] A mind in which no experience is going on is as impossible as a physical object without extension simply because there is, for Descartes, no real distinction between (a) the concrete existence of the attribute of thinking and (b) the concrete existence of thinking substance. His root — radical — idea about the nature of the subject of experience or soul is that it is somehow wholly and literally *constituted* of experience, i.e. of conscious experiencing: that is what *res cogitans* — a soul — is. A Cartesian soul is nothing like an immaterial soul as traditionally conceived. Rather, its whole being is experience, a matter of occurrent experiencing conceived of as some sort of inherently active phenomenon — so it obviously can't exist when there isn't any.[50] In the *Principles* Descartes talks of 'our soul or our thinking' as if the two terms were strictly interchangeable (1644, p. 184). Later he writes, seemingly unequivocally, that 'thinking', in being the essential attribute of thinking substance, 'must be considered as nothing else than thinking

---

the particular modes that its attributes are exemplifying at that time, than there is between a substance and its attributes *tout court*. Its being, if I may say so, is its being — of interest, perhaps, to 'trope theorists'.

[49] This is how Locke polemically chooses to treat it (1689, p. 108 (2.1.10)).

[50] I record this point about Descartes in M1994, pp. 124-7; for a better statement see Strawson 2007, §9. The view raises puzzles that I can't address here; it seems that the process of conscious experiencing that constitutes the mind is also to be conceived as a 'potentiality'.

substance itself..., that is, as mind' (1644, p. 215). In his *Notes against a Certain Broadsheet*, reiterating his official doctrine of the real distinction between mind and body in the face of Regius' most unwelcome exposure of his actual baseline view (see p. 214 below), he treats being a thing and being an attribute as effectively the same, saying of the attributes of extension and experience 'that the one is not a mode of the other *but is a thing, or attribute of a thing*, which can subsist without the other' (1648a, p. 299; my emphasis picks out two expressions that are offered as equivalent). Questioned on the point by Burman, he confirms that his view is that 'the attributes [of a substance], when considered collectively, are indeed identical with the substance' (1648b, p. 15).

Contrary quotations can be found — at one point in his conversations with Burman Descartes also speaks of substance as a 'substrate' — but his basic commitment is quite clear, and the matter is usefully adjudicated by Cottingham.[51] Descartes, as Broad remarked, 'was a man of genius with an extreme dislike of anything misty and confused' (Broad 1944, quoted by Kemp Smith 1952, p. 190), and when he seems equivocal it must be remembered that he is concerned not to wake the Church and the philosophers of the Schools by explicitly denying the existence of entities to whose existence they are committed ('I do not deny that...' is a recurring phrase). The trick, for Descartes, is to do one's philosophy using the conventional terminology, but without making any real substantive appeal to these entities, knowing that one's intelligent readers will see that this is what one has done. 'I wish above all that you would never propose any new opinions', he wrote to Regius in 1642,

> but, while retaining all the old ones in name, only offer new arguments. No one could object to that, and anyone who understands your new arguments properly will conclude immediately from them what you mean. Thus, why did you need to reject substantial forms and real qualities explicitly?[52]

The standard picture of the immaterial soul or self, then, is nothing like Descartes's. According to the standard picture P1 there is [i] some

---

[51] Descartes 1648b, p. 17, Cottingham 1976, pp. 17, 77-9. Nadler agrees, noting that Descartes's 'considered position ... is that while there is a conceptual distinction between substance and attribute ... there is not a real distinction between them. Substance and attribute are in reality one and the same' (2006, p. 57). He goes on to point out that Spinoza also holds this view.

[52] Quoted by Clarke 2006, p. 224, see Descartes 1619-50, p. 205). Descartes had particular reason to ask Regius to be more circumspect because Regius was already publicly identified with the Cartesian cause.

sort of immaterial soul-substance or soul-stuff that is [ii] the *ground* or *bearer* of conscious experiencing and that [iii] can continue to exist even when there isn't any going on, and that therefore [iv] has some nature other than conscious experiencing.

Everyone agrees that Descartes rejects [iii], in holding that a mind or subject must always be experiencing, but his claim that 'each substance has one principal property which constitutes its nature or essence... and thinking *constitutes the nature of* thinking substance' (1644, p. 210, my emphasis) is often read as allowing, as in P2, [iv] that the soul has *some* other necessary manner of being that is not conscious experiencing. This reading is, however, very problematic, because to claim that something constitutes the nature of something is to claim that nothing else does, even if it has other essential attributes like temporal duration (and ultimate dependence on God).

THREE PICTURES OF THE IMMATERIAL SELF

[A] *continuously existing immaterial soul or self or subject represented by thick continuous line*

[B] *gappy process of consciousness (allowing, e.g. for dreamless sleep) represented by thin gappy line*

[C] *continuous stream of consciousness represented by thin continuous line*

[P1] the standard picture: [B] going on in ontologically distinct [A]

[P2] possible picture of Descartes's view: [C] going on in ontologically distinct [A]

[P3] proposal about Descartes's fundamental idea: [C] = [A]

Diagram 1

Descartes, then, rejects all of [ii] to [iv], and accepts [i] only inasmuch as he takes it that there is no real distinction between a substance and its attribute. His picture is more like [P3]. There is 'no real distinction, in the Cartesian sense, between a thing and its properties' (Clarke 2003, p. 215). To that extent, Descartes is an attribute dualist or

[6] a *property dualist.*

Insofar as we seek 'to keep the peace with the philosophers' (1637b, p. 268) we employ the language of 'inherence' and 'substance'; we speak of thought or extension as 'inhering in a substance'; we agree that it is obvious that properties or 'attributes … must inhere in something if they are to exist', for to say this is simply to say that they have to be somehow concretely instantiated, if they are to exist, 'and we *call* the thing in which they inhere a substance' (1641, p. 156, my emphasis). 'Inherence in a substance' is, however, and to repeat, a dummy phrase used simply to express the fact that the properties or attributes in question are concretely instantiated ('exist'). The word 'substance' does no separate ontological work; the clear implication of *Principles* 1.11 is that 'substance' simply means 'not nothing', 'existent' (1644, p. 196).

It may be added that being able to exist on one's own (God apart) is a sufficient condition of being a substance, on Descartes's view. Thus he notes, strikingly, that those who think that 'the heaviness of a stone is a real property distinct from the stone' deny that heaviness is a substance, but 'in fact they conceive of it as a substance because they think that it is real and that it is possible … for it to exist without the stone' (letter of 29 July 1648, 1619–50, p. 358).

This point will be important later on. For the moment, the proposal is that Descartes is a property dualist in any sense in which he is a substance dualist, and that to say that he is a substance dualist is, on his terms, to add nothing to the claim that he is a property dualist. Descartes is not, therefore, a 'Cartesian dualist', as this term is usually employed. The word 'substance' is furthermore subsidiary to the word 'property' or 'attribute', being a mere placeholder that does no more than express concrete existence. Descartes's dualism is completely and more clearly expressed simply as the claim that the attribute of experience can possibly be concretely instantiated without the attribute of extension being concretely instantiated.

It is even better expressed as the claim that there can be experiential *being*, or experiential *reality*, without there being extended being, or extended reality; for this way of putting it avoids words like 'property' and 'attribute' and the conventional metaphysics that they force on most people's minds. The word 'property', in particular, has proved lethally dangerous in recent discussion of the mind-body problem, and is best avoided altogether.

If either of two things can be concretely instantiated without the other being concretely instantiated then there is as noted a real distinction between them, in Descartes's terminology. Now the relation of *being really distinct from* is evidently symmetrical, so that although Descartes concentrates on arguing that the experiential can exist without the physical, he also accepts the converse, including, of course, the possibility that

> [P] there could exist beings who were physically qualitatively identical to experiencing human beings but who were not fully qualitatively identical to them because their existence involved no experience

a view defended in our day by philosophers like Chalmers, whose 'zombie' argument against physicalism is Descartes's argument run in reverse — the two of them make exactly the same sort of conceivability claim — with the restriction described in [P].

Goff reverses Chalmers, arguing that the experiential could exist without the physical, and in particular for the possibility that

> [Q] there could exist beings who were experientially qualitatively identical to experiencing human beings but who were not fully qualitatively identical to them because their existence involved nothing physical[53]

and is on the whole a classical Cartesian. It is, however, arguable that he goes further than Descartes on this point, because Descartes thinks that the truth of [Q], as opposed to the truth of the weaker claim that

> [R] experiential reality can exist without anything physical existing

requires the existence of a supernatural power (whether it be an omnipotent god or a *malin génie*) putting in some special work. On Descartes's view, the experience or thinking of an immaterial mind unconnected with a body will be restricted to operations of pure reason — things like logic and mathematics. It will need specifically targeted supernatural assistance to have sensation, personal memory, imagination and so on, for all these things essentially involve the

---
[53] Or indeed anything else. See Goff 2006.

PANPSYCHISM? 209

brain, on Descartes's view, so far as our actual daily existence is concerned. But perhaps Goff thinks the same.

Many want to say that they are property dualists but not substance dualists, and call themselves 'property dualists' precisely in order to distinguish themselves from 'substance dualists'. As remarked in RMP (p. 28), I think that this position is incoherent if it amounts to anything more than the claim that there are two seemingly very different sorts of properties, and Descartes agrees. The term 'property dualism' has been used in many ways, and has in consequence been rendered more or less useless, but the key question for property dualists (I am assuming that they are dualists with respect to the experiential and the extended, the conscious and the physical) is perhaps this:

> Are the two property-types essentially mutually exclusive in such a way that the being (the concrete realization) of one of them cannot be the same thing as the being (the concrete realization) of the other?

(triangularity and trilaterality provide us with an immediately apparent example of properties where the concrete realization of the one is the concrete realization of the other), or perhaps

> Is there nothing the full being of which can be both the being of extension (or physical existence) and the being of experience (or consciousness)?[54]

The question for those who call themselves property dualists, in other words, is whether they are *ontological* property dualists.

In public Descartes never stopped arguing for ontological property dualism in this sense.[55] That said, he never found a reply to what was perhaps the most vivid objection ever put to him, by Princess Elizabeth of Bohemia[56] when she famously wrote:

---

[54] Note that use of the word 'being', while not fudging anything, crucially avoids the distinction between object/substance and property, which cannot properly stand up in philosophical contexts like the present one.

[55] It seems that most of those who call themselves property dualists today are also ontological property dualists when they say, for example, that experiential (consciousness) properties are non-physical properties. It follows, on the present view, that they are therefore also substance dualists.

[56] 1618-80. Her mother was a Scot, and she was the niece of King Charles I of England, then engaged in civil war.

> I have to admit that it would be easier for me to attribute matter and extension to the soul than to attribute to an immaterial thing the capacity to move and be moved by a body[57]

and it is striking how keen he was on his empirical argument against materialism,[58] which aims to show that a wholly material thing, a mere machine, could never respond to an potentially infinite number of previously unencountered statements in the way that even the 'dullest of men' can.[59] I think he was right, in his time, to find this argument attractive, but it would have been pointless if it had been clear that his *a priori* arguments worked.[60] Worse, it would have been a tactical error to propound it, insofar as to do so is to suggest that the *a priori* arguments are not conclusive. But then, the *a priori* arguments were at the time thought to be inconclusive: the key argument was challenged in every one of the seven sets of Objections to his Meditations, published in 1641–2.

9. Real Descartes [2]

I have set out the sense in which 'substance' is for Descartes an empty term. One might say that it attaches an existence operator (an 'it exists' operator) to a property conceived in the abstract, and does absolutely nothing else. And yet we may I think add something important to it, for when Descartes writes to Henry More he says that he conceives 'incorporeal substances … as sorts of powers or forces' — i.e. as something essentially active, in a large sense of 'active' (letter of 5 February 1649, 1619–50, p. 361). There is no space for a detailed discussion of this, but one might say that Descartes conceives of thinking — conscious experiencing — as a kind of 'powerful process' in some sense of the word 'process' that does not require that there be a

---

[57] Letter of 20 June 1643, quoted in Descartes 1619-50, p. 220 n. In his letter to More of 15 April 1649 Descartes, still without an adequate reply, says rather touchingly that 'it is no disgrace for a philosopher' to think that an incorporeal substance can move a body (1619-50, p. 375).

[58] I use 'physicalism' and materialism' interchangeably.

[59] 1637a, p. 140. Note that Descartes says only that 'it is *for all practical purposes* impossible for a machine to have enough different organs to make it act in all the contingencies of life in the way in which our reason makes us act' (1619-50, p. 365, my emphasis). Note also that the argument aims only to show that a material thing could not have reason. Descartes did not think he could rule out the possibility that a machine, e.g. a non-human animal, could have sensations or experience.

[60] It is interestingly discussed by Chomsky, whose work helped to undermine it once and for all (see e.g. Chomsky 1968). It is a question whether Descartes would also have thought the argument undermined by the fact that there are 10^{11} neurons, with many more interconnections than that. The answer, I think, is 'Very probably'.

substance distinct from the process in which the process can be said to go on or occur, any more than it requires any fundamental ontological or 'real' distinction between concretely existing attributes and substance, or indeed between concretely existing attributes and processes.

I think Descartes has his basic metaphysics exactly right, just as he has almost everything else right, in spite of his central, unsuccessful *a priori* argument for the real distinction in the *Meditations*, which is the only thing holding back his otherwise utterly comprehensive — one might say ruthless — physicalism.[61] I am inclined to go a step further, for the phrase quoted in the last paragraph occurs in a passage that can be read as a well hidden acknowledgement by Descartes of the possible truth of (real) materialism, given the analogy it employs. He says that he conceives incorporeal substances — which, remember, are themselves nothing but thinking or experience (1644, p. 215) —

> as sorts of powers or forces, which although they can act upon extended things, are not themselves extended — *just as fire is in white-hot iron without itself being iron*'.[62]

Now this is only an analogy, but it bears comparison with a proposal in a letter to Arnauld written a few months previously, in which, not for the first time, he uses heaviness as an analogy for the ability for incorporeal mind to move body. Speaking of those who seek (vacuously) to explain the movement of a falling stone towards the centre of the earth by appeal to the property of 'heaviness', he says that

> it is no harder for us to understand how the mind moves the body than it is for them to understand how such heaviness moves a stone downwards (1619–50, p. 358)

and here too one may find a concealed acknowledgement, in someone who was, for all his extreme circumspection, truly and utterly devoted to expressing the truth, that the mind may not in the end be a 'real quality' relative to body, i.e. may not in the end be something that can

---

[61] I have in mind the basic argument that can be variously sliced — as the Conceivability Argument, the Argument from Clear and Distinct Ideas, and the Argument from Doubt — and whose unsoundness was as remarked asserted in every one of the seven sets of Objections to the Meditations published in 1641–2. (I put aside the almost universally derided Argument from the Indivisibility of the Mind, for although there is a sense in which it is correct on its own terms, it is well answered by Kant in his second Paralogism, and by others before him, insofar as it is supposed to offer any support for immortality.)

[62] 1619–50, p. 361, my emphasis. Descartes is here replying to a passage in which More is pressing the same point as Princess Elizabeth.

exist without body.[63] Descartes the indefatigable dissector was in every other pore of his philosophy a physicalist (a real physicalist, in my terms), as just remarked, and had a few months earlier conceded to Burman, directly contrary to his official position, that 'we cannot claim to have adequate knowledge of anything, including even bodies, and that we are obliged to work within the limitations of our concepts even if we recognize those limits' (Clarke 2006, p. 385); from which it appears to follow that we cannot claim to be able to rule out the possible corporeality of mind.[64]

My proposal, then, is that in his letters *Descartes acknowledges the possibility that physicalism, i.e. real physicalism, may be true* — an idea that was already familiar in his time. The case for the proposal may seem weak, but it cannot be otherwise, for Descartes was extremely cautious in what he wrote down and would certainly not have wished it to be otherwise.[65]

He had used the analogy of heaviness before, in a letter to Princess Elizabeth. She had replied robustly, by telling him (in Kenny's enjoyable words) 'that she was too stupid to understand how the discarded idea of a falsely attributed quality could help us to understand how an immaterial substance could move a body, especially as Descartes was about to refute the notion of heaviness[66] in his *Physics*' (letter of 20 June, 1643, 1619–50, p. 220 n.). There are I think three possible readings of her response. First, and wholly implausibly, given that Descartes was himself responding to her earlier and famous protest that she could not conceive how an immaterial thing could move or be moved by a material thing, Elizabeth simply did not register the most striking implication of the analogy, to wit that mind, too, like heaviness might not in the end be a 'real quality' capable of existing apart

---

[63] See the passage about heaviness quoted on p. 211. It is well concealed because the argument is at this point *ad hominem* against the Schools.

[64] For an excellent discussion of 'adequate' as 'opposed to 'inadequate' knowledge, and the associated but different distinction between 'complete' and 'incomplete' knowledge, see Yablo 1990, pp. 158–77. Effectively the same distinctions — and claims — are in play in the current discussion of the mind-body problem.

[65] His caution was not without reason, in Europe in the first half of the seventeenth century, although he was perhaps also a little paranoid, if only in the looser sense of the term. For a good account of his caution, his reaction to the Inquisition's condemnation of Galileo in 1633, his awareness of those whom the Inquisition had put to death, his concealment of and frequent changes of his address, his efforts to keep in with groups likely to be offended and alienated by his overtly stated ideas, see Clarke 2006. Writers at that time had exactly the same sorts of skills of concealment and interpretation as those that evolved under Communist rule in the twentieth century (and at many other times and places in human history).

[66] To refute the idea that heaviness was a 'real quality', i.e. a property capable in principle of existing apart from body. See again the passage quoted on p. 211.

from body. Second, she did register it, but was not prepared to let Descartes get away with the subterfuge. Third, and most plausibly, she saw it quite clearly, and was in her own equally ingenious way making it quite clear to him that she had seen it. The second and third reading are not strictly incompatible, and Descartes's reply can in turn be seen either as an evasion or as a similarly veiled acknowledgement of her point:

> I did not worry about the fact that the analogy with heaviness was lame because such qualities are not real, as people imagine them to be. This was because I thought that Your Highness was already completely convinced that the soul is a substance distinct from the body (1619–50, p. 228).

This, one might say, is a bit rich, and Descartes goes on to close down the discussion in the next paragraph, remarking that it is 'very harmful to occupy one's intellect in frequently meditating upon' metaphysical matters. He wants the exchange to go no further, and it doesn't.

The proposal, then, is that Descartes may be at bottom a real physicalist, in my sense, or rather that he is well aware of this position as a possibility, and suspects (like Locke, who learnt so much from the debates in which Descartes was involved) that it may be true. It is, of course, a possibility that removes a host of difficulties for him, the Princess-Elizabeth difficulty foremost among them, although it is equally obviously incompatible with his official view.

It may be objected that it is really not very helpful to say that Descartes is a real physicalist. More cautiously, less provocatively, more in Descartes's idiom, we may say that he is in the end not a real-distinctionist, i.e.

[7] *not a completely convinced or committed real distinctionist*

i.e. not someone who thinks he really can know for sure that the mind can exist apart from body; and that he is to that extent open to being a

[8] a *realistic monist*

i.e. open to a version of the position to which Spinoza was inexorably led by Descartes's work, and later explicitly adopted, in his own special metaphysical idiom (see §14 below).

The notion of monism is in a sense not helpful, as already remarked (pp. 7, 188), and monism need not be adopted as a positive thesis. It is, to be sure, Occamically agreeable. It is also quite *overwhelmingly* natural as soon as we forget philosophy, religion, fear of death, and so on, and simply consider ourselves, alive now, with our demanding bodies and streaming consciousnesses, so evidently substantially unified, physical, in every aspect.[67] Still, all we need to do at this point is to grant that we cannot know that there is not some good and fundamental sense in which there is only kind of stuff. It is, as Descartes says to Hobbes,

> perfectly reasonable ... for us to use different *names* for substances that we recognize as being the subjects of quite different acts or accidents ..., and ... to leave until later the examination of whether these names signify different things [the dualist option] or one and the same thing [the monist option].[68]

A final piece of evidence, I propose, is provided by Regius's *Broadsheet*. Regius was at one point Descartes's closest intellectual ally, and was well known as a Cartesian, but by 1647 they had fallen out. Regius was irritated by what he saw as Descartes's dishonesty, or lack of courage, and in 1647 published a somewhat careless and provocative broadsheet in which he was bold enough to express something that was perilously close to — indeed was — Descartes's own real view:

> (2) So far as the nature of things is concerned, the possibility seems to be open that the mind can be either a substance or a mode of a corporeal substance. Or, if we are to follow some philosophers, who hold that thought and extension are attributes which are present in certain substances, as in subjects, then since the attributes are not opposites but merely different, there is no reason why the mind should not be a sort of attribute co-existing with extension in the same subject, although the one attribute is not included in the concept of the other. For whatever we can conceive of can exist. Now, it is conceivable that the mind is some such item; for none of these implies a contradiction. Therefore it is possible that the mind is some such item. (1647, pp. 294-5)

He ended his broadsheet in an extremely aggressive way, with a direct quotation from Descartes's own work that directly implied that Descartes was concealing his true views: 'No one acquires a great reputation for piety more easily than the superstitious or hypocritical person' (Regius 1647, p. 296, quoting Descartes 1644, p. 191).

Descartes, understandably, felt very put out, not to say threatened, and replied with a thick version of his official view in order to try to

---

[67] Compare P.F. Strawson's thesis about the 'primitiveness of our concept of a person' (1959, ch. 3). Descartes is also anxious to stress the *extreme* intimacy of the 'substantial union' of mind and body within his official dualistic theory.

[68] 1641, p. 124, my emphasis. Here as so often Descartes talks of 'substance' in the accepted way, without advertising his substance/property identity thesis. Hobbes does, however, pick up on it, and Descartes duly placates him (1641, p. 125; see also Strawson 2007).

cover the damage. I do not, however, think that there is any real doubt that Regius had expressed Descartes's true and sensibly agnostic view. The fact that this is in effect an '*a posteriori* physicalist' view is further illustration of the foolishness of the present-day debate in continuing to ignore the historical debate and turning Descartes into a silly straw man.

In conclusion, then, I suggest that in addition to being [1] a direct realist, [2] an externalist, [3] a certain sort of representationalist, and — this at least no one denies — [4] a real realist about experience, Descartes is also [5] not a substance dualist in any conventional sense of the term, and certainly not in any sense that is incompatible with being a [6] property dualist. Finally, he knows that he cannot actually definitively rule out the possibility that there is no real distinction between mind and body, and indeed suspects that this may be the case. To that extent he is [7] not a truly convinced real-distinctionist with respect to mind and body, and is to that same extent [8] a realistic monist, or at least someone who does not rule it out.

I am aware that textual evidence can be brought against this proposal. It is after all part of the present proposal, and is in any case independently well established, that Descartes had reason to conceal certain of his views, and did so,[69] and was in many respects very evasive in his dealings with others (see, again, Clarke 2006). What this means practically, though, is that it is no good simply piling up contrary quotations against the current interpretation.[70] What one has to do is to feel one's way into the intellectual heart of this man who went to the butcher's for brains when he wanted to understand the mind, and was so very fiercely concerned with reaching the truth whatever the cost, even if it required considerable public circumspection, even deception.[71]

---

[69] Stopping the publication of his book L*e monde*, for example, on hearing of Galileo's condemnation by the Inquisition.

[70] This is what usually happened in the so-called 'Hume wars', as orthodox Hume scholarship responded to the unorthodox interpretation of Hume put forward most visibly by Wright (1983), Craig (1987) and myself (1989). There was very little attempt to confront any of the central arguments behind the new interpretation.

[71] It's worth noting that Descartes did not think one could prove the immortality of the soul, and when the second edition of the Meditations appeared he took care to delete the words 'in which the immortality of the soul is demonstrated' from the subtitle, for they had been added in the first edition without his knowledge or consent. See e.g. Clarke 2006, pp. 202–3. Clarke also quotes Martin Schook, who expresses a common contemporary opinion, if with more than usual force, when he writes in 1643 that Descartes, 'while giving the impression of combatting atheists with his invincible arguments, … injects the venom of atheism delicately and secretly into those who, because of their feeble minds, never notice the serpent hiding in the grass' (2006, p. 235).

But let me end this section with the more moderate words with which Desmond Clarke concludes his book *Descartes's Theory of Mind*:

> Descartes's dualism was an expression of the extent of the theoretical gap between a science of matter in motion, within the conceptual limits of Cartesian physics, and the descriptions of our mental lives that we formulate from the first-person perspective of our own thinking.... The properties that feature in these very different perspectives

are not, however,

> sufficient to justify the conclusion that it is impossible, in principle, to develop an explanation of human thought by including new theoretical entities in one's concept of matter. The underlying support for Descartes's property dualism was not a metaphysical theory of substances, or a plausible argument about the distinctness of properties, but an impoverished concept of matter ....
>
> Was Descartes, then, a substance dualist? Yes and no. He was not a substance dualist if that means that one explains the human mind by reference to a non-material substance. For Descartes, substances as such are non-explanatory. We speak about different substances in the same way as we speak about ... properties that are theoretically irreconcilable. Descartes acknowledged that he had no theory about the way in which thinking might be caused or explained by the known properties of matter, and he was persuaded that such a theory was most implausible. For that reason he was a property dualist. However he also argued unconvincingly in the *Meditations* that the implausibility of finding a theoretical link between thinking and the properties of matter implied a 'real distinction' between the substances to which such properties belong. Cartesian dualism, therefore, is not a theory of human beings but a provisional acknowledgement of failure, an index of the work that remains to be done before a viable theory of the human mind becomes available (2003, p. 258)

As already remarked, the fundamental objection to his central argument for the real distinction between mental reality and physical (extended) reality occurs in all seven sets of Objections published in 1641–2.

I want now to try to regiment the key claims in the mind-body debate. First, though, a brief note about 'supervenience'; for nearly all those who call themselves physicalists accept something called the 'supervenience thesis', so far as the relation between physical and experiential phenomena is concerned.

10. 'Supervenience'

The supervenience thesis as formulated by Joseph Priestley in 1778 is that

[i] 'different systems of matter, organized exactly alike, ...would feel and think exactly alike in the same circumstances'.[72]

He does not consider the converse supervenience thesis

[ii] if different minds thought and felt exactly alike in the same circumstances, then they would be identical in respect of their material constitution

and most, understandably, think [ii] is obviously false. In fact, when people talk today of the supervenience thesis in the philosophy of mind they usually mean [i] plus the denial of [ii], although the denial of [ii] makes no supervenience claim at all. I avoid the notion altogether when discussing the mind-body problem, because it has never been of any real help, and has been very unhelpful in appearing helpful. It is, however, popular, so let me record my view that although [i] is surely correct, given the truth of physicalism, the denial of [ii] may not be correct: not if we are realistic about the nature of mental reality, i.e. the nature of actual concrete mental contents.[73]

The argument for rejecting [ii] seems impregnable, for you and I both believe that grass is green, and so do speakers of many different languages, and our brains are most certainly not identical in respect of whatever it is about them that makes it true of us that we believe that grass is green. True — but this fails to address the real issue. If you understand [ii] in such a way that this gives a sufficient reason to reject [ii], then you are right to reject [ii]. The mental entities that really matter, however, the mental entities that are actually in play when we get down to the concrete business of real metaphysics and raise the question of the truth or otherwise of [ii], are not things like beliefs, or any other such dispositional phenomena. Nor are they particular occurrent conscious phenomena like individual sensations, if these are considered in any sort of isolation from the total experiential fields of which they are a part. We can slice reality in many ways in thought and language, but the mental realities that we have to do with when we are being metaphysically serious are total experiential fields, total

---

[72] 1778, p. 47; the same idea may be said to be succinctly expressed by Spinoza: 'as the body is, so is the soul' (c. 1662, p. 96).

[73] John Heil set me off on this line of thought in 1993. See Heil 1992, pp. 133-4, M1994, p. 49.

occurrent conscious experiential states considered at any given moment, the precise details of whose contents far outrun any possible human description. These are the items that we must consider when we examine the supervenience thesis as a thesis about the relation between the mental and non-mental (experiential and non-experiential) being of the brain. These are the only mental items that are actually to be found when mental reality, i.e. occurrent mental reality, is considered independently of any intellectual abstraction (any 'conceptual distinction', in Descartes's terms) that allows us to consider it, or a segment of it, in a merely partial manner. The question is then this: could two human beings X and Y really be in identical total experiential states, qualitatively speaking, between times t_1 and t_2, and still be in qualitatively different brain states? Could they in other words be identical in their 'E features', between t_1 and t_2 and yet differ in their 'B features'?[74]

Yes, plainly, for their brains are involved in a great deal of activity that has nothing to do with their current experience, or so we may assume. This, though, is uninteresting, and we can tighten the question by putting aside all those things about their brains that have nothing to do with their current experience, together with all those things that are merely causally antecedent to their current experience. Having done so, we consider only those B features that are — how to put it? — directly constitutive of? — 'realizers' of? — well, how else to put it? — their E features between t_1 and t_2. We may call these the B* features, and then ask again whether X and Y can really be identical in their E features, between t_1 and t_2 and yet differ in their B* features. Many, I think, will still confidently answer Yes, rejecting [ii]. Plainly, they will say, X and Y can possibly have exactly the same E features while differing in the precise nature of their B* features. But it is by now far from clear that this answer is obviously right.

On one view (worth noting because it illustrates the treacherousness of the terms in which the mind-body problem is discussed) it can't be right. This is because the tightening of the question has produced a version of the *identity* theory about the relation between B* features and E features. But if a Yes answer is to be possible on the terms of the identity theory, then the theory must involve the assumption that B* features involve something more or other than E features. But this assumption is necessarily false, if the theory is really an

---

[74] I take B features to include E features, so that the question whether their B features can differ from their E features arises only given the standard assumption that their B features include features that are not E features. One could call the relevant B features 'non-E features', but it is not necessary.

identity theory, by the basic logic of the identity relation. This is an old point that must surface whatever terms one uses: where can the gap be, for any supposed identity theorist, between radical eliminativism and panpsychism?

I am going to put this objection aside, however, because many will think that X and Y can plainly differ in their B* features while being identical in their E features even after the tightening has been carried out. (It follows that they can't be identity theorists, if they are real physicalists who are neither radical eliminativists nor panpsychists; I'd rather not enquire further into what they are.) It is, they will say, plainly possible that X and Y should differ in their B* features, even those B* features that are directly constitutive of their E features,[75] while being identical in their E features. Many of us have been brought up on this 'token-identity' possibility and find it very comfortable, and do not think it is threatened by the restriction of attention to colossally complex total experiential fields that is required by a serious metaphysical approach to the question.

Is it a real possibility? Before trying to answer this question, let me first note that the claim is restricted to physical beings. We do not need to rule out the possibility that two creatures in different universes made of fundamentally different kinds of substantial stuff could be in qualitatively identical experiential states. If we knew what God knows about what sorts of substantial constitution are possible for minds, we might see that even this two-universe possibility is not in fact a real possibility; but we don't need to pursue the matter.

Let me also note that I have, in talking of X's and Y's experience, put panpsychism aside. I have put aside all the 'microsubjects' that constitute their brains according to my kind of panpsychism. I am sticking to the point of view of the 'top' subjects, as it were, 'macrosubjects' of experience of the kind that you and I, say, experience ourselves as being in our ordinary everyday experience. Perhaps there are still other subjects of experience in our brains between the macrosubjects and the microsubjects, relatively high-level but still subsidiary subjects, but we may put these, too, aside. The present question is whether subjects of experience X and Y can be E-identical and B*-different.[76]

One might first reply that it may not be physically possible even if it is logically possible — that it is at the very least an open question

---

[75] These are the only ones that we are considering, since tightening the question.

[76] I am not here worried about the precise nature of these 'top' subjects of experience, or whether they have sharp identity conditions.

whether it is physically possible. In reply to that, it will be said that it must be possible for X to have one electron (or indeed a million) in a different place, as compared with Y, between t_1 and t_2, without X and Y being E-different — in which case there can indeed be B*-difference without E-difference. But we must then go back to the notions of realization and direct constitution. Is this electron indeed itself directly constitutive of, realizatory of, any E features? If it is not it is irrelevant; if it is, then it is again not clear whether X and Y can be E-identical and B*-different.

There is more to say about this question, but I am going to leave it here and conclude by recording my suspicion that if we approach the question of supervenience realistically, and as real physicalists, in the way I have sketched, then we have good reason to suppose it to be two-way. If we can avoid incoherence in setting up an interesting, substantive account of what features of the world make supervenience claim [i] true, then the most plausible — if not the only — realistic supposition will be that [ii] is also true. It will not only be true, in other terms, that

[i] physical qualitative identity entails experiential qualitative identity

but also that

[ii] experiential qualitative identity entails physical qualitative identity (in the relevant parts of the brain).[77]

'But the panpsychist supposition that the existence of a human brain between t_1 and t_2 necessarily involves the existence of other — many — subjects of experience requires one to say that there is, irreducibly, more than one experiential field in question, and therefore no such thing as the "total experiential field".' Reply. The total experiential field in question is the experiential field of the 'top' subject. More generally, what is in question is the total quantum of experientiality associated with a brain, and the question is whether two brains could ever have the same total quantum of experientiality and be physically different in any way. The proposal is that the answer is No even if we stick to relatively high-level subjects of experience. If we go on to include the microsubjects (the experientiality of the ultimates) the No cannot be less secure, but the present claim is independent of any such version of panpsychism.

---

[77] It should be unnecessary to note that physical qualitative identity entails identity of physical law.

11. The Basic Framework [1]

I want now to try to make a clear case for realistic monism by putting everything on the table in full view. My aim is to mark out a framework in which to consider the objections to RMP, although I cannot hope to reply to all the points raised by the other contributors to this book. I will begin with a list of theses about this universe (the only object of my concern), some of which have already been mentioned and some of which have no neat name. The list has a somewhat hectic air, because the linked theses constantly jostle each other and overlap, and the real action begins in §13.[78]

The first thesis is

Stuff Monism
[1] There is a fundamental sense in which there is only one kind of stuff in reality

a thesis that I accept for purposes of argument without further comment.[79] [1] contrasts with

Thing Monism
[2] There is a fundamental sense in which there is only one thing in reality

a thesis that I put aside, at least for now and for purposes of argument, aware that there are some rather good reasons for thinking that [2] may be the best thing to say at the level of fundamental ontology.[80]

The next thesis is an *a posteriori* certainty

The Experience thesis
[3] There is experiential reality, experiential being[81]

---

[78] There is no need to try to remember the names of the theses, many of which I will not use again. For another brief but searching survey of the geography, see Nagel 1986, pp. 28–32, 46–53.

[79] The trouble with 'substance monism' is that 'substance' has both a mass-term use, which gives [1], and a count-noun use, which gives [2] below. In §8 I use 'substance dualism' in a standard way to contrast with 'property dualism', and will continue to do so when it seems most natural, but on the present terms 'substance dualism' is properly called 'stuff dualism'.

[80] A compelling present-day version of this view is the view that spacetime is the only thing that exists, and that all particles, for example, are to be 'explained as various modes of vibration of tiny one-dimensional rips in spacetime known as strings' (Weinberg 1997, p. 20). On this view, then, we are made of spacetime, and recent work on loop quantum gravity also appears to support this idea. (If one's conception of spacetime makes it seem implausible, or even impossible, then one needs to rethink one's conception of spacetime.)

[81] I give an argument for this — insofar as one is needed, or possible — in RMP (p. 7 n. 7) and M1994 (pp. 51–2).

THE BASIC FRAMEWORK AT A GLANCE

Metaphysical theses

[1] There is a fundamental sense in which there is only one kind of stuff in reality (*Stuff Monism*)
[2] There is a fundamental sense in which there is only one thing in reality (*Thing Monism*)
[3] There is experiential reality or being (*Experience thesis*)
[4] There cannot be experience without a subject of experience (*Subject thesis 1*)
[5] There cannot be a subject of experience without experience (*Subject thesis 2*)
[6] There is physical reality or being (*Physicality thesis*)
[7] All reality is either experiential or physical (*Only-Experiential-and-Physical thesis*)
[8] There is only one kind of stuff. It has experiential reality and physical reality and no other kind of reality (*Realistic Monism*) [from [1], [3], [6], [7]]
[9] Experiential reality can exist without any physical reality existing (*Real Distinction thesis 1*)
[10] Physical reality can exist without any experiential reality existing (*Real Distinction thesis 2*)
[11] Experiential reality can exist without any extended reality existing (*Real Distinction thesis* 1*)
[12] Extended reality can exist without any experiential reality existing (*Real Distinction thesis* 2*)
[13] Experiential reality *of any kind* can exist without any physical (extended) reality existing (*Real Distinction thesis$^+$ 1*)
[14] Physical/extended reality can exist *in any arrangement* without any experiential reality existing (*Real Distinction thesis$^+$ 2*)
[15] Experiential reality cannot possibly *be* physical reality (*Cartesian Intuition 1*)
[16] Physical reality cannot possibly *be* experiential reality (*Cartesian Intuition 2*)
[17] There is only physical reality (*Physicalism*)
[18] All experiential being is physical being (*Experiential Physicalism*) [from [3], [17]]
[19] All physical being involves experiential being (given that physicalism is true) (*Weak Panpsychism*)
[20] All physical being is experiential being (*Pure Panpsychism*)
[21] At least some ultimates are experience-involving (*Micropsychism*)
[22] At least some ultimates are (wholly a matter of) experiential being (*Micropsychism* strengthened)
[23] All physical reality is (at bottom) the same stuff (*Homogeneity thesis*)
[24] There is non-experiential being (*Non-experientiality thesis*) [compare [6]]
[25] All (concrete) being involves experiential being (*Weak Panpsychism* restated)
[26] All being is experiential being (*Pure Panpsychism* restated)

Metaphysical theses (cont.)

[27] All reality is (at bottom) the same stuff (*Homogeneity thesis* restated)

[0!] All non-experiential reality is experiential reality and conversely (*The Experiential = Non-experiential Thesis*)

[28] Experiential reality cannot be non-experiential reality (*Experiential ≠ Non-experiential thesis 1*)

[29] Non-experiential reality cannot be experiential reality (*Experiential ≠ Non-experiential thesis 2*)

[28] Experiential reality cannot be wholly non-experiential reality (*Experiential ≠ Non-experiential thesis 1* emphatic)

[29] Wholly non-experiential reality cannot be experiential reality (*Experiential ≠ Non-experiential thesis 2* emphatic)

[[3]+[24]] There is experiential reality, there is non-experiential reality (*Fundamental Duality thesis*)

[30] Experiential reality cannot emerge from wholly and utterly non-experiential reality (*No-Radical-Emergence thesis 1*)

[31] Non-experiential reality cannot emerge from wholly and utterly experiential reality (*No-Radical-Emergence thesis 2*)

[32] All physical reality is — in itself, in its fundamental nature — wholly non-experiential reality (*Mindless Matter thesis*)

[33] All facts are (fully) determined by facts about ultimates (*Smallism*)

[34] All reality is either experiential or non-experiential (*The Only-Experiential-and-Non-experiential Thesis*) [compare [7]]

[35] There is only one fundamental kind of stuff. It has both experiential reality and non-experiential reality. There is no other kind of reality (*Fundamental-Duality Monism*) [compare [8]]

[36] Reality is substantially single. All reality is experiential and all reality is non-experiential. Experiential and non-experiential being exist in such a way that neither can be said to be based in or realized by or in any way asymmetrically dependent on the other (etc.) (*Equal-Status Fundamental-Duality monism*)

Epistemological theses

[37] I am acquainted with the essential nature of experience generally considered—i.e. with whatever all possible experiences have in common just insofar as they are indeed experiences—just in having experience (*General Revelation Thesis*)

[38/39] In the case of any particular experience, I am acquainted with the essential nature of that particular experience just in having it (*(Local) Revelation Thesis*)

[40] In the case of any particular experience, I am acquainted with the whole essential nature of the experience just in having it (*Full Revelation Thesis*)

[41] In the case of any particular experience, I am acquainted with the essential nature of the experience in certain respects, at least, just in having it (*Partial Revelation Thesis*)

to which we may immediately subjoin the necessary truths discussed in §6

The Subject thesis
[4] There cannot be experience without a subject of experience
[5] There cannot be a subject of experience without experience[82]

and a standard assumption

The Physicality thesis
[6] There is physical reality, physical being.

I make this assumption myself except where I explicitly question it, and I question it only for terminological reasons, i.e. because the way in which the word 'physical' is standardly used may well mean that I would do better not to use it as I do.[83]

My next assumption, for convenience of argument, and alongside Descartes (but not, say, Spinoza), is

The Only-Experiential-and-Physical Thesis
[7] All reality (or being) is either experiential or physical.

[1], [3], [6] and [7] entail

Realistic Monism
[8] There is only one kind of stuff and it has both experiential reality and physical reality and no other kind of reality.

Consider next

The Real Distinction Thesis
[9] Experiential reality can exist without any physical reality existing
[10] Physical reality can exist without any experiential reality existing.

If you hold that physical reality is necessarily extended,[84] or indeed, with Descartes, that being physical or material is just a matter of being

---

[82] [5] is a necessary truth given the definition of 'thin subject' in §6.

[83] Plainly [6] is just as much of an *a posteriori* certainty as [3] if one takes 'physical' as a pure natural-kind term that applies to anything 'real and concrete', as in RMP, p. 3. Here, however, I have a different argumentative purpose, and do not need to make that move. Stoljar holds a strongly convergent view about the two uses of 'physical'. See his paper 'Two Conceptions of the Physical' (2001).

[84] I do this in RMP in assuming, if only for argument, that 'the universe is spatio-temporal in its fundamental nature' (p. 9). Note that this leaves the question of the intrinsic nature of spacetime wholly open.

extended (a view that seems far from foolish, given Descartes's understanding of what it is to be extended), then you can re-express this as

*The Real Distinction Thesis**
[11] Experiential reality can exist without any extended reality existing
[12] Extended reality can exist without any experiential reality existing

The thesis can also be made more specific as follows:

The Real Distinction Thesis$^+$
[13] experiential reality *of any kind* can exist without any physical (extended) reality existing
[14] physical (extended) reality can exist *in any arrangement* without any experiential reality existing.

Descartes rejects [13] so far as the ordinary running of the universe is concerned, holding that its possible truth requires a special intervention from God. Chalmers, I take it, is an example of someone who accepts [14], but one can accept [12], when thinking of a stone, say, while rejecting [14] when thinking of a living brain.[85]

Like most people I believe that Descartes's *a priori* arguments for the Real Distinction thesis fail (as do his *a posteriori* arguments), but the same appears to be true of all recent *a priori* arguments for the falsity of standard physicalism, many of which are not importantly different from Descartes's.[86] As for Descartes's central intuition — the intuition that there is a fundamental gulf between physical reality and experiential reality of such a kind that experiential reality cannot possibly be physical reality and physical reality cannot possibly be experiential reality —, that is still found intensely compelling, and is still accepted by most participants in the present-day debate:

---

[85] I reject [14]. One reason why Chalmers accepts it may be that he subscribes to the idea that one can in counterfactual speculation suppose that one is talking about qualitatively the same physical objects when one varies the physical laws that govern them. It seems plain to me that the laws of physics are constitutive of the nature of the physical in such a way that one cannot do this. It is I think equally questionable (but this is much more controversial) to think one can take oneself to be talking about numerically the same physical objects when one varies the laws in counterfactual speculation.

[86] Descartes's *a priori* arguments fail in spite of some lovely ingenuity (e.g. his elaboration of what it is to conceive of something as a 'complete thing'). For a very rewarding discussion of his Conceivability Argument see Yablo 1990. Examining Arnauld's (1641) main objection, Yablo notes how powerful it is and expresses exactly my feeling when he writes: 'How wonderful then that Descartes had the chance to hear it and respond' (p.159).

The Cartesian Intuition
[15] Experiential reality cannot possibly *be* physical reality
[16] Physical reality cannot possibly *be* experiential reality

It is striking that the Cartesian intuition receives its most passionate endorsement from those self-styled physicalists who are most dismissive of Descartes — those who, accepting that everything is physical, are thereby led to deny the existence of experience (experiential reality) altogether. These physicalists are so strongly committed to the Cartesian intuition that when they couple it with their belief that physicalism is true they are prepared to deny the existence of the most certainly known thing there is — experiential reality. Descartes was right when he wrote that 'nothing can be imagined which is too strange or incredible to have been said by some philosopher' (1637a, p. 118), but the denial of the existence of experience shows that he was more right than he could ever have imagined. That said, he also noted that a philosopher will take the more pride in his views 'the further they are from common sense ..., since he will have had to use so much more skill and ingenuity in trying to render them plausible' (1637a, p. 115), and I don't suppose he assigned any upper bound to human pride. What he may have underestimated (but no doubt he makes the point somewhere) is the extent to which people can be blinded, rationality and cognitively, by presuppositions that they can't renounce or properly see as such.[87]

I reject the Cartesian Intuition as stated. I accept for purposes of argument that physicalism may be true, but I reject the Cartesian Intuition because I take it that all experiential reality is physical reality — if indeed physicalism is true.[88] If, that is, one accepts

Physicalism
[17] There is only physical reality

then

Experiential Physicalism
[18] All experiential reality is physical reality

follows, given the indubitability of [3].[89]

---

[87] Claims of this sort are risky, because one opens oneself to accusations of exactly the same sort. For my vulnerability in this regard see in particular Goff p. 60, Stoljar pp. 175–6, Coleman p. 47, and pp. 265–6 below.

[88] This conditional is always in place; see §2 above.

[89] [3] and [17] conjoined make up the position I called 'realistic physicalism', or '**RP**', in RMP.

In RMP I also subscribe to the more difficult and panpsychist view that physical reality essentially and constitutively involves experiential reality — the view that any portion of physical reality essentially involves[90] experiential reality — although I do not need to for most of my purposes:

(Weak) Panpsychism
[19] All physical being involves experiential being (given that physicalism is true).

I will call the conjunction of these last two theses

(Weak) Panpsychist Physicalism
[18] All experiential being is physical being (given the truth of physicalism)
[19] All physical being involves experiential being.

Note that [19] invites a stronger restatement as the converse of [18] as follows:

Pure Panpsychism
[20] All physical being is experiential being.

[20] is plainly a stronger version of panpsychism than [19], which still allows for the existence of non-experiential physical being, as [20] does not, at least given standard logic,[91] and if we go down to the level of ultimates, both are plainly stronger than

Micropsychism
[21] At least some ultimates are experience-involving

which was discussed in RMP (pp. 24–5). Micropsychism may in turn be restated more robustly, in line with [20], as follows:

Micropsychism
[22] At least some ultimates are (wholly a matter of) experiential being.

I noted in RMP (p. 25) that micropsychism seems more secure than panpsychism, but went on to assume for argument — as I will here — that panpsychism is likely to be true if micropsychism is. Let me dignify the basis for this assumption with a title

---

[90] A word chosen for its vagueness. Stapp is right that I am not interested in any argument for this that appeals to the rôle of the observer in quantum-mechanical physics.
[91] Many panpsychists would I think say that pure panpsychism is the only kind there is.

The Homogeneity Thesis
[23] All physical reality is (at bottom) the same kind of stuff.

This goes further than, e.g., Coleman's suggestion that 'it goes with our conception of, for example, sub-atomic particles that, *if* they can constitute brains, then any suitably arranged set of them will do',[92] because some of the ultimates recognized in the standard model of physics — e.g. neutrinos — are not fitted to be constituents of brains, as Seager points out (p. 137), and Coleman's formulation duly takes account of this. [23], by contrast, explicitly endorses the strong view that *any* physical stuff could in principle be arranged or situated so as to be a constituent of a brain.[93]

I am still using the word 'physical', but it is plain that it reaches breaking point in [20], at least in modern ears (as noted by Macpherson and McGinn, among others), because it now no longer rules out the possibility that there is no non-experiential being at all.[94] And this makes it sensible to put on the record the result of replacing 'physical' by 'non-experiential' in [6]:

The Non-experientiality Thesis
[24] There is non-experiential being.

More generally, it is time to consider dropping the word 'physical' altogether, as qualifying our stuff monism, and rewriting [19], [20] and [23] as

[25] *Weak Panpsychism*
All (concrete) being involves experiential being

[26] *Pure Panpsychism*
All being is experiential being

[27] *The Homogeneity Thesis*
All reality is (at bottom) the same kind of stuff.

---

[92] p. 48; I take it that Coleman would allow the substitution of 'ultimate' for 'sub-atomic particle'.

[93] Jeremy Butterfield tells me (private communication) that all particles in the standard model are excitations in a quantum field, 'as a wave is an excitation of the sea', and that 'since quantum fields are mathematically very special, there is a fundamental commonality' to them. Against that, he notes that there are of course various fields recognized in today's physics (eg quark, electron, neutrino) and that it remains unclear whether tomorrows physics will find them all to be aspects of one field.

[94] See e.g. RMP p. 8, M1994, p. 74, M2003a, p. 52. Note that the pure panpsychist use of the word 'physical' didn't bother Eddington and Russell at all, although they were in McGinn's terms 'in flagrant violation of common usage' (p. 91); as also, no doubt, was Whitehead.

[25] is compatible with [24], if, that is, it is coherent to suppose that all being could involve both experiential and non-experiential being. But [26], of course — pure panpsychism — is not.[95] Plainly we cannot replace 'physical' with 'non-experiential' in [20], and couple it with its converse to get

[0!] *The Experiential = Non-experiential Thesis*
All non-experiential reality is experiential reality
All experiential reality is non-experiential reality

— not unless Graham Priest can help us out with his 'dialethic' logic. And this is a very good measure of the problem that faces us.[96]

Pure panpsychism as just characterized in [26], the revision of [20], is accepted by Coleman and Skrbina, among others (the word 'physical' having been discarded), and the fundamental challenge for all serious and realistic participants in the current mind-body debate who wish to reject [26], and therefore hold out for [24], the irreducible existence of non-experiential reality, is to show how this can be done. This is, and always has been, the heart — the real heart — of the mind-body problem. A number of the contributors to this book assume that I am in RMP committed to [24], and so to impure panpsychism at best (many of McGinn's objections depend on this assumption, and also on a non-Cartesian metaphysics of object and property that I reject), and it is true that I mostly work with the assumption that [24] is true, because it is almost universally accepted, and so best kept in play when beginning to put the case for panpsychism. From M1994 on, however, I draw attention to the fact that it is an assumption, and I make this point repeatedly in RMP in order to keep it clear that it is an assumption that must be in doubt (see for example pp. 7, 8, 9 n., 17–18, 24, 26). I think, in fact, that the non-experiential can be retained only if it is literally identical with the experiential in some Spinozan way. If that is as impossible as it sounds, then it is impossible.

The adjective 'realistic', as applied to 'physicalism', may be reasonably thought to require not only that one acknowledge the irreducible existence of experiential reality but also that one acknowledge the

---

[95] Note that pure panpsychism is not idealism. I avoid the word 'idealism' because conventional idealism — the claim that reality consists entirely of ideas or experiences — is blatantly incoherent (given [4]) in assuming [a] that the subject of experience is in some way ontologically over and above its experiences and [b] that the subject of experience is not itself a mere idea. One might put this by saying that it is a very bad name for nearly all the positions — e.g. Berkeley's immaterialist position — that it is usually used to denote. This issue arises in Seager's paper; see also M1994 ch. 5, 'Mentalism, Idealism, and Immaterialism' and M2003b.

[96] Priest 2006; see also his article 'Dialethism' in the *Stanford Encyclopedia of Philosophy*.

irreducible existence of non-experiential reality. I am going to argue that the second requirement cannot in the end have the same degree of force as the first, but I will for the moment continue to understand the 'realistic' in 'realistic physicalism' (and in 'realistic monism') in this way.

12. The Basic Framework [2]

Although I reject the Cartesian intuition, I accept something related to it, something that we may surely take to be a necessary truth (but see below): that experiential reality, considered just as such, cannot be non-experiential reality, considered just as such, and conversely that non-experiential reality, considered just as such, cannot be experiential reality, considered just as such.

> *The Experiential ≠ Non-experiential Thesis*
> [28] Experiential reality cannot be non-experiential reality
> [29] Non-experiential reality cannot be experiential reality

The 'considered just as such' may need a little attention, and it may be that *the whole 'mind-body' problem lies in whatever it is that needs attention*, but the claim as it stands — I will call it 'the [E ≠ NE] thesis' for short — is just an instance of the necessary truth that if something is F then it cannot also be *not-F* although it may possibly be G, H, J, and so on. Descartes appealed to this truth when every one of the seven sets of published objections to his *Meditations* asked him to prove conclusively that the physical or corporeal, call it G, could not possibly be experiential or mental, call it F. For he argued, precisely, that for this F and G, F entailed *not-G*.

Someone may say that human beings are real and have experiential properties and also and at the same time have non-experiential properties, and are in this sense both F and *not-F*. This, of course, is not ruled out by the [E ≠ NE] thesis (any more than a Strawsonian acknowledgement of the primitiveness of our ordinary concept of a person is), for the [E ≠ NE] thesis claims only that experiential being itself cannot also be (wholly) non-experiential being.[97]

It may help to be more emphatic in the following way:

---

[97] This objection provides a further small instance of why it is wise to avoid using the word 'property' at all, when discussing the mind-body problem, so far as it is humanly possible to do so.

The [E ≠ NE] thesis
[28] Experiential reality cannot be wholly non-experiential reality
[29] Wholly non-experiential reality cannot be experiential reality.

That, at least, seems reasonably safe.[98]

Consider next the key claim that there is indeed both experiential reality and non-experiential reality

The Fundamental Duality Thesis[99]
[3] There is experiential reality
[24] There is non-experiential reality.[100]

The Fundamental Duality thesis has of course been rejected, both by certain sorts of mentalists or (misnamed — see n. 96) 'idealists', and by materialists who are eliminativists with respect to experience (whether overtly or covertly), and it seems prudent to say (to repeat) now that if it cannot in the end go unchallenged then it is of course the non-experiential, not the experiential, that will have to yield, because [3], the existence of the experiential, is beyond doubt. This is why I've always felt obliged to be formally agnostic about the existence of non-experiential reality, even as I have always assumed its existence for purposes of argument in taking it to be an essential feature of anything that can reasonably be called 'physicalism' (see §2 above).

The key question, now, given the key claim, i.e. the Fundamental Duality thesis, is this: Is the Fundamental Duality thesis combinable with a generally monist position? This is, I think, the key question for almost all of us, and the best hope for the combination is perhaps some kind of Spinozistic dual-aspect or dual-attribute view of things.

I will come to this. The next thing I want to lay down is related to the [E ≠ NE] thesis, although most think that it is not so secure. This is the claim, argued for in RMP, that experiential reality cannot 'come from' or 'emerge from' physical reality that is in itself wholly non-experiential (and conversely).

The No-Radical-Emergence Thesis
[30] Experiential reality cannot possibly emerge from wholly and utterly non-experiential reality

---

[98] So too wholly experiential reality cannot be non-experiential reality, and conversely.

[99] I am taking it that a *duality* (or plurality) thesis is not automatically a *dualistic* (or pluralistic) thesis, and hoping for Spinozan support.

[100] [24] is not the same as [6], the Physicality thesis, with which [3] was originally contrasted, and the transition is important.

[31] Non-experiential reality cannot possibly emerge from wholly and utterly experiential reality

I argued for this claim — exercised this intuition — in RMP, and here I am just unblushingly owning up to it. It aligns me (and e.g. Coleman) quite closely with those, like Jackson, who are now called '*a priori* physicalists', and I am not going to defend it further in this paper.[101] On the face of it, it is plainly different from the [E ≠ NE] thesis, but I think that denying it requires one in effect to reject the [E ≠ NE] thesis — a higher price, I suppose, than any emergentist would wish to pay — although there are many more or less frilly notions of emergence in the air that make it seem that this is not so.

It may be that it is right, in the end, to reject the [E ≠ NE] thesis. But if this is so it is so for reasons that render any Radical Emergence thesis entirely unnecessary.

I am nearly finished with metaphysical theses, but I should now add

The Mindless Matter Thesis
[32] All physical reality is — in itself, in its fundamental nature — non-experiential reality

which I called 'N-E' in RMP (p. 11) and reject.[102] Some, I know, are unable to renounce [32] (which may be expressed in linguistic mode as the idea that it is true by definition of 'physical' that physical reality is non-experiential reality), although it is a view to which physics in its abstractness gives no support. I mention it here in order to point out that one need not feel a conflict between the [E ≠ NE] thesis ([28] and [29]) and Panpsychist Physicalism ([18] and [19]) unless one also endorses [32] the Mindless Matter thesis.

Finally: on p. 9 above I explicitly assumed [i] that 'there is a plurality of ultimates (whether or not there is a plurality of types of ultimates)' and [ii] that 'everything physical (everything physical that there is or could be') is constituted out of ultimates of the sort we actually have in our universe. I took [ii] to entail [iii] that everything's being as it is at any given time just is the ultimates being as they are at that time, this being a simple matter of identity, although I did not explicitly say this. To that extent I endorsed the thesis that I will follow Coleman in calling 'smallism', taking it to be part of the

---

[101] Against Stoljar's doubts, for example, or Lycan's. McGinn is right that it is only given this claim (and the Homogeneity thesis) that I can say that 'physicalism entails panpsychism' (p. 93); that was my rhetorical choice.

[102] One can wonder about the exact force of 'in itself' and fundamental', and something is said about this in RMP. Here I'm using a broad brush.

conditional or *ad hominem* physicalism (p. 186) that forms my basic platform. Smallism, in Coleman's words, states that 'all facts are determined by the facts about the smallest things, those existing at the lowest "level" of ontology' (p. 40), and can be expressed equally well as follows:

Smallism
[33] All facts are (fully) determined by facts about ultimates

or, distinguishing 'ultimate facts' (i.e. facts about ultimates) from 'non-ultimate facts' (i.e. facts about non-ultimates)

Smallism
[33] All non-ultimate facts are (fully) determined by ultimate facts.

The word 'determine', however, is slippery,[103] and while I want to record smallism explicitly here as an extra thesis, and am untroubled by any commitment to the truth of determinism that it may bring with it,[104] I am not sure that I endorse it if the determination in question amounts to anything more than the determination that is necessarily involved in constitutive identity (if X wholly constitutes Y, if X is constitutively identical with Y, then X of course 'determines' Y). To say that all facts are determined by facts about ultimates is (at least) to say that if you fix all the ultimate facts then you fix all the non-ultimate facts. It also, however, suggests that the converse is not also the case, and although perfectly good sense can be given to this asymmetry claim (it is a one-way 'supervenience' or 'multiple realizability' claim) it is not as if there is some kind of metaphysically real *determinative flow* going from the ultimate facts to the non-ultimate facts. The actually existing ultimate facts do not *do* anything to the actually existing non-ultimate facts in any sense in which the converse is not also the case; and in fact neither kind of fact does anything to the other. They are, in reality, the same thing: there is, as Descartes would say, no real distinction between them, only a 'distinction of reason' or conceptual distinction (p. 203). One might try to convey this impressionistically by saying that if one fixed all the *non*-ultimate facts that actually are the case now, and fixed them exactly as they actually are, in all the full richness of their being, one would (of

---

[103] The trouble is that it has both a metaphysical and an epistemological use and, in the hands of some philosophers (especially empiricists), slides disastrously from one to the other.

[104] Especially given 't Hooft's recent work. This, however, is not a central issue (it should always be borne in mind that the thesis of determinism is — provably — both unverifiable and unfalsifiable).

course) equally have fixed all the ultimate facts. This is trivial as stated: my aim is to convey that the asymmetry as applied to what actually exists 'outside our thought' is at bottom just a matter of difference — and fineness — of description. I think that a truly enormous amount of confusion has been created in philosophy by treating these conceptual distinctions as if they were real distinctions.

13. Fundamental-Duality Monism

So far I have listed thirty-one theses — a mix of necessary truths, intuitions and more or less natural assumptions that stand in assorted relations of entailment, mutual exclusion and logical independence. My hope is that they will allow me to say more clearly what it would be to be a genuine realistic monist.

On p. 226 I rejected the Cartesian intuition that experiential reality cannot be physical (extended) reality and that physical (extended) reality cannot be experiential reality. But this, by now, is just a matter of words, a matter of how one uses the word 'physical'. It's time to drop the word 'physical' once and for all, as McGinn recommends (p. 92), and speak instead of non-experiential reality. After all, and as Stoljar says, when my opponents use the term 'physical' 'they mean, near enough, "non-experiential"' (p. 175).

So [6], the thesis that there is physical reality, gives way to [24], the thesis that there is non-experiential reality, and [7], the only-experiential-and-physical thesis that all reality is experiential or physical, gives way to

The Only-Experiential-and-Non-experiential Thesis
[34] All reality is either experiential or non-experiential.

The Cartesian intuition ([15] and [16]) accordingly becomes the

The [E ≠ NE] thesis
[28] Experiential reality cannot be (wholly) non-experiential reality
[29] (Wholly) non-experiential reality cannot be experiential reality

and all those whom I have castigated for accepting the Cartesian intuition — like Dennett, and, in this book, as I understand them, Lycan, Rey, Rosenthal, and Smart (I am uncertain about Jackson) — turn out to be quite right in insisting on what after all appears to be an unassailable necessary truth. They go wrong, given that they are stuff-monists who accept the [E ≠ NE] thesis, only in taking the existence of non-experiential reality for granted. It seems utterly reasonable for them to do so, of course, but it forces them to reject the existence of experiential reality — real experiential reality — altogether, and that is more unreasonable than anything is reasonable.

The argument can be laid out as follows:

[i]   There is only one fundamental kind of reality (premiss =[1], stuff monism)
[ii] There is reality of the non-experiential fundamental kind (premiss =[24])
[iii] All reality is non-experiential (lemma from [i] and [ii])[105]
[iv] Non-experiential reality can't also be experiential reality (premiss =[29])
[v]   There is no experiential reality (conclusion from [iii])

But since [v] is knowably false, the only genuinely realistic option for monists, given the [E ≠ NE] thesis, and [34], the Only-Experiential-and-Non-experiential thesis, is to reject non-experiential reality altogether and embrace pure panpsychism.

[i]   There is only one fundamental kind of reality (premiss =[1])
[ii] There is reality of the experiential fundamental kind (premiss =[3]; obvious)
[iii] All reality is experiential (lemma from [i] and [ii])
[iv] Experiential reality can't also be non-experiential reality (premiss =[28])
[v]   There is no non-experiential reality (conclusion from [iii]).

This seems, of course, an intolerable result. But the only thing left to do, given the certainty of [3], is to give up [1], stuff monism, and to become a stuff dualist, i.e. a good old fashioned substance dualist.

Well, this is an old and well oiled merry-go-round. The question is whether one can get off it. I am still trying to defend what I am still calling 'realistic monism', if only to find out whether or not it is coherent, now that the word 'non-experiential' has been substituted for 'physical'.

The result of this substitution is that realistic monism ([8]) has become the thesis that there is only one kind of stuff, that it has both experiential reality and non-experiential reality, and that there is no other kind of reality. The term 'realistic' in 'realistic monism' has also outlived its usefulness, however, and I propose now to call this position, not 'realistic monism', but

---

[105] I take a lemma to be an intermediate conclusion.

Fundamental-Duality Monism
[35] There is only one fundamental kind of stuff. It (all of it) has both experiential reality and non-experiential reality. There is no other kind of reality.[106]

It is a virtue of this name that it puts the intrinsic difficulty of the position so clearly on view. The word 'realistic' needs to be put aside because although it can be used unrestrictedly when used to indicate full acceptance of the reality of the experiential, it cannot in the end be used equally unrestrictedly to indicate full acceptance of the reality of the non-experiential. There is a sense in which it is always realistic to believe the truth, but the truth, quantum-mechanical or otherwise, has a tendency to go beyond what we find it realistic to believe, and if it turns out that nothing can somehow be both experiential and non-experiential in the way that fundamental-duality monism requires, then, given the above terms, it is no longer realistic for a stuff monist to believe in the non-experiential, and panpsychism is unavoidable. This cannot happen the other way round.

Descartes was well aware of the position I am calling 'fundamental-duality monism'. It is after all the original experience-affirming materialist position,[107] and he was of course well aware of the materialist position.[108] The only reason that it needs a new name now is because the meaning of the words 'materialism' and 'physicalism' has been utterly changed in the last fifty or sixty years. Up to this point I've tried to cope with this change by talking of 'real materialism' and 'real physicalism', but the terms 'materialism and 'physicalism' are I think too far gone in the present-day philosophical ear. This is why I opt here, however clumsily, for 'fundamental-duality monism'.

Descartes was aware of this position, as just remarked, and he was also aware, much more generally, of the possibility that one might, faced with two *seemingly* very different kinds of properties, ultimately conclude that there was in fact only one stuff or substance in question. He makes this clear when replying to Hobbes in the passage quoted on p. 214:

---

[106] [8] is the product of [1], [3], [6], and [7]. [35] simply replaces 'physical' with 'non-experiential', i.e. [6] with [24] and [7] with [34].

[107] Prior to the twentieth century, *all* materialists or physicalists were real physicalists in my terms, in fully acknowledging the existence of experiential reality.

[108] It was popular in Rome at the time, interestingly, and was canvassed as a possibility in all seven sets of objections. And then, of course, there was Hobbes.

> ... we do not come to know a substance immediately, through being aware of the substance itself; we come to know it *only* through its being the subject of certain acts [or accidents]. Hence it is perfectly reasonable, and indeed sanctioned by usage, for us to use different *names* for substances which we recognize as being the subjects of quite different acts or accidents, and ... *to leave until later the examination of whether these different names signify different things or one and the same thing'* (1641, p. 124; my emphasis).

The idea is that we have two apparently fundamentally different kinds of properties, *F* properties and *G* properties. These are our data. We posit some thing or substance *a* as the thing that has *F* properties and we posit some thing or substance *b* as the thing that has *G* properties, and we leave open till later the question whether *a* and *b* are different things or substances or the same single thing or substance.

Our position as enquirers today is exactly the same as Descartes's.[109] We make the fundamental-duality assumption, as he does: we take it that there are two very different kinds of being, experiential and non-experiential. The question for us, as it was for him, is whether or not these two seemingly very different kinds of being can possibly be ways of being of something that we can have good reason to think of in a stuff-monist way as the *same* something, both when it has experiential being and when it has non-experiential being. Descartes concluded that we could not in the end do this, that thinking and extension are just too different, that stuff dualism — i.e. substance dualism in his sense — is correct, although he retained, at the very least, an open mind on this question (as argued in §8).

The choice, in any case, *given* the fundamental-duality assumption, is, as always, between stuff dualism and fundamental-duality monism. The question, for most of us, is how to avoid stuff dualism. The challenge, for committed monists attached to the non-experiential, is to avoid pure panpsychism.

If we try to express fundamental-duality monism using the terms of the arguments set out on page 235 we get the following set of statements:

[i] There is only one fundamental kind of reality (stuff) ([1])
[ii] There is reality of the experiential fundamental kind ([3])
[iii] There is reality of the non-experiential fundamental kind ([24])

---

[109] Except that we now have a rich and mysterious conception of the physical that is intuitively far less hostile to the idea that the physical might somehow be experiential than Descartes's conception of the physical as simply the extended. See M2003a, p. 66.

[iv] Experiential reality can't also be non-experiential reality ([28])
[v] Non-experiential reality can't also be experiential reality ([29])

and this gives the measure of our problem. [ii] and [iii] are incompatible with [i] given [iv] and [v], but [i] is presumably definitive of stuff monism, [ii] is non-negotiable, [iii] is non-negotiable for the purpose of this argument, and [iv] and [v] jointly constitute the [E ≠ NE] thesis, which appears to be a necessary truth of the first water. Neutral monism offers a quick solution by treating [ii] and [iii], experiential and non-experiential reality, as non-fundamental relative to the fundamental kind of reality referred to in [i], but we have already ruled out neutral monism, because we can't downgrade [ii], we can only downgrade [iii] — which leads to panpsychism.[110]

At this point in the paper we cross a bar. What follows is 'without prejudice' in the technical sense, because I haven't had time to think it through, and don't know enough.

14. Spinoza

When I first stated the [E ≠ NE] thesis on p. 230

> experiential reality, considered just as such, cannot be non-experiential reality, considered just as such, and conversely that non-experiential reality, considered just as such, cannot be experiential reality, considered just as such

I said that it could be taken simply as an instance of the necessary truth that if something is *F* then it cannot also be *not-F* (although it may be *G, H, J,* ...). I added that the 'considered just as such' might need a little attention, and that the whole mind-body problem might lie coiled up in whatever it is that needs attention. This is what we now come to. Can we defend fundamental-duality monism in any form, or is it at bottom a contradiction in terms? (And if it is a contradiction in terms, is that the end of it?)

We might start by asking what 'fundamental' means, but perhaps this is superfluous, for whatever exactly it means, fundamental-duality monism seems to be ruled out for someone who, like myself, agrees with Descartes in rejecting conventional substance/property

---

[110] Whatever '*a posteriori*' physicalists think they're doing, they can't plausibly reject [ii]. More generally, the *a posteriori* physicalist project of dissolving the metaphysical issue into an epistemological issue seems doomed. See further pp. 263–5 below.

metaphysics. Why? Because it requires in effect that one be a 'property dualist' (about fundamental, intrinsic properties or attributes) and at the same time a stuff or substance monist. Such a position is generally taken to be at least coherent, but it is plainly incoherent given the (correct) Cartesian view of the substance/property relation; for there is, as Descartes rightly says, no real distinction between a substance and its properties. We may allow (as Descartes did not) that an entity can have two different fundamental properties ('attributes') F and G which are utterly different, but it can't be true to say of it that it has fundamental property F and the no less fundamental property of being not-F unless it is in fact two substances, for there is no real distinction between substances and their properties.[111]

We owe an enormous debt to Descartes for making this so clear to everyone in his time (I don't know why it has since become so unclear), if only because one of the people to whom he made it so clear was Spinoza, who, speaking of 'substances, or what is the same, their attributes' fully agreed with Descartes that there is no real distinction between substance and attribute (between a thing and its properties).[112] Spinoza, for all that, made it his task to be a stuff monist, in spite of everything Descartes had said,[113] and I had enough of a sense of Spinoza's project by the time I started this paper in July 2006 — fresh from re-reading Descartes and Clarke and suitably predisposed by various premonitions *de longue date*[114] — to reason as follows:

> None of us is as clever as Spinoza. So if there is any chance that anyone is going to succeed in being any kind of monist without being a pure panpsychist it is Spinoza. So if one wants to avoid pure panpsychism — and one may not — one should look for a Spinozistic solution.

I knew very little about Spinoza, but two weeks later an advance review copy of Nadler's admirable book *Spinoza's Ethics*, to which I am considerably indebted, fell happily under my hand.

---

[111] This argument could be disputed, but I am happy to leave it undefended here.

[112] 1677, p. 411 (1p4d); see Nadler 2006, pp. 57–8. Note that when Spinoza says that 'substance' and 'attribute' are two names for the same thing, just as 'Jacob' and 'Israel' are two names for the same person (in the Bible), he is not treating 'Jacob' and 'Israel' as pure proper names. Jacob was so called 'because he had seized his brother's heel' (1663, p. 196 (letter 9)) — the name derives from the Hebrew for 'heel' — while 'Israel' means 'he who has wrestled God'.

[113] Although not in a way entirely unanticipated by Descartes, who missed nothing.

[114] See e.g. M1994, pp. 97–8.

Spinoza does not deny the [E ≠ NE] thesis insofar as it is indeed a necessary truth, but he does deny the Cartesian intuition that [15] experiential reality cannot possibly be physical/extended reality and that [16] physical/extended reality cannot possibly be experiential reality. He holds that

> a mode of thought [e.g. a hope or sensation] and a mode of extension [e.g. a concrete entity with shape, size, position, motion, and so on] are but one and the same thing expressed in different ways (Nadler 2006, p. 144)

The difficulty, of course, is contained in the word 'expressed', but the claim is otherwise clear, and, as addressed to Descartes, it strikingly resembles a claim close to the heart of those today who call themselves '*a posteriori* physicalists':

> even if two attributes may be conceived to be really distinct (i.e., one may be conceived without the aid of the other), we cannot infer from that that they are two beings or two different substances.[115]

Nadler continues his gloss of Spinoza as follows:

> the human mind and the human body are not two ontologically distinct things. They are two different expressions — incommensurable and independent expressions, to be sure — of one and the same thing

and then quotes Spinoza himself:

> The mind and the body are one and the same individual, which is conceived now under the attribute of thought, now under the attribute of extension (Nadler 2006, p. 144, Spinoza 1677, p. 467 (2p21).

Now this quotation will immediately lead many to think that Spinoza is after all a neutral monist of some variety who is treating thought and extension in some subjectivist fashion as merely appearances of some more fundamental phenomenon that is in itself neither thought nor extension. It seems plain, however, that this is not his position, although the point needs careful expression (see e.g. Nadler 2006,

---

[115] Spinoza 1677, p. 416 (1p10s). Having recently learnt the terms '*a priori* physicalists' and '*a posteriori* physicalists' I think both may fit the RMP position, though it may be better called '*a priori* and *a posteriori* monism'. The Jacksonian apriorism lies in the rejection of any sort of radical emergence (compare Jackson, p. 63). The a posteriorism appears (once Spinozan ESFD monism — see the next page — has been put aside) in the idea that the radical difference between experiential and non-experiential is mere seeming, a distinction of mind (a 'conceptual distinction' in the very largest sense of Descartes's term), rather than an irreducible ontological reality (a 'real distinction' in Descartes's terms, a distinction in the things themselves). It differs from all standard versions of a posteriorism, as represented here by (e.g.) Carruthers and Schecter and Papineau, because it is held alongside the view, discussed further in §15, that we are acquainted with the essential nature of experience, at least in certain respects, in having experience as we do.

p. 127ff). His position is not *neither-nor* (i.e. *ultimately* neither-nor), but *both-and* (*ultimately* both-and). Thought and extension both really exist and are both really the same thing.[116]

We can continue to take this thesis as I have been taking it, as a rejection of the Cartesian intuition ([15] and [16]) that none the less respects the [E ≠ NE] thesis. If, however, we suppose that Spinoza took it for granted that the physical, the extended, had at least some irreducibly non-experiential aspect — and this, from now on, is thoroughly speculative — we can put things more dramatically by saying that it is the [E ≠ NE] thesis itself that is being rejected. The view now up for inspection is in fact equal-status monism, which I introduced in §3, and which can be restated more briefly as follows:

> reality is irreducibly experiential and non-experiential and substantially single. The experiential and non-experiential coexist in such a way that neither can be said to be based in or realized by or in any way asymmetrically dependent on the other; or if there is any sense in which one can reasonably be said to be dependent on the other, then this sense applies equally both ways

and further adjusted in what I am now taking to be a Spinozistical fashion as follows:

> *Equal-Status Fundamental-Duality monism*
> [36] Reality is substantially single. All reality is experiential and all reality is non-experiential. Experiential and non-experiential being exist in such a way that neither can be said to be based in or realized by or in any way asymmetrically dependent on the other (etc.)

I will call this 'ESFD monism' for short. I think it is what many people really want (although they are likely to deny it), because it is the only

---

[116] It may be said that a proposed solution to the present problem along these lines cannot be anything more than 'Spinozistical', because it retains [34], the strictly dualistic 'only-two' thesis that all reality is either experiential or non-experiential, and this is something that Spinoza explicitly rejected, for he took it that reality involved all possible ways of being, and that it therefore had not two, but an 'infinite' number of ways of being. This objection is too fast, however, if the two terms we employ are 'experiential' and non-experiential', for these are as a matter of logic exhaustive of the field of possibilities. Even if there is a way of counting fundamental kinds of being given which there is an 'infinite' number of them, it seems that they must all classify as either as either experiential or non-experiential. Are we at a point where even these certainties must be doubted? I am going to rely on this one, whatever its calibre before God (as it were), if only for my remaining purposes of argument.

option short of pure panpsychism and radical eliminativism.[117] Note the explicit addition of 'all' in 'all reality is experiential and all reality is non-experiential'.

The question is whether experiential being and non-experiential being can fail to be identical, given ESFD monism. Spinoza's answer (given the present supposition) is that they are indeed in some metaphysically immoveable sense identical, but that maintaining this is compatible with maintaining some sort of real fundamental duality and not falling back into some sort of 'neutral monism' that finds only one fundamental way of being, call it X-being, and reduces the experiential and the non-experiential to mere subjective aspects or appearances of X-being.[118]

This is clearly an answer of the right calibre, an answer with the right degree of difficulty and excitement. It amounts to an abandonment of the [E ≠ NE] thesis as it stands, and returns a firm No to the question put to dualists[119] on p. 209:

> Are the two forms of reality (now characterized simply as *experiential* and *non-experiential*, rather than as *experiential* and *physical*, or *experiential* and *extended*) essentially mutually exclusive in such a way that the being (the concrete realization) of one of them cannot be the same thing as the being (the concrete realization) of the other?[120]

I think that it has to be right, in some version, if any form of ESFD monism is true. If it is incompatible with fundamental principles of thought and language, then (if ESFD monism is true) this is just one more proof of the limitations of human understanding.[121] We know — we have copious proof, both in science and in philosophy — that

---

[117] It is the only option given [27], the Homogeneity Thesis that all physical reality is at bottom the same stuff, and [30], the No-Radical-Emergence Thesis that experiential reality cannot possibly emerge from wholly and utterly non-experiential reality.

[118] I am putting aside Spinoza's view that being has all possible attributes, not just thought and extension.

[119] It doesn't matter whether they call themselves 'property dualists' or 'substance dualists'.

[120] Macpherson may well be right that the position I work with in RMP counts as a property dualist position, given the ordinary understanding of property dualism. My acceptance of this position (the position that supposes there to be irreducibly non-experiential being) is always provisional, however, and I end up rejecting it with Eddington in 1928 (Skrbina points out on p. 154 that Eddington had already rejected it in in 1920, and did so again in 1939). What Macpherson's argument shows, I think, is that the mainstream terminology (in particular the uncritical use of the substance/property distinction, so useful in many areas, so disastrous in this area) has substantive philosophical consequences, and makes the expression of fundamentally important metaphysical positions seem quirky or implausible.

[121] The proposed metaphysical identity is perhaps not much harder to swallow than the metaphysical identity between an experience E, the (thin) subject of E, and the content of E

PANPSYCHISM? 243

many of our ways of thinking of reality are quite hopelessly inadequate to reality as it is in itself,[122] and the perceived difficulty of the mind-body problem — vividly marked by the fact that it has led some people to make the silliest claim ever made in the history of humanity — is just one more proof (if indeed ESFD monism is true).

That said, let me say again that I agree with all pre-twentieth-century psychologists and philosophers and almost all twentieth-century and almost all living philosophers that the reality of the experiential is not in doubt, and that we can and do know the intrinsic nature of the experiential, at least in certain respects, simply in having it (see further §16 below). It follows that if the identity claim is not compatible with the fundamental duality then it is the non-experiential, not the experiential, that must give way.

Giving way would not involve any sort of conventional reductive 'idealism' with respect to the non-experiential (see n. 96). It would be, rather, the Eddingtonian (Russellian) view, best represented by Coleman in this book, that the energy-stuff that makes up the whole of reality is itself something that is experiential in every respect.[123] The universe consists of experience (and hence also subjects of experience — but this is in the end a conceptual distinction, not a real distinction) arrayed in a certain way. This experience must not for a moment be conceived as some sort of 'mere experiential content', where this is in some way passively conceived. Experience — experientiality, the experiential — is itself something intrinsically active, energetic, as Descartes also supposed (see pp. 204, 210). One must not slip into thinking that the *experientiality* of experience, considered strictly, is just a matter of experiential content conceived as somehow just inertly ontologically given, and go on to infer that there must as a matter of metaphysical necessity be something more to experience than its experientiality — some sort of energy substratum that is not itself strictly speaking experience. The inference is valid, but the premiss is unsound. Energy is experientiality; that is its intrinsic nature.

The metaphysic, stated brutally (contrary to Descartes's advice — see p. 205), is this. All that exists is substance (substance properly

---

(internalistically construed) that I argue for in M2003b — although this may be no recommendation.

[122] I defend the entirely respectable and effectively indispensable notion of 'as it is in itself' in Strawson 2002, §2 (revised version, Strawson 2008b).

[123] It seems plain that the priority must be accorded to Eddington, at least on the local (i.e. early twentieth-century) stage; see Skrbina (p. 154). Lockwood (1981, 1989, 1992) is the best recent exponent of Russell, for whose views see e.g. Russell 1927a, 1927b, 1948, 1956, all of which were quoted in M2003a.

conceived, 'substance/attribute' substance). Or, in plural form, all that exist are substances (properly conceived, 'substance/attribute' substances). All subjects (insofar as they are properly conceived as plural) are substances. All substances (insofar as they are properly conceived as plural) are subjects. Equally, all substance is experientiality. Equally, all substance is energy, for substance is essentially active. The fundamental definition of 'substance' is not Aristotle's or Descartes's, but Leibniz's: to be a substance is *to act*.[124] Various further conclusions can be drawn from these claims.

'The universe consists of experience arrayed in a certain way.' Plainly this view involves no logical contradiction. It solves, albeit in a very general way, the greatest problem in physics (the problem of the existence of experience). It is not ruled out by anything in current physics (as has been known for a long time). It may also be said to be particularly hospitable to the correct account of the rôle of experience in present-day physics stressed by Stapp — whatever exactly that account is.[125]

It may now be objected that *space* — *extension* — cannot then really be anything like we ordinarily suppose it to be. But we have known this for a long time, and the pressure that has been exerted on anything resembling the ordinary idea of space by recent developments in science goes way beyond the quite extraordinary pressure that has already been put on it by its absorption into the spacetime of Einstein's theory of relativity.[126] 'Space', we may say, is a natural-kind term. It refers to a certain concrete reality, and we cannot have any confidence that we know its nature, considered as a concrete reality, over and above whatever we can know about its mathematical characteristics or 'abstract dimensionality' (M2003a, pp. 58–9). If we assume, as we surely must, that the concrete real has dimensionality of some sort, we may say the following: existing conceptions of space or spacetime are conceptions of the nature of the dimensionality of the concrete real that fit with existing conceptions of the nature of the

---

[124] I am going to talk freely of substances, although in the final analysis I share Nagarjuna's doubts about the notion.

[125] I take it that the information transfers treated of in quantum mechanics do not necessarily involve the experience of experimenters, and so on, although they do indeed involve experience — they are a matter of experience — given micropsychist or unrestricted panpsychism. See pp. 270–1 below.

[126] See e.g. M2003a, pp. 57-60. See also, more excitingly, Greene 2004, ch. 16. In M2003a I am still clinging in the face of Lockwood's and Eddington's scepticism to the idea I have since abandoned: the idea that we can be confident that we get something fundamental right about the intrinsic and non-structural nature of space when we hold up our hands in a Moorean fashion and consider, not them, but the space between them.

concrete real as non-experientially propertied physical stuff. To suppose instead that the nature of the concrete real is experientiality is to suppose accordingly that the nature of the dimensionality of the concrete real is something that fits with the nature of the concrete real conceived of as experientiality.[127]

Similar things may be said about the phenomenon of causation. For what is 'physical causation', the only sort we acknowledge insofar as we call ourselves 'physicalists', meaning 'real physicalists'? It is an old thought that there is a fundamental sense in which we do not know its intrinsic nature, although we know it exists. All we know of causation is regular succession, constant conjunction, as Hume said — something that does not (as he insisted) capture its intrinsic nature at all.[128] Even if it is allowed that we may also know something of its nature in experiencing pushes and pulls, and so on,[129] there remains a clear sense in which we do not thereby know its intrinsic nature.

This is why panpsychists may if they choose call the laws of physics — taken to be the actual principles of working that inform the whole of concrete reality, rather than any conceptual or linguistic items — 'the laws of experience', or (equivalently) replace 'physics' by 'psychics'. They will not in so doing change a jot or tittle of the laws of physics as currently rendered by the science of physics, for physics is psychics, in the sense that experience is (as Eddington and Whitehead supposed) its whole concrete subject matter. So far as our laws are concerned, conceived of as linguistic items, the panpsychist hypothesis is merely a hypothesis about the nature of the things referred to by the referring terms in those linguistic items.

These issues are of great importance in a full discussion, but it is enough for my present purpose just to log them.[130] Certainly no worthwhile objection to ESFD monism or pure panpsychism can be based on appeal to the notion of space, or that of causation. For many in the current debate the main task, in fact, is to get to the point where it becomes apparent that the issues under discussion here are indeed the

---

[127] This line of thought has affinities, perhaps, with Newton's speculation that space is God's 'sensorium'. We may treat this God as Spinoza's God, an entity that has nothing to do with established religion (other, perhaps, than the Romantics' 'Religion of Nature').

[128] Hume holds that his definitions of cause are 'imperfect', even though they are the best possible. The trouble is that 'we cannot remedy this inconvenience, or attain any more perfect definition, which may point out that circumstance in the cause, which [actually] gives it a connexion with its effect' (1748, 7.29). Hume was not misunderstood on this point for some time (not until the nineteenth century, as far as I know).

[129] See e.g. Anscombe 1971, P.F. Strawson 1985.

[130] I have said elsewhere (in M2003a) the little that I feel competent to say about space, and that I now doubt.

central issues. Those who are prepared to take panpsychism seriously, on the other hand, must not get so carried away that they no longer have the resources to account for simple facts like the facts of reproduction and evolution, whose undeniability I will as a naturalist (a panpsychist naturalist) take as my benchmark, with the crucial proviso that their undeniability *does not legitimize anything like our ordinary conception of space*.

The dialectical situation, in conclusion, is as follows. Assuming [27], the Homogeneity thesis that all reality is at bottom the same kind of stuff (pp. 231–2), and the No-Radical-Emergence thesis (p. 228), which I have not further defended in this paper—except to say that I suspect that to reject it is in the end to reject the [E ≠ NE] thesis, a move that immediately renders any Radical Emergence thesis superfluous—, we must as monists choose between two things: radical eliminativism and pure panpsychism.[131] Radical eliminativism is ruled out for all those who approach the mind–body problem in a serious fashion. That leaves pure panpsychism. The only alternative is to abandon the [E ≠ NE] thesis and embrace some form of ESFD monism. Since many judge that ESFD monism is not in the end coherent, on the grounds that it involves abandoning the law of non-contradiction, they are left with nothing but pure panpsychism, or at least 'micropsychism'—the view that experiential properties are among the fundamental properties. So be it. For my part, I am fond of ESFD monism.[132] For now, though, I will now focus mainly on pure panpsychism of the specifically smallest, micropsychist variety, and on the difficulties that arise from the view that the macro-experiential arises from ('emerges from', in the legitimate sense of the term, see p. 27) the micro-experiential. I will drop the word 'pure' for the most part, and it will be plain that I am not always talking about pure panpsychism, e.g. when discussing the views of others.

15. Panpsychism and the 'Compounding of Consciousness'

RMP is a schematic and exploratory paper, and gives no precise definition of panpsychism, as remarked by several contributors to this book (see e.g. Stapp, pp. 163–4). Now, however, the focus is on pure

---

[131] Abandoning [27] allows one to dilute pure panpsychism to micropsychism.

[132] What we need, Nagel says (using 'mental' and 'physical' where I use 'experiential' and 'non-experiential'), is a 'psychological Einstein' who will build on the work of a 'psychological Maxwell' and show 'that the mental and the physical are really the same' (1986, p. 53).... Alternatively, we can think that Spinoza (as interpreted here) is right even though we don't think that a psychological Einstein will ever come along.

panpsychism — arguably the only respectable kind of panpsychism — and the basic idea, at least, is easy to state. For pure panpsychism has only one kind of thing in its fundamental ontology: subjects of experience in the 'thin' sense expounded in §4, subjects of experience each of which is at the same time an experience, an experiencing, i.e. literally identical with an experience or experiencing.[133]

When I speak of subjects of experience, then, I mean only 'thin' subjects. I propose to call them 'sesmets', for reasons that are not relevant here.[134] I take it that there is a fundamental sense in which there is more than one of them at any given time (this corresponds to assumption [1] in RMP, p. 9), and I also take it that while some pluralities of sesmets constitute further numerically distinct sesmets, others do not. I take it, in other words, that not every plurality of sesmets constitutes a further sesmet, without claiming to know this with certainty.[135]

The term 'ultimate' may be taken as before (RMP, p. 9) to correspond to terms like 'particle' 'string', 'loop', 'simple', 'preon' (whether simple or 'braided'), 'field quantum'. We may then say that all ultimates are sesmets (for sesmets are the only things that exist), and that many many sesmets are ultimates. It may be, in the end, that the best notion of an ultimate will include things that count as composite relative to the apparent plurality of field quanta, strings, and so on, for it is not obvious that 'ultimate' must entail 'simple or non-composite', although we standardly assume that ultimates are non-composite, and often take it to be true by definition.[136] It may be, in other words, that certain sesmets — certain experiential[137] field quanta — may be truly ultimate, in some metaphysically fundamental sense,[138] even if they are analysable as composite for certain purposes; and in that case we will be able to say not only that all ultimates are sesmets, but also that all sesmets are ultimates (if such proposals feel logically unacceptable, bear in mind that they are just the kind of thing that physics

---

[133] This second feature flows from the Subject Thesis ([4] and [5], p. 224) and a true Cartesian-Spinozan metaphysics of object and property (one may also say that substance is all that exists, and all substances are subjects). I argue for the strict metaphysical identity of subject and experience in M2003b, as remarked in note 121.

[134] The acronym stands for 'subject of experience that is a single mental thing'; see Strawson 1999b.

[135] One no longer has any right to be impressed by 'spatial separation' (whatever the ultimate nature of space), and we are taught that particles light years apart may be 'entangled' in such a way as to put their real or ontological distinctness in question (especially once we have a correct metaphysics of object and property).

[136] van Inwagen uses the count noun 'simple' where I use 'ultimate'.

[137] The word is redundant.

[138] It is of course also a physically respectable sense, given a complete physics.

encourages). Alternatively, to the same effect, it may be that the correct notion of an ultimate is simply the one that has the consequence not only that all ultimates are sesmets but also that all sesmets are ultimates — whatever other intuitions we have about the large and the small, the composite and the non-composite. The plurality of 'particles' or field quanta may not always be what it seems — a point anticipated, in effect, by the later William James.

However this may be, I take it that I now have to face what I will call the *Composition Problem*:[139] the problem of how pluralities of sesmets can jointly compose or constitute distinct and 'larger' single sesmets; the problem of how 'microsesmets', e.g. electron sesmets or string sesmets, can possibly compose single macrosesmets, e.g. human sesmets. It is a problem faced by any 'smallist' (p. 233) panpsychist, and I note its difficulty, and James's well known formulation of it, in RMP (p. 26). In this book it is forcefully pressed by Goff and Carruthers and Schecter, among others.

In 1890 James argued that Composition was impossible, and that its impossibility was fatal to any plausible (hence smallist) panpsychism — even while appearing to hold, in the same chapter, that some version of panpsychism must none the less be right, so that Composition could not really be impossible. In the 1900s he worked himself closer to the view he had condemned as impossible — it is as if he knew from the start that Composition could not really be impossible, but could not see how — and by the time of his Oxford Hibbert lectures, published in 1909 as *A Pluralistic Universe*, he is comfortable with a not-rigidly-particulate, field-quanta-friendly form of Composition, as Skrbina observes — partly on the cheerful Fechnerian ground that 'we know it's actual, so it must be possible'. He still doesn't 'logically see how a collective experience of any grade whatever can be treated as logically identical with a lot of distributive experiences' (p. 204), and so finds himself 'compelled to *give up the logic*, fairly, squarely, and irrevocably' that rules it out (Skrbina, p. 156, quoting James 1909, p. 212). Once again I think there is support for this in physics, but the idea remains the same even if this is not so: given smallism, we have to accept some form of the view 'that states of consciousness, so- called, can separate and combine themselves freely, and keep their own identity unchanged while forming parts of simultaneous fields of experience of wider scope'.[140]

---

[139] Compare van Inwagen 1990.

[140] James 1909, p. 181. It is not clear that we have to suppose that 'smaller' states of consciousness remain qualitatively unchanged when becoming parts of 'larger' ones — if,

In *A Pluralistic Universe*, and in particular in the remarkable chapter 'The Compounding of Consciousness', James is often concerned with the Problem of Composition as it arises on the largest scale: the problem, pressing for many of the gravely bearded philosophers of his time, of how many minds with diverse contents, e.g. human minds, can all somehow be parts of the single mind of the universe.[141] He is, however, very clear on the point that this is just one possible case of the phenomenon that concerns him — which is, as he says, the entirely general phenomenon of 'collective experiences ... claiming identity with their constituent parts, yet experiencing things quite differently from these latter' (pp. 203–4) — , and my concern is only with a smaller-scale case. It is only with the question of how a human experience as one knows it in one's own person can possibly be somehow composed of many micro-scale experiences/subjects of experience. The problem lies at the heart of smallist panpsychism, and no one is saying it is easy, least of all James:

> Sincerely and patiently as I could, I struggled with the problem for years, covering hundreds of sheets of paper with notes and memoranda and discussions with myself over the difficulty. How can many consciousnesses be at the same time one consciousness? How can one and the same identical fact experience itself so diversely?... I found myself in an *impasse*....[142]

Goff addresses just this question, and begins by quoting James's 1890 argument:

> Take a hundred of them [feelings], shuffle them and pack them as close together as you can (whatever that may mean); still each remains the same feeling it always was, shut in its own skin, windowless, ignorant of what the other feelings are and mean. There would be a

---

that is, our conception of their numerical identity conditions is such as to allow them to persist through qualitative change.

[141] This problem also arises, of course, for Spinoza. I say 'the mind of the universe', rather than 'God', or (perhaps more respectably than 'God') 'Brahman', because I'm not sure that either Spinoza or William James believes in God in any sense in which Dennett does not also believe in God. I'm sure I don't. Spinoza and James have much harder naturalistic noses than Dennett, and I follow them. What distinguishes us and many others, including most of the gravely bearded philosophers, from Dennett on this issue is rather that we are more struck by — philosophically concerned with — the unignorable idea of the universe considered as a whole. Dennett sometimes suggests that when people like myself issue such denials they are really smuggling in some kind of mysticism under cover of a pretence of atheistic naturalistic orthodoxy. He cannot be argued out of this position, because it is indefeasible, but he is of course quite wrong, and mysticism as he understands it is a tame thing compared with current physics and cosmology.

[142] James 1909, pp. 207–8. It is sad that James's later work appears so flaky, on first reading, to those like myself who have been brought up in the analytic tradition — so that they do not persist with it. These things take time.

hundred-and-first feeling there, if, when a group or series of such feelings were set up, a consciousness *belonging to the group as such* should emerge. And this 101st feeling would be a totally new fact; the 100 feelings might, by a curious physical law, be a signal for its *creation*, when they came together; but they would have no substantial identity with it, nor it with them, and one could never deduce the one from the others, or (in any intelligible sense) say that they *evolved* it. (James 1890, 1.160)

Goff argues in his turn that even if Composition could happen it would be 'unintelligible ..., as brute and miraculous as the emergence of experiential properties from non-experiential properties, [so that] Strawson's panpsychism is itself committed to the very kind of brute emergence which it was set up to avoid' (p. 54).

My first reply, as in RMP, is that unintelligible experiential-from-experiential emergence is not nearly as bad as unintelligible experiential-from-non-experiential emergence.[143] I stand by this reply, although Goff tries to block it by turning one of my own arguments neatly against me (p. 60). It is, however, no defence against Goff's next move. For he now simply allows for the sake of argument that Composition is possible, in order to object that it cannot be the way I need it to be even it is possible, because my basic definition of experience simply rules it out.

16. Revelation

I have avoided epistemological issues as far as possible, and I am going to overfly criticisms that raise epistemological questions that are for all their popularity (e.g. among *a posteriori* physicalists) irrelevant to the present discussion.[144] I cannot avoid them entirely, however, for I not only take the very stuff of existence — experience — to involve knowing or acquaintance; I also hold, in a passage from M2003a picked up by Goff, that

> we are acquainted with reality *as it is in itself*, in certain respects, in having experience as we do...the having is the knowing.[145]

Plainly many who are not panpsychists can agree with this claim. Equally plainly, it has particular force for a panpsychist who holds

---

[143] Unintelligible' means not understandable, and, in particular, not understandable by us. It does *not* mean incoherent (a crucial point when studying Hume, who uses the word regularly).

[144] See, though, pp. 263–5 below. The *a posteriori* physicalists' main error (as Coleman intuits, p. 47) is their focus on phenomenal *concepts* — which derives, no doubt, from their philosophy-of-language upbringing.

[145] p. 55, quoting M2003a, p. 54. This claim does not feature in RMP.

that all that exists, substantially speaking, is experience (*sive* experiencers, *sive* sesmets).

We may re-express it as the claim that in having experience we know its essential nature, and analyse this in turn into two parts:

The General Revelation Thesis[146]
[37] I am acquainted with the essential nature of experience generally considered — i.e. with whatever all possible experiences have in common just insofar as they are indeed experiences — just in having experience

and

The Local Revelation Thesis
[38] In the case of any particular experience, I am acquainted with the essential nature of that particular experience just in having it.

I think, though, that we may take it that [38] entails [37], and also drop the word 'local' from [38], restating it as

The Revelation Thesis
[39] In the case of any particular experience, I am acquainted with the essential nature of that experience just in having it

a thesis that I accept (subject to an imminent qualification). But when Goff rephrases the original claim as follows: 'in introspecting one's conscious experience, one perceives that metaphysical reality "*as it is in itself*"' (p. 57), I do not accept the rephrasal, although I am in general quite happy with the expression 'as it is in itself'. The first of my two problems with Goff's rephrasal is that the original claim makes no use of the second-order notion of introspection. It is resolutely first-order. It is that 'the having is the knowing'.[147] If one engages in

---

[146] 'Revelation' began as Mark Johnston's term for the (Russellian) view that colour properties are to be understood as 'properties [whose] whole and essential nature can be and is fully revealed in sensory ... experience given only the qualitative character that that sensory experience has' (Strawson 1989, p. 224, commented on in Johnston 1992). Since then the use of the term has expanded, and Stoljar uses it in the mind–body debate as I do here (following him), although not in his contribution to this book. In particular, Stoljar uses it for the thesis that 'if one has a conscious experience, then one knows the essence or nature of experience' (2006, p. 96; this is General Revelation in the present terms), and also for the thesis that if you have an experience 'you know *all* the essential properties of the experience' (see e.g. 2006, p. 221; this is Full Revelation — see below — in the present terms). Goff speaks instead of 'transparency', a metaphor that fits well in the discussion of perception, but not here.

[147] 'When we claim (with Russell) that to have an experience is *eo ipso* to be acquainted with certain of the intrinsic features of reality, we do not have to suppose that this acquaintance

the higher-order operation of introspecting one's experience, then the acquaintance is no longer direct, and it is a commonplace that taking one's experience as an explicit object of knowledge precludes knowing of it as it in itself in the direct acquaintance of having it.[148]

The second problem is that Goff takes the original claim to amount to 'full disclosure' — to

The Full Revelation Thesis
[40] In the case of any particular experience, I am acquainted with the whole essential nature of the experience just in having it

— and all his main criticisms of smallist panpsychism hang on this reading. But although it is a natural reading of [39] as it stands, and a possible reading of the original version of the claim, so is

The Partial Revelation Thesis
[41] In the case of any particular experience, I am acquainted with the essential nature of the experience in certain respects, at least, just in having it.

The original claim, after all, is that 'we are acquainted with reality as it is in itself *in certain respects* in having experience as we do', and it is followed three lines later by the observation that this claim 'is fully compatible with the view that there may also be fundamental things we don't know about matter considered in its experiential being', where these are not just 'facts about experience in sense modalities we lack, or (e.g.) about the brightness-saturation-hue complexity of seemingly simple colour-experience, but also, perhaps, murkier facts about its composition, and also, perhaps, about the "hidden nature of consciousness" postulated by McGinn'.[149]

---

involves standing back from the experience reflectively and examining it by means of a further, distinct experience. It doesn't. This picture is too cognitivist ... The having is the knowing' (M2003a, p. 67). I make this time-honoured point against Papineau on pp. 263–5 below. In the recent mind–body debate it has been stressed in a different idiom by Kripke (1972) among others.

[148] A similar point is often made about the subject of experience's attempt to know itself as it is in that very act of knowing. For a doubt about whether this is really impossible, see Strawson 1999b, section X.

[149] M2003a, p. 54 and n; McGinn 1989. Experimenting, I find that the quality of *having hidden aspects* is part of what I feel I experience, when I attend to my experience rather than just having it unreflectively. It may be an illusion, or have an explanation of the sort offered by C. O. Evans (1966), and I rest nothing on it — but my experience certainly feels to me as if it is essentially something that is ontologically more than what is revealed to me in having it in the present sense of 'reveal'.

It seems, then, that I can as a 'smallist' panpsychist who endorses [41] suppose that one of the hidden facts about the nature of my experience — whose essential nature is partly revealed to me simply in my having it — is that it is somehow constituted — composed — of many other experiences. Nothing exists other than experience, on this panpsychist view, and in having an experience I am *ipso facto* acquainted with the essential nature of my experience; but it does not follow that I know the whole experiential nature of the event that occurs when I have that experience, just in having it. Acquaintance with something need not involve exhaustive knowledge of its nature, any more than direct contact with something need involve direct contact with all of it. Nor need I have any inkling of the Laws of Experiential Composition by which the existence of many small experiential fields somehow constitutes the existence of my own phenomenologically unified experiential field. Nor do I, if smallist panpsychism is true.

Goff may object that the natural reading of the claim that 'the having is the knowing' is [40], the Full Revelation thesis, a thesis to which he himself subscribes. In reply let me note and put aside an unclarity about Full Revelation, and then record — with a proviso — a sense in which I do accept [40].

The unclarity has to do with the notion of attention. If it turns out that what a defender of Full Revelation takes to be fully revealed at any given time is just what is in the focus of attention at that time, then the doctrine is falsified as soon as it is allowed, seemingly most plausibly, that something can be genuinely part of the overall content of one's awareness at a given time without being in the focus of attention at that time — in such a way that it truly exists as part of the content of one's experience before (or whether or not) one turns one's attention to it. It seems plain, however, that a Full Revelationist need not favour the focus of attention in this way, and can hold that the *whole* nature of the experience is indeed fully revealed just in the having of it, even though it is actually impossible for what is known and revealed by this kind of acquaintance-knowledge to be brought whole into the focus of attention.

In what sense do I accept [40]? Well, there's a sense of the expression 'my experience' given which I agree that I do — necessarily — know the whole essential nature of my experience when it occurs. It cannot be otherwise, in fact, given that 'the having is the knowing' — given the sense in which (in James's words) 'as a psychic existent *feels*, so it must *be*'.[150] The proviso is that this acceptance of [40] must sit alongside a sense in which [40] is too strong.

I will try to clarify this. First, though, note that the phrase 'the having is the knowing' is becoming increasingly uncomfortable, because both 'having' and 'knowing' suggest a metaphysical separateness I reject. As for 'having': it suggests the distinctness of possessor and possessed that I reject insofar as I take it (in M2003b) that there is in the end no real distinction between experience and experiencer; perhaps one might better say 'the being is the knowing'. As for 'knowing': it suggests a distinction between the knowing subject and the thing constituted as object of knowledge by the act of knowing that I also reject as inapplicable to acquaintance-knowledge; perhaps one might better say 'the knowing is the being'.

From 'the being is the knowing' and 'the knowing is the being' one can presumably derive 'the knowing is the knowing' and 'the being is the being' — which seem unhelpful. I like tautologies, but we can put them aside, because the present proposal is simply that one can admit a sense in which [40] is true without ruling out smallist panpsychism. One might try to put the point by saying that it does not follow, from any sense in which it is true that I know the whole essential nature of *my experience, e_1*, when it occurs, that I know the whole essential nature of *the event, E_1, that occurs when I have an experience* — even if it is true, as it is by panpsychist hypothesis, that E_1 consists in nothing but experience, and even though the occurrence of e_1 consists in the occurrence of E_1.

This is intended to sound problematic, and I will say more about it in the next section. For the moment the proposal is that for all that has been said so far, E_1 may involve the existence of many 'small' experiences e_2-e_n with which I have no 'from-the-inside', being-is-knowing acquaintance (necessarily so, given that they are experiences had by subjects of experience that are numerically distinct from me) even though they somehow jointly constitute my experience e_1. This bears comparison with the fact that the centre of gravity of a composite thing is distinct from the centres of gravity of all its constituents, although it is wholly constituted by them. (A centre of gravity, however, is not a real concrete item.)

Goff questions the coherence of this proposal on several grounds, and reasonably so, given that he attributes Full Revelation to me.[151]

---

[150] 1890, 1.162, quoted by Goff (p. 54).

[151] He might also have cited M1994, p. 46: 'experiential phenomena are those phenomena that are entirely constituted by experiences' having the experiential character they have

First, and independently of Full Revelation, he endorses the earlier James claim that 100 subjects can't combine to make a 101st. Second, he takes it that it follows from Full Revelation that my experience can't be 'constituted of the experiential being of billions of micro subjects of experience' without this fact being revealed in the content of my experience (p. 57); for nothing is hidden, given Full Revelation. Certainly nothing experiential is hidden, and experience is all there is, given panpsychism.

Third, suppose we assume for argument that such constitutive Composition is possible — that experiences can somehow compose a further numerically distinct experience even while their individuality is somehow preserved. In this case we are really no better off, Goff says, because Full Revelation dictates that constitutive Composition cannot be anything other than merely *additive*, a mere summation of the constituent experiential parts. There can be no *blending* analogous to blue and yellow blending to make green, say. For just as nothing can be hidden, nothing can be lost, in the sense in which blue and yellow are lost when they blend to make green. Nor can there be any less obvious or as one might say *chemical* principles of combination that allow complex human thoughts — about philosophy, say — to be constituted of components that in no way resemble them, in the way that smallist panpsychism must surely suppose. In fact, as Goff says, it seems that 'the experiential being of a higher-level subject of experience [cannot be] significantly qualitatively different from the experiential being of the lower-level subjects of experience of which it is constituted' (p. 57). My severe pain, then, BIG PAIN, can only be made out of LITTLE PAINS, at best, and not, for example, LITTLE RED-EXPERIENCES. But even this won't work, Goff says, because 'what it feels like to be LITTLE PAIN 1 is not part of what it feels like to be BIG PAIN. LITTLE PAIN 1 feels slightly pained, BIG PAIN does not. The phenomenal character of LITTLE PAIN 1's experience, i.e. feeling slightly pained, is no part of the phenomenal character of BIG PAIN's experience, i.e. feeling severely pained' (p. 58; McGinn makes similar points). As for the constituent parts of the thought that *Hume needed to admit the respectability of transcendental arguments in his discussion of personal identity*, go figure.

This is I think a devastating refutation of Full Revelation smallist panpsychism, on at least one natural understanding of it; so I am glad that I do not hold such a view. I think there are many reasons for a

---

for those who have them as they have them', but this (taken without qualification) is the claim I have shifted away from in M2003a.

panpsychist to reject a Full Revelation thesis of this sort, in fact, and I will finish my main discussion in the next section by considering some of them in a little more detail.

What follows is rough, and it is written in the awareness that there must be a great deal of sophisticated discussion of these issues that I don't know about. The interim conclusion, in any case, is that anyone who thinks that Full Revelation as just expounded is a necessary part of panpsychism must abandon smallism, even when smallism is taken in a realistic — a later-Jamesian, quantum-field-theoryish, not crudely or brutally atomistic — way.[152]

17. Panpsychism and Duality

In the last section I claimed that my experience e_1, with which I necessarily (by definition) have direct, from-the-inside, being-is-knowing acquaintance, may be somehow constitutively composed of many 'small' experiences e_2-e_n with which I have no such direct from-the-inside acquaintance (equally necessarily, for they are the experiences of numerically distinct subjects). This is how it must be, I think, if any realistic version of smallist panpsychism is to stand up, for we are trying to give an account of our own experience, and in having an experience we have no experience of ourselves as somehow being many subjects of experience. I cannot avoid this difficulty in the way that Coleman can (pp. 48–50), by proposing that an experience of mine may be somehow composed of many experiences whose existence does not essentially involve subjects of experience, because I not only accept [4], the Subject Thesis according to which there cannot be an experience without a subject of experience, but also the ultimate identity of experience and experiencer. I find no difference between panexperientialism and panpsychism, as remarked on p. 189.

I have used the expression 'from-the-inside'. It is not entirely stable, but it is very natural in this context, and it offers one way of making a distinction that must be made, if realistic smallist panpsychism is to have any chance of being true. It may also lead us forward in a crucial way, because it may give us a first intimation of how panpsychist monism can allow some sort of fundamental and all-pervasive *duality* to existence (a glimmering of the possibility that ESFD monism may be intelligible after all). And this, perhaps, is just

---

[152] Chalmers notes that the seeming difficulty of the problem of how experiences may constitutively compose other experiences 'may well arise from thinking of experiential composition along the lines of physical composition, when it might well work quite differently' (1997, p. 43).

as well, for it is extremely natural to think that we cannot in the end do without some such duality.[153] It cannot be a betrayal of naturalistic panpsychism to require this, if naturalistic panpsychism is to have any chance of being true. (We must, as remarked, accommodate the facts of reproduction and evolution.)

A first, inadequate way to put the idea just canvassed is that while an experience, a sesmet, which is energy-stuff, necessarily has an 'inside', a being-is-knowing inside, which is its essential nature, it must also, as energy-stuff, have an 'outside', which is no less part of its essential nature. 'Inside' and 'outside' are likely to mislead, however. It is not as if any sort of non-experiential stuff is being introduced, because there is on this view a fundamental sense in which the inside of an experience or sesmet like e_1, i.e. its experiential nature, is its whole essential nature, its whole essential being. e_1's outside is not something ontologically extra.

What is it, then? Well, two main issues arise, when it comes to providing for a duality of 'inside' and 'outside': the issue of causation and the issue of constitution. With respect to causation, we may say that e_1's outside is just a matter of how e_1 is disposed to interact with other sesmets, other parts of (experiential) energy-stuff, given its inside. With respect to constitution, we may say that it is a matter of how e_1 is constituted of numerically distinct sesmets e_2-e_n. At bottom these are two aspects of the same thing, the given reality.[154]

Mysterious, you may say; but the proposal about causation returns us to a crucial point, one that first surfaced in the discussion of Descartes (p. 204; see also pp. 210–11): experience cannot be thought of as just passive content, in any plausible (reproduction-and-evolution-allowing) panpsychism, but must always be understood to be active substance.[155] I think that this, too, is a difficult idea for us, and that effective grasp of it requires considerable acclimatization, but the basic smallist picture remains plain for all that. Many believe that it is legitimate to think of our actual world, conceived of as involving

---

[153] Skrbina remarks that my position is one of 'dual-aspect monism ... an approach that dates back at least to Spinoza ... and strongly urges one toward panpsychism' (p. 153).

[154] I refrain from saying that e_1's inside, i.e. its experiential nature, is wholly non-relational, for I take it that its experiential nature will be partly a function of its interactions with other sesmets. The effect that e_1 has on e_{n+1} will indeed be wholly a function of its experiential nature, but its experiential nature may be partly a function of how it is being affected by e_{n+2}. Great complications lie here, no doubt, about which I have said nothing (Mach's famous principle — that everything in the universe is affected by everything else — comes to mind).

[155] All substance is active, as Leibniz says (activity does not imply any sort of intentional agency).

non-experiential substance (substance that is not experience), as in some sense composed wholly of energy, in various forms, and the present (and so far still Eddingtonian) suggestion, which is becoming increasingly familiar in philosophy, is simply that the intrinsic nature of that energy is experience, i.e. something whose essential nature is fully revealed to us, at least in part, just in our having it. Everything that exists (including of course reproduction and evolution) is left in place by the panpsychist hypothesis, then. Panpsychists can fully agree with Papineau, in the old idiom, that 'pains are one and the same as C-fibre firings' (p. 102), and at the same time reject his suggestion that 'straightforward physicalism is strongly backed by causal-explanatory considerations' (p. 101) in any way that gives it an advantage over panpsychism.

Perhaps the main reason why it is difficult to think of what I am calling 'experience' as a stuff or substance (apart from the prevalence of the bad picture of the object/property or substance/attribute relation) is that we have, as previously noted, a tendency to think of it as 'just' content, experiential content conceived of as something passive, content contained in a container. And it is of course perfectly workable to think of it in this way in many contexts. In the present context, however, it is highly obstructive, and I will try to offset the obstruction by sometimes speaking instead of *experiencing*.[156]

My first impulse is to add immediately that 'experiencing' in this use is not just a verbal noun denoting an activity, but a noun denoting a certain sort of substance. This, however, obscures the deeper point, which is that the activity in question is the substance in question. The same idea animates the notion of a sesmet, and the claim that (thin) experiencers are in the final (Cartesian) analysis identical with experiences.

With this in place, consider another line of thought that focuses on the issue of causation.

[1] Naturalistic panpsychism is true [premiss]

[2] Naturalistic panpsychism requires causation (including reproduction and evolution) [premiss]

[3] If there is to be causation, substances must affect each other[157] [premiss]

---

[156] It's useful, when reading Hume on personal identity, to replace 'perception' by 'perceiving'.

[157] Substances are all that concretely exist.

[4] Substances are experiencings (i.e. subjects-having-experiences, sesmets) [premiss]

∴ [5] Experiencings must affect each other. ([1]–[4])

[6] There is a fundamental *respect* in which experiencings are wholly 'closed' to each other, as I will say — a fundamental respect in which experiencings, conceived of as things of which it is true that the being is the knowing, are necessarily separate from each other. One way to put this point is to say that experiencings are 'logically private' or have, in Searle's phrase, a 'first-person ontology'.[158] [premiss]

So far, I think, so good.

[7] Things that are metaphysically wholly closed to each other in any respect cannot affect each other in that respect [premiss]

∴ [8] That in virtue of which experiencings are able to affect each other cannot be that in respect of which they are wholly closed to each other. ([6], [7])

∴ [9] There is something about experiencings other than that in respect of which they are wholly closed to each other. ([8], [5])

[10] That in virtue of which experiencings are wholly closed to each other is that in virtue of which they are correctly said to be fully revealed [premiss]

∴ [11] There is something about experiencings other than that in respect of which they are correctly said to be fully revealed. ([9], [10])

∴ [12] Full Revelation is false. ([11])

I think the argument goes wrong in [7], but I am going to leave it in place, with the unclear term 'metaphysically wholly closed' unexamined, because I think it may be suggestive when it comes to understanding the sense in which Full Revelation must be false if panpsychism is true. The basic idea is that all 'first-person-ontology' phenomena must exist in such a way that they are also 'third-person-ontology' phenomena, i.e. phenomena that have causal reality in the third-person-ontology reality (we know that this is how things are

---

[158] See e.g. Searle 1980. Compare James: 'the breaches between ... thoughts ... belonging to different ... minds ... are the most absolute breaches in nature' (1890, 1. 226).

on our ordinary picture of the world, and there is no reason to think it should be different on the panpsychist picture). In other terms again, one might say that while there are many 'perspectival' realities that are indeed real realities, irreducibly real, any such perspectival reality is also and nonetheless part of a reality that is not that perspectival reality (remember that experience is all that exists, on the current view). The difficulty in this idea is perhaps no greater than the difficulty in the idea that the space depicted in a painting of an imaginary landscape is no part of the space in the world in which the museum in which the painting hangs is located, although there is also a sense in which its whole ontological being is entirely included in the larger space.

This may or may not be helpful, but there is, on any account, causation. Where does this causation (and constitution) take place? We have to assume, in line with §14, that the dimensionality of the concrete real, although not understood by us, is something that fits with the nature of the concrete real conceived of as experientiality in the same general way as the way in which space or spacetime — which is certainly not understood by us — fits with the nature of the concrete real conceived of as good old fashioned non-experientially propertied physical stuff. The causal effect of anything on anything will have an experiential aspect, will indeed be experiential, and that is why even microsubjects — ultimate sesmets — may be said to have experience, and may even be said to have intentionality and represent things, on many currently favoured accounts of what intentionality and representation are, rather than just having 'bare experientiality'.[159] On these accounts, there is no more difficulty in the idea that ultimate sesmets have experience and intentionality and represent things than there is in the idea that one particle exerts attractive or repulsive force on another — for these are in fact the same thing. Obviously the intentionality will not be explicit conceptual intentionality. Nevertheless the experiential state that is particle a's registering of the repulsive force of particle b may be said to be of or about particle b; not only on any theory of 'intentionality' according to which intentionality does not require experience or consciousness, but also on many theories of intentionality, according to which intentionality does require experience.[160]

---

[159] On this issue see e.g. Coleman (p. 50), Lycan (p. 70), Simons (p. 150), and McGinn (pp. 96–7).

[160] In Strawson 2008a, I consider particles' claim to have intentionality in a context in which I put aside panpsychism. Note that I'm happy to attribute 'animation' to ultimates in Wilson's sense (pp. 181–2).

Might we in the end have to posit a universe-wide sesmet in order to posit the existence of many sesmets existing in a dimension that allows for their interaction? I've been assuming that the answer is No, but I would not be much troubled if it were Yes, first because a universe-wide sesmet would have no more to do with dogmatic religion than the view that there is a single universe, second because of a methodological principle integral to serious naturalism: if one finds oneself pushed towards an apparently extraordinary hypothesis like panpsychism, when one is trying to account for the given natural facts, of which the first and most fundamental is and will always be the fact of experience, one should bear in mind the certainly equal and arguably much greater extraordinariness of many of the hypotheses seriously entertained, and in some cases well supported, in present-day physics and cosmology.

All this needs, to put it mildly, development. The basic proposal is that ultimates — sesmets — experiencings — can be as they are to themselves, and their being as they are to themselves can be what it is intrinsically, compatible with their having causal effects on other sesmets and compatibly with their playing a part in constituting other numerically distinct sesmets (sesmets that are not only numerically but also qualitatively distinct). They have the effects or constituting rôles they have wholly in virtue of their experiential being, which is all the being they have (note the tension with the argument set out on pp. 258–9), and yet when one sesmet or experiencing affects another, in accordance with the Laws of Experiential Nature, whatever they are, or goes to constitute another, in accordance with those Laws of Experiential Nature that are Laws of Experiential Composition, the second obviously will not have access to the from-the-inside nature of the first in the way in which only the first can. Nor is there any more reason to think that the second will take on the experiential character of the first, in some direct way, than there is to think that a positively charged particle will in some direct way take on the character of a negatively charged particle with which it is in interaction — a point independent of the fact that the second of these two phenomena is, on the current view, an instance of the first. In this sense experiential realities may be said to *function* as non-experiential but experience-causing realities for other experiential realities, and to *function* as non-experiential but experience-constituting realities for other experiential realities. Again, it may be said that although there is no non-experiential being absolutely speaking, there is non-experiential being relatively or relationally speaking.

If this is so, [40] — Full Revelation — is false, at least as interpreted here. If the cost of maintaining Full Revelation is retreat into a world without causation, reproduction and evolution, it is too high.[161] If we try to hold onto causation (panpsychist, experiential causation) along with Full Revelation, it looks as if we may have to take on the consequence that Full Revelation must involve full revelation of the causal powers of any sesmet, full revelation of the very nature of causation, indeed, of a sort that renders it fully intelligible to that sesmet. It seems, though, that a theory is refuted rather than supported if it has any such consequence. It is plain that the kind of acquaintance-knowledge that we have of experience just in having it simply does not contain this kind of causal knowledge, even though it is, in being direct acquaintance-knowledge of the actual living of occurrent lived experience, far more than just knowledge of passive experiential content.

If, then, one holds, as I do, that a thing's causal powers are essentially and literally constitutive of its nature, one can conclude that Full Revelation must be false if panpsychism is true.[162] If it follows directly from the conception of experience as active substance that Full Revelation is false, then, *given the assumption* that smallism is true, then, again, Full Revelation is false; for reality is not inert, so if reality is experience-stuff, then experience-stuff is not inert. This result doesn't undermine the characterization of experience as 'the qualitative character that experiences have for those who have them as they have them, where this qualitative character is considered wholly independently of everything else' (M2003a, p. 50); it remains valid as a characterization of the intrinsic nature of experience in virtue of which it has the effects it does.

This may seem like uncontrolled speculation. But it is not entirely uncontrolled, and it is not unwarranted, because I am not defending a thesis that is already crazy and that is now pushing me into further craziness. The dialectical situation is rather this. A hard (genuinely naturalistic) nose for reality obliges one to endorse some sort of panpsychism or (micro)psychism long before any wild speculation has taken place. Given that one then knows that some sort of panpsychism must be true, speculation as to how it could be true is fully licensed,

---

[161] Leibniz, whose fundamental ontology is panpsychist, came to the view that there is no causal interaction between his monads.

[162] This connects with the point (p. 195) that *there is no real distinction, only a conceptual distinction, between a thing's categorical properties and its dispositional properties*. This in turn combines with the point that there is no real distinction between an object and its properties (§8 above) to lay the foundation of a decent metaphysics, insofar as it is possible to have one while retaining the standard philosophical notions of object and property.

and strongly to be encouraged. 'The truth ... must be strange' in this area, as Russell once said (1912, p. 19), and we have to do our best to understand how what must be true could be true. So when Goff says (p. 60) that I have nothing to offer on the question of how macroexperientiality emerges from microexperientiality, only 'faith that *it must happen somehow*', I enthusiastically agree — *given the assumption* that smallism is true — and find the James of *A Pluralistic Universe* by my side. The only argument for the claim that macroexperientiality emerges from microexperientiality is transcendental, and it depends on the assumption that smallism is true, an assumption that can, as Coleman says, be questioned.

18. A Few Further Responses

I have already replied to many of the points made by the other contributors to this book, especially Goff and Coleman, and the endorsement of the Cartesian view of the relation between objects and properties in §§7–9 is in effect, and as noted, a reply to Macpherson. I am particularly grateful to these three philosophers, the youngest contributors to this book. I have many further detailed replies, but space for only a few. There is obviously no correlation between my appreciation of a paper and the length of my response to it, if only because the more I agree the less I am likely to have to say in reply. In this spirit I salute Skrbina and thank him for putting me onto the earlier expression of Eddington's view and for prompting me to read William James's *A Pluralistic Universe*; and also Seager, whose case against ontological 'relationalism' I fully endorse.

I argued earlier, against the *a posteriori* materialists, that when it comes to our knowledge of the nature of our experience, the kind of knowledge that is fundamentally in question is a strictly first-order matter. The having (the being) is the knowing, and our acquaintance with it most certainly does not involve standing back from the experience reflectively and examining it by means of a further, distinct experience. I take it that this point voids many of Papineau's criticisms, for he focuses on the second-order phenomenon of what he calls 'phenomenal thought', such as 'think[ing] about the experience of seeing something red', and the third-order phenomenon of 'reflect[ing] introspectively on phenomenal thoughts' (p. 104). His highly intellectualistic conception of the mind also leads him to express doubt about the very idea of direct acquaintance; he thinks that it 'assumes some mode of thought where objects become completely transparent to the mind', and suspects it to be inspired by a visual model of peer[ing] in at some immaculately illuminated scene' (p. 102). This, however, is a vivid description of exactly what it is not, and of what it is not inspired by, for it is entirely non-perceptual. It is in fact (as Aristotle and Locke well knew) nothing more than the phenomenon of 'what-it's-likeness'.[163]

A posteriori physicalists, Carruthers and Schecter as well as Papineau, will demand further explanation of this acquaintance-knowing, but the right response to this demand is, once again, Louis Armstrong's: 'If you gotta ask, you ain't never going to know'. And the power of this response lies in the fact that of course they do know: the fact that they are still asking shows that they have been led astray by theory.

It is perhaps the focus on phenomenal *concepts*, rather than on the fact of direct experiential acquaintance, that is the central error of the philosophy-of-language grounded *a posteriori* physicalist approach (as remarked in note 144). Papineau makes this apparent when he says, with characteristic honesty, 'I don't recognize any way in which the mind "captures" something, apart from simply referring to it' (p. 106). This is enough to secure the *a posteriori* physicalists in their position, but the reply is simple. This is not how one 'captures' — knows, is acquainted with — the experience of red or pain — or free fall or chili or nausea — when one has it.[164] Coleman gives a sympathetic presentation of the *a posteriori* physicalists' case even as he rejects it (pp. 45–7) and ends by noting that it is their '*phenomenal concepts strategy*' that needs arguing against. Exactly so. But it is not as if anything more needs to be done, because the simple reply just given is devastatingly sufficient for the task. To say any more is to fog things over.

The *a posteriori* physicalists will deny this. They will stonewall and accuse me of doing the same. That's fine by me, because I'm stonewalling on a stone wall: the fact of experience, the fact that we are acquainted with the essential nature of experience, at least in certain respects, simply in having it, because the having is the being acquainted, as Descartes observed in his Second Meditation (1641, p. 19).

---

[163] On Aristotle see in particular Caston 2002, a wonderful piece of work. On Locke see e.g. Strawson 2011b, §7.

[164] 'Capture' is my word, but I do not use it well, and Papineau is right to express doubts about my talk of *terms* fully capturing the nature of things. What I mean is that *we* (not terms or theories) can mentally 'capture', i.e. grasp, the essence of things in certain respects, e.g. in having experience as we do. Papineau shows that he understands my intent when he talks in the quotation just given of 'the mind' rather than a term capturing something.

PANPSYCHISM? 265

This is an infallibility claim, no less, and infallibility claims are often thought to be beyond the pale. So perhaps philosophers who accept the acquaintance view can allow that we don't have any kind of perfect infallibility in the matter (although it is unclear how we could fail to), insisting merely that we are in fact always nearly right, if not necessarily absolutely right, about the essential nature of our experience, at least in certain respects, just in having it. Certainly they can agree that all infallibility claims are immediately voided as soon as one starts to conceive knowing in a more cognitive way as the taking of something as object of intellectual attention, moving to the second-order, or the third-order, and starting to *think* about our experience, and about our thoughts about our experience. To do this, however, is to step back — disconnect — from the real mind-body problem.[165]

All very well — but now the *a posteriori* physicalists (and indeed everyone else) can make against me a move that I make myself when I conjoin claims of the following sort

[a] our fundamental thought categories fail to get the world right when we think about the relation between an object and its properties

[b] we can as philosophers nevertheless just about see how things are

or when I talk airily of the need for 'cultivation of a shift in intuitions, acquisition of the ability to sustain a different *continuo* in place in the background of thought' (M2003b, p. 302). 'Look', they say, 'the same applies to your claim that we know the essential nature of reality in certain respects in having experience, because the having is the knowing. We are all deeply and indeed helplessly committed to this conviction, but we can nevertheless sufficiently see how it may yet be false — how we may be completely wrong about the nature of experience even in the very having of it.'

I have three replies. The first is the reply to Dennett in RMP (pp. 5–6, nn. 6 and 7). The second has also been sufficiently laid out; it is that there is no good motivation for this move, and in particular that it receives precisely *no* support from the success and beauty and 'causal-explanatory power' of the physical sciences (contrary to

---

[165] I can sense the feeling of confidence that many will have that this is an easy objection to meet, something that they don't really have to think about, because I am familiar with the analytical-philosophy problematic that underwrites this confidence (a merely sociological remark).

Papineau and McGinn, among others). The third is 'Fine, we're at the end of argument. I'm happy where I am, if you're happy where you are you should stay there'. For my aim as a pupil of Epictetus is not to convince anyone of anything, only to try to state the truth as far as I can. Note, though, that the fact that a form of argument is reversible doesn't show that the considerations it adduces have no more force one way than another. Rather it shows that neither side can use it to argue the other out of their position. The most they can hope to do is jolt the other side's intuitions by the way in which they use it.

This formal vulnerability in my position — the fact that one of my styles of argument can be turned against me — does not trouble me. Nor does the fact that I am vulnerable in the same sort of way when I claim that the deep reason why people come to endorse the notion of the radical or brute emergence of the experiential from the non-experiential is that it 'marks a position that seemingly has to exist' given their prior commitments. Surely, Goff says, 'we could say the same about Strawson's hypothesis that macro-experiential being emerges from micro-experiential being' (p. 60) — his point being that I have to say this given other things I am committed to. True; and the same can be said about my attempt to argue that radical emergence is impossible, by proposing (for example) that it is as impossible as the emergence of the spatial from the non-spatial. My case for the impossibility of radical emergence is, as I stress in RMP, the exercise and articulation of an intuition, an attempt to open up a scene of thought. I am not seeking to disturb the *a posteriori* physicalists' fortifications, I am outside their walls with no wish to get in.

Smart recruits Occam against me, Occam with his razor. But whatever Occam prompts us to do, he cannot have us cut away the only thing we know for certain to exist — experience — , and the strictest reading of his famous dictum that entities should not be multiplied beyond necessity[166] leads straight to the view that experience is the *only* thing that exists, a view that one can break out of only if some form of Kantian transcendental argument of the following form

[1] Experience exists
[2] Experience cannot possibly exist unless X exists
[3] Therefore X exists

is valid, where X is something that is, knowably, ontologically distinct from experience. It is true that a transcendental argument of this sort

---

[166] When one theorizes, one should not commit oneself to the existence of anything more than one has to.

can establish the existence of a subject of experience, and it is very natural to think that a subject of experience must be ontologically distinct from any experience itself; in which case we can at least establish that something other than experience exists, in spite of Occam's razor. However, even the view that the subject of an experience must be ontologically distinct from the experience itself is open to doubt, given one natural understanding of the notion of a subject of experience — the 'thin' understanding employed in this paper — and I reject it. And even if it were correct, the transcendental argument for the existence of a subject of experience would not permit us to establish the existence of anything like a physical world as ordinarily understood — nor even a world with intrinsically non-experiential features.[167] If that is something you feel you need to do, the best argument is that rare thing, Lockean humour.[168]

I am as much of an identity theorist as Smart, in fact — I would say I am more so —, but an identity theory is an identity theory, and to say that experiential states are brain states of certain sorts (something we may take to be true of all macro-experiential states, at least on this earth) is to say just that: that our experiential states, the only thing we know for certain to exist, are certain sorts of brain states. By the same token, of course, it is to say that brain states of these sorts just are experiential states. As Sprigge remarks — in another book I have been fortunate to be led to in writing these replies — 'anything going for the identity theory is evidence for the truth of panpsychism, as was realized long ago by philosophers such as Josiah Royce'.[169] To identify X with Y is not to say that X does not exist. You can't apply Occam's razor and cut away X (experiential states) without also cutting away Y (brain states); for X is Y.

So while I have great respect, even fondness, for Smart's position, I find that the Occam he appeals to rules against him and in my favour. It is certainly not Smart's fault, but I think it is a shame that his justly famous 1959 paper so eclipsed Herbert Feigl's 1958 paper, which was also once widely known. For Feigl never put the reality of experience

---

[167] Kant's 'Refutation of Idealism' won't do the trick.

[168] 'If any one say, a dream may do the same thing, and all these ideas may be produced in us, without any external objects, he may please to dream that I make him this answer...' (1689, p. 537 (4.2.14)). See also p. 634 (4.11.8): 'And if our dreamer pleases to try, whether the glowing heat of a glass furnace be barely a wandering imagination in a drowsy man's fancy, by putting his hand into it, he may perhaps be wakened into a certainty greater than he could wish, that it is something more...'.

[169] Sprigge 1983, p. 102.

in doubt any more than Ullin Place did in his equally well known 1956 paper.[170]

Rosenthal points out something I hadn't noticed — my argument in RMP is in part a simple reversal of Smart's. For just as Smart takes the general nature of the physical as given and proposes a 'topic-neutral' of experiential phenomena that leaves it open whether or not experiential phenomena are physical phenomena, so I take the general nature of the experiential as given and propose a 'topic-neutral' account of physical phenomena that leaves it open whether physical phenomena are experiential. We are both then free to invoke 'mind-body correlations together with Ockham's razor [to] show that experiences are physical' (p. 123), or as I would put it, although the difference is only one of emphasis (for identity is identity), to show that the physical is experiential.

An important part of Rosenthal's position is that non-conscious perceptual states can be said to have 'qualitative character' (to be 'experiential' in my sense). I'm happy to concede this for argument, and I agree with him that it 'allows for useful theoretical ways to identify and taxonomize mental qualities independently of the way individuals are conscious of them' (p. 118). But I don't believe that it would provide a way in to non-first-person knowledge of the nature of qualitative character. For although I also agree that we could in principle determine that 'two individuals have the same mental qualities' from the third-personal point of view, I deny that this supports the claim that we could thereby know what those qualities were like, qualitatively speaking. For although we could determine the qualitative mental sameness of two persons — if, for example, we could know that they were atom- for-atom identical — we still would not know what those qualities were (unless of course we were one of them).

There is a striking symmetry about Rosenthal's and my positions, for even as he rejects my claim that we have merely structural knowledge of the intrinsic nature of the non-experiential,[171] he offers what is to my ears a purely structural (state-space) account of the intrinsic nature of the experiential. What is for him the keystone of his position

---

[170] Feigl's paper was somewhat ill expressed in terms of 'raw feels' (a notion inherited, perhaps, from Carnap). For a concise and sympathetic criticism of Feigl's use of 'raw feels' that at the same time draws out the respect in which Feigl's instincts were right — and panpsychist — see Sprigge 1983, pp. 100–4.

[171] On the grounds that we have knowledge of properties like mass, spin, and charge (p. 123). On my view, we no more have knowledge of the intrinsic, non-mathematically expressible nature of these properties than we do of the intrinsic nature of space. (Can we at least be said to have some irreducibly non-structural knowledge of the nature of *force*? Not in any way that shows it to be intrinsically non-experiential rather than experiential.)

— in one expression, it is the claim that 'your mental quality of red and mine are automatically the same if our abilities to discern physical color properties are the same' (p. 126) — is for me a *reductio* of it, comparable to saying that any two spatial entities that are topologically identical must have identical shapes (it is arguably a special case of such a claim, once the reference to spatiality is removed). And Rosenthal could of course make the converse move against me! Perhaps our positions are from a topic-neutral point of view the same! To that extent, we are both right about one thing — that the truth can't lie somewhere in between.

Georges Rey and I have always disagreed about experience. We enjoy this. Always he has always told me — the first time in a downpour in Aix-en-Provence — that there is no non-question-begging argument for the existence of experience. Always I have replied that it is question-begging to request a non-question-begging argument for its existence, and have requested his non-question-begging argument for the existence of non-experiential reality. 'You're prepared to go directly against ordinary opinion in rejecting the possibility of radical free will', he has always said, 'so why do you think the rejection of ordinary opinion is absolutely ruled out in the case of experience?' I have never had anything to reply except that the cases are of a completely different order, and that if he can't see that then I don't know what more to say. Now, though, I am tempted to add that panpsychism also goes wildly against common sense, although not as wildly as his eliminative reductionism (for that is the ultimate wildness), and that I'm still in the radical club.

Lycan correctly reports my view that '(1) The nature of (real) experience cannot be specified in wholly non-experiential terms', and grants that my conclusion — that those who seek to reduce the experiential to the non-experiential are denying the existence of the experiential — might 'follow if (1) were analytic. But [he says] (1) is not analytic; it is a highly contentious philosophical claim.' It does not follow, though, from the fact that a claim is contentious, that it is not analytic,[172] and I find on reflection that I am happy to say that (1) can be counted as analytic for present purposes, or at least as a necessary truth. More moderately: it is at least as much a necessary truth as the claim that one could not fully reveal the nature of colour-experience to someone using only means that did not in any way involve their having any sort of colour-experience. If Lycan wishes to hold to the view

---

[172] I'm tempted to say that it is analytic — inescapably contained in the notions of infinity and concrete existence — that there cannot be an actual concrete infinity of things.

that (1) is not a necessary truth, then I think he must at least concede that its converse, (2) the nature of the non-experiential cannot be specified in wholly experiential terms, is not analytic or any sort of necessary truth. With this in place we could cut the cards and reconsider our mutual positions.

Jackson's paper is a shaft of sun in a dark place, but the shaft is too thin. I resonate with the intuitions that lie behind his *a priori* entailment approach (given the way in which I reject radical or brute emergence); my difficulty with his paper in this form is simply that I am not clear what he means by 'consciousness', or what he has in mind when he talks of arrangements of non-experiential items 'causing' or 'generating' consciousness (experience). So I am tantalized, and want more.

I like Simons's stylish and helpful paper very much, and am pleased that he thinks that the argument in RMP actually works, once the case against radical emergence is granted. I should, though, say that panpsychism as I understand and characterize it has nothing to do with 'the idea of electrons making decisions about how to spin, nuclei harbouring intentions to split, or photons with existential Angst' (p. 146), any more than it has anything to do with tables and chairs being subjects of experience (RMP, p. 26); and when Simons writes 'I hope it is clear that adopting panpsychism is of the same order of desperation as denying experience or accepting dualism, because to all appearances there is nothing like experience down among the quarks and leptons' (p. 148) I cannot agree that the 'because' clause does the work he needs it to do — apart from the other reasons that exist for denying the charge of desperation.

Stapp makes some excellent and correct debating points, but I think his central criticism fails. After observing, rightly, that by 'physicSalism' I mean in effect present-day physics, he points out that present-day physics involves ineliminable reference to experience in involving reference to the conscious choices and feedback experiences of scientific experimenters. He concludes that I am wrong to say that what I call 'real physicalism cannot have anything to do with physicSalism unless it is supposed — obviously falsely — that the terms of physics can fully capture the nature or essence of experience' (p. 4); for actual present-day physics does indeed include such terms.

There is much to be said here in reply, and my use of 'fully capture' (p. 4) is indeed unsatisfactory (as Papineau shows, pp. 100 ff.), but I will make only three small points. First, even if this is right, the real problem — the mind-body problem — remains: it just needs reformulation. Second, I take it that the object of physics existed in full before there were any scientific experimenters, in such a way that a full description of its nature could in principle be given independently of any reference to the experiences of scientific experimenters. Third, I also take it that although collapse of the wave function involves a certain sort of transfer of *information*, this need not involve any sort of consciousness or experience of the scientific-experimenter kind, i.e. human-style experience (although it will indeed involve some sort of conscious experience, according to panpsychism).

I have touched on McGinn's paper at several points. Here I will say only that my hair did get very long at the end of the 1960s (see his p. 93), and that I am glad he thinks that panpsychism is — for all the faults he finds in it — 'one of the loveliest and most tempting view of reality ever devised' (ibid). I have already claimed (p. 229) that a correct, Cartesian attitude to the relation between objects and substances and their properties and attributes undercuts many of his detailed objections, and I would like to respond to his fearless proposal of '*extended panpsychism*: experience exists at every point in the spatial universe, whether occupied by matter or not' (p. 97) by fearlessly accepting it as a serious hypothesis, with two provisos. First, I take 'spatial' to be a natural-kind term for whatever the intrinsic dimensionality of reality is. Second, I delete the words 'whether occupied by matter or not', to leave just 'experience exists at every point in the spatial universe', on the scientifically well-attested ground that there is (as Descartes had already intuited) really no such thing as 'empty space'. I forgive McGinn his jokes at the expense of panpsychism, especially now that I have learnt from Wilson that Lucretius had the same impulse over two thousand years ago (p. 177), and he was surely not the first, given the wit of the Greeks.

Central to the present position since M1994 ('Agnostic materialism') has been the Lockean claim that we are profoundly ignorant of the nature of the physical.[173] Macpherson thinks my terminology obscures the Lockean connection and that I don't explicitly recognize it, but I have always done so, and do so again in RMP.[174] It is in any case a claim massively backed by physics, and I'll call it *Ignorance*.[175]

---

[173] Two key passages in Locke's *Essay* are 2.23.28-32 and 4.3.6 — one of philosophy's greatest paragraphs (1689, pp. 311-14 and 539-43).

[174] p. 8. See e.g. M1994, p. 102, M1999a, p. 31, M2003a, pp. 63, 65.

[175] Locke, of course, argues that we are equally ignorant of the nature of the substance of the mind; what he has principally in mind is that it may for all we know be wholly material. One cannot properly understand Locke on substance without a good knowledge of the debate centred round Descartes, as I have recently discovered, and Locke's correspondence with Stillingfleet shows, I think, that Macpherson is wrong to hold (p. 79) that

One thing that follows from Ignorance is that we can't know that 'Australian' zombies are possible even if we allow that whatever is conceivable is possible (Australian zombies are perfect physical duplicates of experiencing human beings that are wholly experienceless beings).[176] Daniel Stoljar agrees, and wonders why I call them 'Australian'. The answer is that I felt that they needed to be distinguished from their more important and long-established cousins, 'classical' zombies, who are '*outwardly* and *behaviourally* indistinguishable from human beings while having unknown (possibly non-biological) insides and [are] accordingly of considerable interest to functionalists and behaviourists' (M2003a, p. 84 n. 115); for it seemed that a whole generation was growing up unaware of the existence of classical zombies. I called them 'Australian' in spite of the fact that I disapprove of them, because they seemed particularly popular on that continent.[177]

Another closely connected thing that I take to follow from Ignorance, as Stoljar notes, is that we do not know enough about the nature of the physical to know that it cannot itself be experiential. Well and good, and again Stoljar agrees. But he then points out, quite correctly, that I also claim that we do know enough to know [30] that the experiential cannot emerge from the wholly and utterly non-experiential. And [30] is plainly inconsistent with Ignorance.

True. I am not a radical Ignorantist. With Socrates, Locke, and, happily, most of the rest of the First XI, I subscribe to Ignorance in general, both in RMP and elsewhere, but within that frame I try to make the case for the intuition that the experiential cannot emerge from the wholly and utterly non-experiential as vigorously as I can, even as I stress that the case for the intuition cannot ultimately rest on argument from generally agreed principles.[178] If someone says that one of the things that follows from Ignorance is that there may be non-

---

Locke thinks that 'no combination of physical properties can produce mentality' (see Locke 1696-9, pp. 459-62, M2003a, p. 82, n. 84).

[176] The point is made in M1999a, pp. 28-9, more fully in M2003a, p. 72, and in RMP, p. 22 n.

[177] I write as a long-standing 'honorary Australian philosopher' (appointed by Frank Jackson, Michael Smith, and Philip Pettit in the last century). Since writing his commentary, Stoljar (private communication) has traced Australian zombies to the 'imitation man' in Keith Campbell's 1970 book *Body and Mind*, and thinks the epithet may after all be appropriate.

[178] After saying that the property of being intrinsically suitable for constituting experience cannot be possessed by 'wholly, utterly, through-and-through non-experiential phenomena' I comment: 'This is the unargued intuition again. Bear in mind that the intuition that the non-experiential could not emerge from the wholly experiential is exactly parallel and unargued' (p. 21). Coleman picks up on this last sentence, claiming that this argument from parity is disingenuous (p. 46). Experiential-from-non-experiential emergence may

experiential facts, unknown to us, that will serve to 'yield' experiential facts, I will not be able to prove that that's not so, but I will not be greatly troubled, for until more is said it amounts to simply dismissing the considerations brought in favour of the intuition that the experiential cannot emerge from the non-experiential, and is I think close to saying (albeit with an added, dynamical flavour) that not-P might entail P. Now one might get away even with this last claim, I think, but only if one endorsed the Spinozan, ESFD-monist view that P and not-P are at bottom the same thing, and Stoljar, of course, does not hold this view. Nor would it help him even if he did, for then — an important point — he would no longer need (or be able) to claim that there was any kind of experiential-from-non-experiential *emergence*.

In sum, I think that the assertion of radical Ignorance is not an adequate or interesting response. The radical Ignorantist position is consistent, indefeasible, safe, but it opts out of the real difficulty — if, that is, its exponents really are real realists about the experiential. If they're *not* really real realists about the experiential — and certainly none of those who think that experiential properties could literally be functional properties, however realized, are real realists about the experiential — they haven't yet got to the real difficulty.

I did not know that I was an 'Anaxagorean panpsychist' until Wilson told me so (p. 177), but it is plain that I am, in rejecting the radical emergence of the experiential from the wholly non-experiential. Her elegant paper does not, however, convince me that Spinoza and Leibniz are not fundamentally on my side, in their central metaphysical commitments, at least when it comes to the matter of panpsychism. As for Spinoza, I disagree with Wilson's suggestion that 'Spinoza is ... best read as an ordinary physicalist who thinks that human and animal minds depend on their brains' (p. 179), for I cannot see how Spinoza could accept any such use of the notion of dependence. It seems to me, furthermore, that everything Wilson cites in support of taking him to be some sort of ordinary physicalist on pp. 178–9 is equally compatible with taking him to be some sort

---

be impossible, he says, but at least it fits the standard model for emergence in being a case of 'higher-level' phenomena emerging from 'lower-level' phenomena. Non-experiential-from-experiential emergence, by contrast, would, he says, be a bizarre case of lower-level phenomena emerging from higher-level phenomena. This objection is based on a misunderstanding, however. In being explicitly couched in terms of emergence, the challenge to those who reject the intuition that the experiential could not possibly emerge from the wholly non-experiential takes for granted the standard model of emergence (the idea that emergence is a matter of 'higher-level' phenomena emerging from 'lower-level' phenomena). The challenge is, precisely, that those philosophers who think it possible that the (real) experiential could emerge from the non-experiential must think it equally possible that the non-experiential could *emerge* from the experiential.

of ESFD monist (an Equal-Status Fundamental-Duality panpsychist monist, see p. 241), and that the view that he is some sort of ESFD monist is better supported on other grounds. I know, though, that this needs a great deal of further discussion.

Wilson thinks that it is harder to show that I cannot claim Leibniz as an ally, because, as she says, he does quite clearly hold that 'everything that is a real, individual thing, and not an entity by convention, has experiences' (p. 180). To that extent he is clearly a panpsychist. However, these metaphysically simple real individual things, these 'monads', are not physical, on Leibniz's view, if only because 'everything the natural scientist encounters in the physical world ... is [on Leibniz's view] complex and divisible', and it follows, Wilson says, that I cannot assimilate them to my 'ultimate physical constituents' (ibid). She goes on to say that my appeal for alliance fails in another way, because the monads 'are supposed to give rise to the phenomenal world of extended bodies, and Strawson says that the emergence of extension from nonextension is incoherent' (ibid).

The first point is plainly right; I can't assimilate Leibniz's monads directly to my ultimates. But this is not because I call my ultimates 'physical', whereas his monads are by definition non-physical; for my stretched use of 'physical' covers whatever is 'real and concrete (p. 8) and therefore comfortably encompasses his monads.[179] The real problem, I think, is that his monads do not interact causally in any way, while my ultimates do. This is a big difference. It's true that allies don't have to have identical views, only common interests, and I claim Leibniz as an ally only in a general way. I also have a strong suspicion that a physics-informed study of the respect in which the universe is a single entity might take us a very long way towards assimilating Leibniz's notion of non-causal 'pre-established harmony' to whatever in that study survived of the notion of cause. The fact remains that the causal/non-causal difference is, on the face of it, a big one.

I reject Wilson's second point because I don't think that the way in which monads are supposed to give rise to the phenomenal world of extended bodies, in Leibniz's scheme, is a real case of the emergence of extension from the unextended. It is not, as I understand it, the emergence of real extension from the unextended, because Leibniz's 'phenomenal world of extended bodies' is just that, a phenomenal world, a *phenomenon bene fundatum*, to be sure, in Leibniz's terminology, but not fundamentally real, that is, not fundamentally real in

---

[179] Plainly my use of 'physical' is confusing, in spite of my explicit provision. I had a specific rhetorical purpose in using it in the way I did in my argument for Ignorance, but I do better now to give it up, and do so in this paper.

the sense that monads are fundamentally real. I cannot support this claim with quotation, though, and the larger lesson I draw from Wilson, very willingly, is that I had better read Leibniz, some of whose other views I suspect I am simply recapitulating.[180]

19. Conclusion

There is, I feel sure, a fundamental sense in which monism is true, a fundamental sense in which there is only one kind of stuff in the universe. Plainly, though, we don't fully understand the nature of this stuff, and I don't suppose we ever will — even if we can develop a way of apprehending things that transcends discursive forms of thought.

The existence of the mind-body problem is the best evidence we have that our understanding will fall short. I think we can see how things might be in a general way (i.e. some version of panpsychism) but on the whole I am happy to stick my neck out into the future and agree with Emile Du Bois Reymond that in this matter *ignoramus et ignorabimus*: we don't know and we won't know, we are ignorant and will remain so.

There are mathematical propositions that are provably unprovable, and metaphysical propositions also ('Determinism is true', 'Determinism is false'). There are metaphysical propositions whose ununderstandability-by-us can be compellingly argued for, and there are many scientific propositions whose truth is well attested although they too remain in a clear sense ununderstandable by us. So be it — full understandability-by-us is no more a condition on metaphysical truth than it is on truth in physics, nor is it any more a condition on reasonable acceptance of propositions in metaphysics than it is on propositions in physics. There are many propositions, about wave-particle duality, for example, or quantum entanglement, or superposition, with which we have made our peace although we cannot claim to have any real sense of understanding of the phenomena in question. We need to cultivate the same attitude in metaphysics, even while we continue to press for greater understanding.

For each of us, perhaps, the sense of the difficulty of the mind-body problem has a special spin. Certain conundra mark the heart of our incomprehension in a particularly vivid way — they are the 'first-rank symptoms' of our affliction. I have always felt this about the so-called 'problem of mental causation', and this has determined my sense of

---

[180] A lesson powerfully reinforced by reading Garber's 2005 paper 'Leibniz and Idealism'.

what it would be to reach a decent stopping point in the mind-body problem. It would be to contemplate a fabulously detailed and exhaustive specification in neurological or particle-physics terms of the causation involved in a line of thought or a practical decision, and to feel no force in the objection that the availability of this specification showed that the mental was epiphenomenal or causally inefficacious. It would be to feel this because one no longer had any intuitive sense of a radical conflict between the physics/neurology explanation and a mental explanation along the following lines. 'She wanted to get to London in good time to give her paper. She wondered whether to go by train or by car. She knew that the fast trains take about two hours, and then suddenly remembered the new Congestion Charge restrictions on driving in Central London … etc.'

I think I've made it. It's some time since I stopped having any intuitive difficulty with the idea that this red-experience, this thing with whose essential nature I am in certain respects fully acquainted just in having it, is just (just is) this patch of complex neural activity, and this sense of there being no great intuitive difficulty has spread from cases like that of having red-experience to more complex cases, cases of perceptual experience, for example, or cases of consciously entertaining and comprehending propositions like 'nobody could have had different parents', or indeed any of the propositions expressed in this paper.

One of the keys, I am sure, is to see that there is a fundamental component to the business of consciously entertaining and comprehending propositions that is just a matter of 'qualitative-experiential character' in every sense in which an experience of red is just a matter of qualitative-experiential character;[181] so that once one has no great intuitive difficulty in the idea that this red-experience is just a patch of complex neural activity one is well on the way to finding no great intuitive difficulty in the idea that this conscious comprehending of this thought is just a patch of complex neural activity, even it still seems that much more puzzling. It takes time, though.[182]

References

Anscombe, G. E. M. (1971/1981) 'Causality and Determination' in *Metaphysics and the Philosophy of Mind* (Oxford: Blackwell).

---

[181] I develop the point in M1994, e.g. pp. 5–13, 182–3, Strawson 2005, pp. 47–53 and 2008a.

[182] I am very grateful to Sam Coleman for reading and commenting on a draft of this paper, and to Anthony Freeman for his help on many fronts.

Armstrong, D. M. (1980/1997) 'Against "Ostrich Nominalism"' in *Properties*, edited by D. H. Mellor & Alex Oliver (Oxford: Oxford University Press).
Arnauld, A. (1641/1985) 'Fourth Set of Objections' in *The Philosophical Writings of Descartes*, Volume 2, translated by J. Cottingham et al. (Cambridge: Cambridge University Press).
Arnauld, A. (1683/1990) *On True and False Ideas*, translated with an introduction by Stephen Gaukroger (Manchester: Manchester University Press).
Broad, C. D. (1925) *The Mind and Its Place in Nature* (London: Routledge and Kegan Paul).
Campbell, K. (1970) *Body and Mind* (New York: Doubleday).
Caston, V. (2002) 'Aristotle on Consciousness' *Mind* **111**, pp. 751–815.
Chalmers, D. (1997) 'Moving forward on the Problem of Consciousness' *Journal of Consciousness Studies* **4**, pp. 3-46.
Chalmers, D. (2003) 'Consciousness and its Place in Nature' in *Blackwell Guide to the Philosophy of Mind*, edited by S. Stich and F. Warfield (Oxford: Blackwell).
Chomsky, N. (1968) *Language and Mind* (New York: Harcourt, Brace & World).
Clarke, D. (2003) *Descartes's Theory of Mind* (Cambridge: Cambridge University Press).
Clarke, D. (2006) *Descartes* (Oxford: Oxford University Press).
Coleman, S. (2006) *In Spaceships They Won't Understand: The Knowledge Argument, Past, Present and Future* (University of London: PhD thesis).
Cottingham, J (1976) 'Commentary' in *Conversations with Burman*, translated with a philosophical introduction and commentary by J. Cottingham (Oxford: Clarendon Press).
Crane, T., and Mellor, D. H. (1990) 'There Is No Question of Physicalism.' *Mind*, **99**, pp.185–206.
Descartes, R (1618-50/1964-76) *Oeuvres*, edited by C. Adam & P. Tannery (Paris:Vrin).
Descartes, R. (1618-28/1985) *Rules for the Direction of the Mind* in *The Philosophical Writings of Descartes* Volume 1, translated by J. Cottingham et al (Cambridge: Cambridge University Press).
Descartes, R. (1619-1650/1991) in *The Philosophical Writings of Descartes* Volume 3, *The Correspondence*, translated by J. Cottingham et al (Cambridge: Cambridge University Press).
Descartes, R. (1637a/1985) *Discourse on the Method* in *The Philosophical Writings of Descartes* Volume 1, translated by J. Cottingham et al (Cambridge: Cambridge University Press).
Descartes, R. (1637b/2001) *Meteorology* in *Discourse on Method, Optics, Geometry, and Meteorology*, translated by P. J. Olscamp (Indianapolis: Hackett).
Descartes, R. (1641/1985) *Meditations* in *The Philosophical Writings of Descartes* Volume 2, translated by J. Cottingham et al. (Cambridge: Cambridge University Press).
Descartes, R. (1642/1984) *Letter to Father Dinet* in *The Philosophical Writings of Descartes* Volume 2, translated by J. Cottingham et al. (Cambridge: Cambridge University Press).
Descartes, R. (1644/1985) *The Principles of Philosophy* in *The Philosophical Writings of Descartes*, Volume 1, translated by J. Cottingham et al. (Cambridge: Cambridge University Press).
Descartes, R (1645-6/1991) letter in *The Philosophical Writings of Descartes* Volume 3, translated by J. Cottingham et al (Cambridge: Cambridge University Press).

Descartes, R. (1648a/1985) *Comments on a Certain Broadsheet* in *The Philosophical Writings of Descartes* Volume 1, translated by J. Cottingham et al (Cambridge: Cambridge University Press).
Descartes, R. (1648b/1976) *Conversations with Burman*, translated with a philosophical introduction and commentary by J. Cottingham (Oxford: Clarendon Press).
Descartes, R. (1649/1985) *The Passions of the Soul* in *The Philosophical Writings of Descartes*, Volume 1, translated by J. Cottingham et al. (Cambridge: Cambridge University Press).
Eddington, A. (1920) *Space, Time, and Gravitation* (Cambridge: Cambridge University Press).
Eddington, A. (1928) *The Nature of the Physical World* (New York: Macmillan).
Eddington, A. (1939) *The Philosophy of Physical Science* (Cambridge: Cambridge University Press).
Evans, C. O. (1970) *The Subject of Consciousness* (London: Allen & Unwin).
Frege, G. (1918/1967) 'The Thought: A Logical Inquiry' in *Philosophical Logic*, edited by P. F. Strawson (Oxford: Oxford University Press).
Feigl, H. (1958), 'The "mental" and the "physical"' in *Concepts, Theories, and the Mind-Body Problem* (Minneapolis, MN: University of Minnesota Press).
Garber, D. (2005) 'Leibniz and Idealism' in *Leibniz: Nature and Freedom*, edited by D. Rutherford and J. Cover (Oxford: Oxford University Press).
Goff, P. (2006) *Should materialists be afraid of ghosts? Arguing for substance dualism* (University of Reading: PhD thesis).
Greene, B. (2004) *The Fabric of the Cosmos* (New York: Knopf).
Griffin, D. (1977) 'Whitehead's philosophy and some general notions of physics and biology' in *Mind in Nature*, edited by J. Cobb and D. Griffin (University Press of America).
Heil, J. (1992) *The Nature of True Minds* (Cambridge: Cambridge University Press).
Hume, D. (1739-40/2000) *A Treatise of Human Nature*, edited by D. Norton and M. Norton (Oxford: Clarendon Press)
Hume, D. (1748/1999) *Enquiry Concerning Human Understanding*, edited by T. Beauchamp (Oxford: Clarendon Press).
James, W. (1890/1950) *The Principles of Psychology*, 2 vols. (New York: Dover).
James, W. (1909/1996) *A Pluralistic Universe* (Lincoln, Nebraska: University of Nebraska Press).
Johnston, M. (1992) 'How to speak of the colors', *Philosophical Studies* **68**, pp. 221-63.
Kant, I. (1781-7/1933) *Critique of Pure Reason*, translated by N. Kemp Smith, London: Macmillan.
Kant, I. (1781-7/1996) *Critique of Pure Reason*, translated by W. S. Pluhar, Indianapolis: Hackett.
Kant, I. (1781-7/1999) *Critique of Pure Reason*, translated by P. Guyer and A. Wood (Cambridge: Cambridge University Press).
Kemp Smith (1952) *New Studies in the Philosophy of Descartes: Descartes as Pioneer* (London: Macmillan).
Kripke (1972/1980) *Naming and Necessity* (Oxford: Blackwell)
Lewis, D. & Langton, R. (1996/1999) 'Defining "intrinsic"' in D. Lewis, *Papers in metaphysics and epistemology*, (Cambridge: Cambridge University Press).
Locke, J. (1689-1700/1975) *An Essay Concerning Human Understanding*, edited by P. Nidditch (Oxford: Clarendon Press).

Locke, J. (1696-9/1964) 'The Correspondence with Stillingfleet' in *An Essay Concerning Human Understanding*, edited and abridged by A.D. Woozley (London: Collins).
Lockwood, M. (1981) 'What *Was* Russell's Neutral Monism?' in *Midwest Studies in Philosophy* **VI**, 143-158.
Lockwood, M. (1989) *Mind, Brain, and the Quantum* (Oxford: Blackwell).
Lockwood, M. (1992) 'The grain problem' in *Objections to Physicalism*, edited by H. Robinson (Oxford: Oxford University Press).
Loux, M. (2002) *Metaphysics: a contemporary introduction*, second edition (London: Routledge).
McGinn, C. (1989) 'Can We Solve the Mind-Body Problem?' *Mind* **98**, pp. 349–66.
McGinn, C. (2000/2004) 'Solving the Philosophical Mind-Body Problem' in *Consciousness and its Objects* (Oxford: Clarendon Press).
Montague, M. (2007) 'The content of perceptual experience' in *Oxford Handbook in the Philosophy of Mind* edited by A. Beckermann and B. McLaughlin (Oxford: Oxford University Press).
Nadler, S. (2006) *Spinoza's Ethics: an Introduction* (Cambridge: Cambridge University Press).
Nagarjuna (c150/1995) *The Fundamental Wisdom of the Middle Way*, translated with a commentary by Jay Garfield (Albany, NY: SUNY Press).
Nagel, T. (1979) 'Panpsychism' in *Mortal Questions* (Cambridge: Cambridge University Press).
Nagel, T. (1986) *The View From Nowhere* (New York: Oxford University Press).
Nietzsche, F. (1885-8/2003) *Writings from the Last Notebooks*, translated by Kate Sturge, edited by Rüdiger Bittner (Cambridge: Cambridge University Press).
Pascal, B. (c1640-62/1965) *Pensées* revised edition, translated by A.J. Krailsheimer (London: Penguin).
Pascal, B. (1656-7/2004) *Provincial Letters*, in *Thoughts, Letters and Minor Works* (Whitefish, MT: Kessinger Publishing).
Place, U. T. (1956) 'Is consciousness a brain process?' *British Journal of Psychology* 47, pp. 44–50.
Priest, G. (2006) *In Contradiction*, second edition (Oxford: Oxford University Press).
Priestley, J. (1777-82/1818) *Disquisitions Relating to Matter and Spirit* in *The Theological and Miscellaneous Works of Joseph Priestley*, volume 3, edited by J. T. Rutt (London).
Pyle, A. (2003) *Malebranche* (London: Routledge).
Ramsey, F. (1925/1997) 'Universals' in *Properties*, edited by D. H. Mellor & Alex Oliver (Oxford: Oxford University Press).
Regius (1647) *An Account of the Human Mind...* in *The Philosophical Writings of Descartes*, Volume 1, translated by J. Cottingham et al. (Cambridge: Cambridge University Press).
Reid, T. (1785/2002) *Essays on the Intellectual Powers of Man* (Edinburgh: Edinburgh University Press).
Russell, B. (1912/1959) *The Problems of Philosophy* (Oxford: Oxford University Press).
Russell, B. (1927a/1992a) *The Analysis of Matter* (London: Routledge).
Russell, B. (1927b/1992b) *An Outline of Philosophy* (London: Routledge).
Russell, B. (1948/1992c) *Human Knowledge: Its Scope And Limits* (London: Routledge).
Russell, B. (1956/1995) 'Mind and Matter' in *Portraits from Memory* (Nottingham: Spokesman).

Searle, J. (1980), 'Minds, Brains, and Programs', *Behavioral and Brain Sciences* **3**, 417–24.
Shoemaker, S. (1986/1996) 'Introspection and the Self' in *The First-Person Perspective And Other Essays* (Cambridge: Cambridge University Press).
Skrbina, D. (2005) *Panpsychism in the West* (Cambridge, MA: MIT Press).
Smart, J. J. C. (1959) 'Sensations and Brain Processes', *The Philosophical Review* **68**, pp. 141–56.
Spinoza, B. (c 1662/1984) *Short Treatise* in *The Collected Works of Spinoza* volume 1, edited and translated by E. Curley (Princeton, NJ: Princeton University Press).
Spinoza, B. (1663/1984) 'Letter 9' in *The Collected Works of Spinoza* volume 1, edited and translated by E. Curley (Princeton, NJ: Princeton University Press).
Spinoza, B. (1677/1984) *Ethics* in *The Collected Works of Spinoza* volume 1, edited and translated by E. Curley (Princeton, NJ: Princeton University Press).
Sprigge, T. L. S. (1983), *The Vindication of Absolute Idealism* (Edinburgh: Edinburgh University Press).
Stoljar, D. (2001) 'Two Conceptions of the Physical', *Philosophy and Phenomenological Research* **62**, pp. 253–81.
Stoljar, D. (2006) *Ignorance and Imagination* (Oxford: Oxford University Press).
Strawson, G. (1986, reprinted with corrections 1991) *Freedom and Belief* (Oxford: Clarendon Press).
Strawson, G. (1989) 'Red and 'Red'.' *Synthese* **78** pp. 193–232.
Strawson, G. (1994) *Mental Reality* (Cambridge, MA: MIT Press).
Strawson, G. (1997/1999) '"The Self"' in *Models of the Self* edited by S. Gallagher & J. Shear (Thorverton: Imprint Academic).
Strawson, G. (1999a) 'Realistic Materialist Monism' in *Towards a Science of Consciousness III*, edited by S. Hameroff, A. Kaszniak & D. Chalmers (Cambridge, MA: MIT Press).
Strawson, G. (1999b) 'The Self and the Sesmet' in *Journal of Consciousness Studies* **6** pp. 99-135.
Strawson, G. (2002) 'Knowledge of the world' in *Philosophical Issues* **12** pp. 146–75
Strawson, G. (2003a) 'Real materialism' in *Chomsky and his Critics*, edited by L. Antony and N. Hornstein (Oxford: Blackwell).
Strawson, G. (2003b) 'What is the relation between an experience, the subject of the experience, and the content of the experience?' in *Philosophical Issues* **13** pp. 279–315.
Strawson, G. (2005) 'Intentionality, terminology and experience' in *Phenomenology and Philosophy of Mind*, edited by D. Smith and A. Thomasson (Oxford and New York: Oxford University Press).
Strawson, G. (2007) 'Selves' in *Oxford Handbook in the Philosophy of Mind* edited by A. Beckermann and B. McLaughlin (Oxford: Oxford University Press).
Strawson, G. (2008a) 'Real intentionality 3' in *Real Materialism and Other Essays* (Oxford: Oxford University Press).
Strawson, G. (2008b) 'Can we know the nature of reality as it is in itself?' in *Real Materialism and Other Essays* (Oxford: Oxford University Press).
Strawson, G. (2011a) *The Evident Connexion: Self, Mind and David Hume* (Oxford: Oxford University Press).
Strawson, G. (2011b) *Locke on personal identity* (Princeton, NJ: Princeton University Press).
Strawson, P. F. (1959) *Individuals* (London: Methuen).
Strawson, P. F. (1974/2004) *Subject and Predicate in Logic and Grammar*, reprinted with a new introduction (Aldershot: Ashgate).

Strawson, P. F. (1985) 'Causation and Explanation' in *Essays on Davidson: Actions and Events*, edited by B. Vermazen and M. Hintikka (Oxford: Clarendon Press).
't Hooft, G. (2006) 'The mathematical basis for deterministic quantum mechanics' www.arxiv.org/quant-ph/0604008
van Inwagen, P. (1990) *Material Beings* (Ithaca, NY: Cornell University Press).
Weinberg, S. (1997) 'Before the Big Bang' in *The New York Review of Books* (June 12).
Yablo, S. (1990) 'The real distinction between mind and body' *Canadian Journal of Philosophy* supplementary volume **16**, pp. 149–201.

Postscripts

Sam Coleman

A Panqualityist Manifesto

A visual field could in a congenitally blind person just lie there — like a hidden pool in a corner of the mind: multi-coloured, unchanging, unremarked, perhaps never to come to the attention, forever dormant. This highly unnatural state of affairs is a real possibility, and that it is opens up an all-important space between consciousness and the visual field, even though the visual field of its nature lies open to consciousness.   Brian O'Shaughnessy, *Consciousness and the World*, p.502

We can begin at once with the peculiar predilection of scientific thinking for mechanical, statistical, and physical explanations that have, as it were, the heart cut out of them . . . [P]eople ceased trying to penetrate the deep mysteries of nature as they had done through two millennia of religious and philosophical speculation, but were instead satisfied with exploring the surface of nature in a manner that can only be called superficial . . . the great Galileo Galilei, always the first to be mentioned in this connection, eliminated the question of what were nature's deep intrinsic reasons for abhorring a vacuum and consequently letting a falling body penetrate space after space until it finally comes to rest on solid ground, and settled for something far more common: he simply established how quickly such a body falls, the course it takes, the time it takes, and what is its rate of downward acceleration . . . However disconcerting it may sound nowadays to speak of someone as inspired by matter-of-factness, believing as we do that we have far too much of it, in Galileo's day the awakening from metaphysics to the hard observation of reality must have been, judging by all sorts of evidence, a veritable orgy and conflagration of matter-of-factness! But should one ask what mankind was thinking when it made this change, the answer is that it did no more than what every sensible child does after trying to walk too soon; it sat down on the ground, contacting the earth with a most dependable if not very noble part of its anatomy, in short, that part on which one sits.   Robert Musil, *The Man Without Qualities*, pp.326–27

Introduction

A monistic theory of the mind, including consciousness, and how it relates to the physical body is implicitly a theory of all Reality — of the existing concrete universe and its nature. For what is required of this theory is to fit the conscious mind into the physical universe that, one way or another, provided the conditions for the formation of the body which carries the mind, and within which that minded body is deeply embedded, enmeshed — the universe which nourishes that body and mind, is perceived by them, acted upon by them, and so on. That's why the mind–body problem is the world-knot.[1]

There is reality *inside* one's mind — by which I mean individual medium macro-minds like ours — and reality *outside* one's mind. These two smaller realities which together make up all Reality are, in at least some ways, different from one another: all realist theories acknowledge this, since the only view that denies this discontinuity is solipsism, on which there is no reality outside one's mind. It is this difference between inner and outer reality that makes the mind–body problem so difficult. For, since a monistic theory of the mind is also a theory of all Reality, it must strike a sweet spot that is exceedingly hard to hit: it must posit a ground for mentality, including consciousness, that is at once a ground for the physical world. It must do justice to both, at the expense of neither. Justice here means *full justice*.

In 2006 Galen Strawson helped to put panpsychism back on the map.[2] The variety of panpsychism he defended was what I then called 'smallist'[3] — it conceived of our world as more akin to a bucket of shot than a bucket of jelly, in Russell's phrase, with the facts and features pertaining to the smallest entities determining those of the entities they compose. Strawson's panpsychism populated reality with micro-instances of consciousness, whose composites, in the right arrangements, included conscious minds like ours. Strawson's argument for panpsychism placed great emphasis on inner reality, for his reasoning was that unless physical matter has consciousness built in it is in no shape to constitute conscious minded bodies. Claiming that this was possible was physicalism's big error, he maintained — even attributing to physicalists a sneaking eliminativist attitude toward the mind.

Since 2006, in some ways recapitulating much older debates, many authors have highlighted problems for Strawson-style constitutive smallist panpsychism — *constitutive* in that it aims to explain how conscious macro-minds are intelligibly formed from micro-things, as opposed to emerging from them. Notable among these were 'combination problems', and most notable among the combination problems

---

[1] Schopenhauer's term, see Griffin (1998). Cf. Nagel (1986), pp. 52–3.

[2] See also Nagel (1979), Chalmers (1996), Strawson (1994), Griffin (1998).

[3] Coleman (2006), Goff (2022).

was the problem of understanding how many conscious subjects could come together and, in and of themselves, make up a larger subject without achieving their own annihilation — something composite *and yet* single in respect of consciousness.[4] With Strawson's contribution of impetus panpsychism was propelled high and wide. But then, faced with these problems, many, perhaps most, advocates have come back down to ground and abandoned smallist panpsychism, essaying instead emergentist, or non-smallist 'cosmopsychist' versions.[5] These face their own problems. Some have moved to or explored neighbouring theories, non-standard (some might say oxymoronic) 'Russellian' forms of physicalism, neutral monism, idealism.[6] Amidst all this activity — early promise slowly giving way to frustration — eliminativist physicalism has once more, unsurprisingly, seen its own rise.[7]

After setting out some groundwork I will make a case for one of these alternatives: *panqualityism*.[8] I will do this by showing that panqualityism, of all the theories at our disposal, and especially as compared with panpsychism, provides the most comprehensive vision of inner and outer reality, hence of Reality.

1. Terms and Theories

Noting first that people are free to define terms and names for theories as they see fit, let me engage in some discussion of how I see fit to define some relevant terms and names for theories, before getting into more substantial matters.

Panpsychism, if we are guided by etymology, has it that mind is everywhere. But by contemporary usage it is, more strictly, the claim that *consciousness* is everywhere, and that it is fundamental.[9] Now, panpsychism is often ridiculed by allegations that it makes rocks, trees, cars, etc. — *all* clumps of conscious particles — themselves each into a consciousness. Some panpsychists embrace this consequence, but many try to avoid it, saying that only *special organizations* of the

---

[4] For a taxonomy of combination problems see Chalmers (2016). Cf. Coleman (2016). See also Goff (2006), Coleman (2014), James (1890/1981), Roelofs (2019).

[5] Brüntrup (2016), Seager (2016), Mørch (2018), Goff (2017), Shani (2015), and arguably Strawson himself, who places less emphasis on the constitutive explanatory virtues of panpsychism these days (see e.g. his 2016).

[6] Montero (2015), Stoljar (2001), Coleman (2016), Chalmers (2019).

[7] Frankish (2016), Kammerer (2021).

[8] Advocates include Mach (1897/1959), James (1904), Russell (1927), Feigl (1975), Coleman (2015), and Cutter (2018).

[9] Goff (2022).

micro-consciousnesses generate macro-consciousness. Which special organizations, though? The best candidates are of course brains, with their capacities for representing, cognizing, information processing,[10] and so on. So that's why brains are conscious and coffee cups are not.

The first, promiscuous, kind of panpsychism is false if there are any pockets of non-consciousness in the universe, among which would figure, notably for my purposes, pockets of *unconscious mentality*. But the second, more chaste, panpsychism also faces a problem regarding unconscious mentality. The problem is this. If the special organization — the recipe — that combines micro-conscious items into macro-consciousness has to do with the brain's distinctive material arrangement and consequent *capacities for representation*, information processing etc., how could *those* capacities and *that* organization obtain without consciousness, as they would do by hypothesis if there is such a thing as unconscious mentality — unconscious beliefs, cognition, desires, perceptions, proprioceptions, and such? The special organization chaste panpsychism invokes as grounding macro-consciousness is just that sort of organization which, more widely, is usually taken to ground *mentality per se* — cognition, representation, and so on. But then unconscious mentality and cognition would seem to be impossible by the lights of even chaste panpsychism, hence for panpsychism of any stripe. Where we would find mentality we would *ipso facto* find consciousness. If the brain is so organized as to produce mental content or representations of any kind, then that would seem to have to be conscious content or representations, given panpsychism — for how could a mental content or representation formed of conscious ingredients fail, itself, to be conscious? This problem is related to the fact that panpsychism nowadays equates to pan-consciousness, and, additionally, that panpsychists, at least tacitly, equate mentality with consciousness. That makes unconscious mentality very hard, if not impossible, for them to compass.

Idealism has been defined as the view that the universe is fundamentally mental, or that all concrete facts are grounded in facts that exclusively involve mental properties.[11] Contemporary panpsychists seem to take it that instantiating consciousness is *eo ipso* instantiating a mental property. But the basic, 'diminished', instances of experience

---

[10] Tononi's Integrated Information Theory associates consciousness with informational complexity, and would seem as such especially vulnerable to the criticism I present just below, though it is somewhat unclear whether IIT actually implies full-blown panpsychism (see e.g. Tononi and Koch, 2015).

[11] Chalmers (2019).

which panpsychists attribute to basic physical particles hardly seem to qualify as genuine mindedness. Minds as we know them are rather complex things: they involve sensitivity to the environment, representation, perception, thought, memory. These are the minds we know, and the sort of mentality we attribute to God, when we think about God. As Russell said, being mental is more similar to being a harmony — it is not something a point particle can be. Since panpsychists explicitly deny that electrons perceive, have worries, thoughts, memories, and so on, I think they should at most claim that *ultimate particles have experience of some kind but lack minds*. Particles have, then, an ingredient of mindedness — because, again, all the minds we know about are conscious at some time, and their capacity for consciousness is an important part of them. In sum, contemporary panpsychism — including Strawson's — can be equated with *panexperientialism*: the ubiquity of experience, if not of mindedness. Idealism, by contrast, places the metaphysical emphasis squarely on mindedness as such. Consider Berkeley's version, on which Reality consists of minds and their ideas, and ultimately all that exists is contained within God's mind. That is a truly mental universe.

Now a crucial point: minds are not limited to what is in consciousness. There is such a thing as *unconscious mentality*, and it would be mistaken to construe this as talking simply in terms of dispositions to conscious mentality.[12] There exists *occurrent unconscious mentality* — thoughts, beliefs, desires, memories. These interact, actively, with conscious states when we do things as simple as reason, converse, react fearfully to things that long ago scared us, manifest anger with colleagues and friends which we do not wish to surface, work out what we fancy for lunch, and fall in or out of love with someone.

Given this point, idealism can be quite opposed to panpsychism. Idealism is compatible with the notion that large chunks of Reality consist of God's *unconscious ideas*, for example, or with a universe of dreamlessly sleeping Leibnizian monads. Panpsychism, on contemporary usage, is not so compatible. Nor is its neighbour, cosmopsychism, which can construe the universe as comprising a massive single mind: that mind is still supposed to be conscious through-and-through, which is why cosmopsychists expend time arguing *against* unconscious sensory states and the like.[13]

---

[12] Which is not to deny that there are dispositions to conscious mentality. Below I argue that these should not be classed as mental just in virtue of their manifestations. This is a surprisingly controversial stance.

[13] Shani (2022).

Next we come to panqualityism. It can be picked out by comparison with a claim made by panpsychism, namely:

Reality is populated with properties of the same broad class as those of our acquaintance, which comport consciousness.

By this the panpsychist means *experiential* or *phenomenal* properties like perceptual consciousness of red, experiencing pain, feeling anger, and so on, except they don't say that properties of *exactly* these kinds are had by particles. Rather there is an extension, by analogy, to conceptions of far simpler sorts of experience supposedly had by the smallest entities.[14]

Panqualityism, which is also a smallist constitutive theory, makes a closely related claim:

Reality is populated with properties of the same broad class as those of our acquaintance, which do not comport consciousness.

The difference between panpsychists and panqualityists, then, is that panpsychists see consciousness or experience as *built into* qualities (or qualitative states) like perceptual redness, pain, and anger; the kinds of quality we are often conscious of. Panqualityists disagree. They think that not only can we conceive of such properties — qualities — as existing without conscious awareness, but that we have good reason to believe they actually do so exist — and often. Perceptual redness by itself, on this conception, is not a phenomenal property — not a property the having of which by itself constitutes an experience. It is more like the *possible content* of an experience or episode of awareness, a content that can exist without our being aware of it. The properties of the 'ultimates' are qualities in the sense, then, of being conceivable contents of experience for some subject, however small.[15] Awareness, or consciousness, itself, is thus construed as something distinct, so that a full-blown episode of experience involves these two factors: consciousness and qualities.[16]

People profess to find the notion of unexperienced qualities very perplexing, even incoherent. I am consistently surprised by this attitude. My reason is that most philosophers seem to have little trouble making sense of *naïve realism* about perception — or, at least, if they fail to find naïve realism sense-making that is not on account of its having to do with unexperienced qualities. But naïve realism has *everything* to do with unexperienced qualities. A naïve realist says the

---

[14] Goff (2022), Rosenberg (2004).

[15] Lockwood (1989).

[16] A distinction Kriegel (2009) accepts in conceptual terms but rejects as metaphysically impossible. Cf. Rosenthal (2005).

red colour quality I experience on being perceptually conscious of a ripe tomato is contributed by the tomato's redness itself; as the slogan goes, qualia ain't in the head. Fine. But I can close my eyes on what's not in my head. Does the tomato's redness disappear, cease to exist as such, when I close my eyes (or we all close our eyes — or aren't around in the quad where the tomato sits)? No — naïve realists don't say that, because *they are not Berkeleian idealists*. They don't hold that *esse est percipi*. But, then, when the tomato's redness — remember the redness *is* the quality, quale, which I experience with my eyes open — exists unseen *it is an unexperienced quality or quale*. So if naïve realism even makes sense, regardless of its truth or falsity, then unexperienced qualities make sense. Panqualityists tend to — but need not — hold that the qualities of which we intermittently become aware exist inside the head.[17] But otherwise the view can be remarkably similar to direct realism — panqualityist Michael Lockwood labels his view 'inner direct realism',[18] and maintains, with Russell, that what we are immediately aware of in consciousness are qualities belonging to our brains — painted brains, agonized brains, happy, smelly, loving brains.

I am not at all ashamed to say such things, and, as I'll now argue, panqualityism is preferable because of all theories in the ballpark it makes best sense of reality inside and out, hence of Reality. My focus will be on comparing panqualityism with Strawsonian panpsychism.

2. reality Inside

Panpsychism struggles with unconscious mentality. Strawson (2006) is rightly up in arms about standard physicalism's proclivity for demeaning or eliminating consciousness — often eliminating it in effect, just by giving a glaringly inadequate account of it. But we should be up in arms, too, about any theory that demeans or eliminates the unconscious — that is, unconscious *mentality*. There is an unconscious side to the mind — perhaps the larger side. We have this from Freud. We have this from modern cognitive science. *We have this from common sense.* People do not lose all their beliefs and desires when dreamlessly asleep, nor do sleepers cease to be *people*, qua mental beings. We are awoken by pains; but pains which bring us to consciousness — to feeling — cannot themselves have been being felt. Migraines can last for days — mercifully, one need not feel them all

---

[17] See Coleman (2015, 2016), Russell (1927), Cutter (2018).

[18] *Op. Cit.*

the time. When we talk, when we infer, when we ponder, act, these activities involve real chains of content that dip in and out of consciousness — for it is absurd to hold that everything that is relevant to one's drawing a certain conclusion, or coming to a certain course of action, or resolution, flows in the stream of consciousness. That stream is fast, but rather narrow, by all accounts.[19] We can be angry all day without feeling it — and it takes someone else to tell us, with whom we *have been being angry*. It can take therapy to realize that one is depressed, or loathing. As the therapy concludes the depression or loathing does not appear *ex nihilo* to consciousness — it surfaces. It was there, waiting to be uncovered.

I argued above that panpsychism is on the face of it incompatible with unconscious mentality. What do panpsychists and their kin say, in practice, about unconscious mentality? None allow that the very qualities we are aware of in thinking, perception, emotion, and so on, could exist unconsciously as such — hence from the get-go they proffer some sort of bifurcated picture of the mind. Freud, too, denied there were unconscious qualities — but he had the sense of propriety to acknowledge that the conscious mind had better be *continuous in nature* with the unconscious mind. So he made the whole thing neural in nature — qualities for him are just the pleasing, if illusory, and wholly epiphenomenal, way we represent the brain to ourselves.[20] This is little short of illusionism — but at least it keeps the mind in one piece. Those of the contemporary panpsychist ilk do not want to forgo the insight that qualities matter to the conscious mind: what we think, feel emotionally, how we act, infer, perceive; all that which in respect of intentionality and causality turns on qualities. But they bar qualities from the unconscious. So what do they replace them with? Neural dispositions to conscious qualities — the Searlian/Strawsonian solution from the 1990s.[21]

In truth this is an old idea: ironically, younger Freud also held that in the unconscious there exist, at most, dispositions to consciousness, but he did this *expressly to eliminate unconscious mentality*. The Searles and Strawsons, and those who now follow their lead,[22] purport to give an account of unconscious mentality. But their theories really

---

[19] See Strawson (2009).

[20] Wakefield (2018).

[21] Searle (1992), Strawson (1994).

[22] E.g. Kriegel (2011), Mendelovici (2018), Smithies (2019). But see Pitt (2016) for refreshing candour about this strategy — he tellingly observes that a dispositional sensation is not a sensation; just so, 'dispositional mental content', especially on a phenomenal intentionality account, is *not* mental content. Cf. Coleman (2022).

give this aspiration the lie. Searle affirms, revealingly, that it's merely a terminological matter whether we say that someone unconsciously believes that *p*, or just that there's a neural state in them poised to produce the conscious belief that *p* if they are ever prodded by the question whether *p*. Strawson thinks we have unconscious mental states in the sense in which there is music on a CD. But there *is* no music on a CD! Dispositions to x are not x, only potential x. Potential x is not x. So, potential qualitativity, potential mentality, is no qualitativity, no mentality at all. That would be like calling potential pain real pain, believing that wishes literally made one rich, or saying that a dormant bomb, with the fuse perfectly intact, is exploding *right now*, 'in a sense', thanks to its disposition to explode if lit.

Despite advocates' protestations of realism,[23] this eliminativism is, we can see, properly in keeping with panpsychism: if one conflates mentality and experientiality, it is going to be a struggle, and a pointless one, to accommodate genuine unconscious mentality.

And this is where panqualityism really earns its corn, because it offers us a realist account of (occurrent) unconscious mentality that is continuous with conscious mentality, and which thus makes best sense of its existence and their interaction. Mentality is the organization — a Russellian harmony — of qualities, by brains; an organization apt to carry content, to represent Reality: to recall it, to think about and act upon it. But these qualities can exist unconsciously. So unconscious mentality is real, occurrent, not-merely-as-if or dispositional. Panqualityism and its associated doctrine of the qualitative nature of unconscious mentality can make the mind all of one piece — doing justice to mental life as a whole. Panpsychism cannot. Crucially, here, panqualityism can give a univocal account of mental *content*, as qualitative; hence of the contentful interactions of conscious and unconscious mental processes. If, therefore, panqualityism does just as well with outside reality as panpsychism, as I shall briefly show below, it is preferable overall.

Epistemic humility is congenial to Strawson at choice moments — e.g. when it comes to the combination problem that threatens to derail panpsychism. We should be more humble about consciousness itself. We do not *know* that it accompanies every possible instance of qualities. Introspection cannot tell us *that*. All we know is that consciousness comes with such qualities *when we are conscious of them*. This is hardly a surprising or earth-shaking datum. And it no more licenses

---

[23] Though Kriegel and Mendelovici are fairly frank about the eliminativist trajectory of their accounts, in fairness.

the belief that unconscious qualities cannot exist, than cats are licensed to believe that milk cannot exist outside their bowls — which is, simply, where they find it. Saying qualities *must* be conscious might be like saying that water must be liquid, potable, and fall from the sky. Kripke taught us that when it comes to a kind's nature, its most salient properties — those prominently associated with our encounters with it — may not be in its deep essence.[24] Consciousness could just be a reference-fixing property, quite contingent, of qualitative characters. We do not know whether this is true or not directly, since the only samples of qualities we get hold of are conscious. What we need to do is see where the best, most coherent, most satisfying theory takes us and what it says.

But how qualities appear is how they are, so this is no mistake!

Panqualityists do not, or need not, deny this. But how qualities appear is *as qualitative*. To say that, in appearing via consciousness — in appearing at all — qualities' appearing, their being conscious, is itself part of the appearance, is to beg a weighty question. People can, if they want, define qualities into consciousness — as qualia — but then that's a definition that only begs the question, too. Definitions, perhaps surprisingly, *can* beg the question — when they latch onto a real-world phenomenon and frame it as we wish it to be, as opposed to how it is. Just look up the definition of democracy.

3. reality Outside

Panqualityism shares the *considerable* virtues of pansychism when it comes to making sense of reality outside the mind. It offers a picture of the world as continuous in nature with the mind. Russell pointed out how such a continuity thesis was helpful for comprehending the process of perception — how the world enters the mind. This also helps to explain how, when a small portion of the world is rearranged in the right way, qualitative mentality is the product. Panqualityism and panpsychism, further, both offer answers to the question — again influentially posed by Russell — of what there is to matter other than that part physics reveals, a part that falls manifestly short of providing a sufficient basis for mentality.

4. Reality

But, now, panpsychists, in equating consciousness with mentality, and in saying that consciousness is everywhere, wish to imply that

---

[24] Kripke (1980).

mentality is everywhere. I have said that consciousness *simpliciter* does not suffice for mentality. Consciousness is an *ingredient* of mentality, mindedness — having a mind: but an ingredient of mentality is no more mentality than an ingredient of lasagne is lasagne. But panpsychists ought not to hold that reality outside the mind is mental in any case. There is just no need for this, in order to make the world into one in which macro-minds can arise. If consciousness, phenomenal qualities, is the carrier of mental content — if to be intentional is to be phenomenally-qualitative, as Strawson additionally holds, then every particle in the universe is a bearer of intentionality. Unless we *want* to be pantheists, there is just no need to say this. That is not duly to respect the difference between Reality inside and outside. A theory that hits the sweet spot will respect everyday intuitions sufficiently to be able to say that outside reality is non-mental, whilst nevertheless beautifully suited for mentality in organization. It will make inner and outer reality at core continuous, but will have the resources to say that the inner is intentional and mental, some of it conscious some of it not, while the outer is physical and non-mental/intentional. That sweet-spot theory is panqualityism. Call this the 'sweet spot argument' for panqualityism.

But what of consciousness? What makes a bunch of qualities really subjectively 'for me'?

We need to be careful about how much of a premium we place on consciousness. Of course there is consciousness, subjective character — of course some qualities are *for me* and not *for you* and some are *for you* and not *for me*. But, as I've explained, if we cram consciousness into base matter we will eliminate part of the mind — the unconscious. So matter, in itself, *must* lack consciousness. Panpsychism, then, is flatly false.

But panpsychism as such does not even explain the much-touted subjective for-me-ness. Let us say that all the qualities, all the enqualitied ultimates, in the universe are aflame with consciousness. Still, what makes one clump of those *for-me*, and another, distinct, clump subjectively *for-you*? Why do you experience *those* ones but I experience *these* ones? What, in other words, is the panpsychist account of individual macro-*perspectives*? The answer cannot be that those ultimates are for you, fall within your perspective, which form 'your' mind — for since the panpsychist conflates your mind with your conscious mind, which ultimates you consciously experience (and which not) and which compose your mind (and which not), and why, is just one and the same issue. Hence the panpsychist, on top of saying all ultimates are conscious, in any case owes us an account of the extra relational property, the metaphysical lasso, that harnesses sets of (already conscious) ultimates into distinct perspectival macro-subjects. The panqualityist, too, must propose some such relational property — this will be the property that confers consciousness on enqualitied ultimates in the first place (for instance, a relation of higher-order representation). But that relation will do double-duty: it will capture consciousness and for-me-ness/macro-perspectivalness in one. And that is as it should be, for, *contra* panpsychism, they *are* as one. So even when it comes to consciousness, panqualityism is to be preferred to panpsychism. And, as I've explained, panqualityism can, where panpsychism seemingly cannot, make room for our unconscious mentality.[25]

References

Brüntrup, G. (2016), 'Emergent Panpsychism,' in G. Brüntrup and L. Jaskolla (eds.) *Panpsychism* (New York: Oxford University Press).

Chalmers, D. J. (2019), 'Idealism and the Mind–Body Problem,' in W. Seager (ed.) *The Routledge Handbook of Panpsychism* (New York: Routledge), pp. 353–373.

— (2016), 'The Combination Problem for Panpsychism,' in G. Brüntrup and L. Jaskolla (eds.) *Panpsychism* (New York: Oxford University Press).

— (1996), *The Conscious Mind* (New York: Oxford University Press).

Coleman, S. (2022), 'The Ins and Outs of Conscious Belief,' in *Philosophical Studies* **179**, pp. 517–48.

— (2016), 'Panpsychism and Neutral Monism: How to Make Up One's Mind,' in G. Brüntrup and L. Jaskolla (eds.) *Panpsychism* (New York: Oxford University Press).

— (2015), 'Neuro-Cosmology,' in P. Coates and S. Coleman (eds.) *Phenomenal Qualities* (Oxford: Oxford University Press) pp. 66–102.

— (2014), 'The Real Combination Problem: Panpsychism, Micro-Subjects, and Emergence,' *Erkenntnis* **79**, pp. 19–44.

— (2006), 'Being Realistic: Why Physicalism May Entail Panexperientialism,' *Journal of Consciousness Studies* **13** (10–11), pp. 40–52.

Cutter, B. (2018), 'Paradise Regained: A Non-Reductivist Realist Account of the Secondary Qualities', *Australasian Journal of Philosophy* **96**, pp. 38–52.

Feigl, H. (1975), 'Russell and Schlick: A Remarkable Agreement on a Monistic Solution to the Mind-Body Problem,' *Erkenntnis* **9**, pp. 11–34.

Frankish, K. (2016), 'Illusionism as a Theory of Consciousness,' *Journal of Consciousness Studies* **23** (11–12), pp. 11–39.

Goff, P. (2022), 'Panpsychism,' in *Stanford Encyclopedia of Philosophy* (available at: https://plato.stanford.edu/entries/panpsychism/)

— (2017), *Consciousness and Fundamental Reality* (New York: Oxford University Press).

---

[25] Thanks especially to Madeleine Cohen for helpful discussion that aided in the preparation of this paper.

— (2006), 'Experiences Don't Sum,' *Journal of Consciousness Studies* **13** (6), pp. 53–61.
Griffin, D. R. (1998), *Unsnarling the World-Knot: Consciousness, Freedom, and the Mind–Body Problem* (Wipf and Stock).
James, W. (1904), 'Does 'Consciousness' Exist?,' *Journal of Philosophy, Psychology, and Scientific Methods* **1**, pp. 477–491.
— (1890/1981), *Principles of Psychology* (Cambridge, MA: Harvard University Press).
Kammerer, F. (2021), 'The Illusion of Conscious Experience,' *Synthese* **198**, pp. 845–66.
Kriegel, U. (2011), *The Sources of Intentionality* (New York: Oxford University Press).
— (2009), *Subjective Consciousness: A Self-Representational Theory* (New York: Oxford University Press).
Kripke, S. (1980), *Naming and Necessity* (Cambridge, MA: Harvard University Press).
Lockwood, M. (1989), *Mind, Brain and the Quantum* (Oxford: Blackwell).
Mach, E. (1897/1959), *The Analysis of Sensations* (Dover Publications).
Mendelovici, A. (2018), *The Phenomenal Basis of Intentionality* (New York: Oxford University Press).
Montero, B. (2015), 'Russellian Physicalism,' in T. Alter and Y. Nagasawa (eds.) *Consciousness in the Physical World: Essays on Russellian Monism* (New York: Oxford University Press), pp. 209–223.
Mørch, H. H. (2018), 'Is the Integrated Information Theory of Consciousness Compatible with Russellian Panpsychism?,' *Erkenntnis* **84** (5), pp. 1065–1085.
Nagel, T. (1986), *The View from Nowhere* (New York: Oxford University Press).
— (1979), 'Panpsychism,' in his *Mortal Questions* (Cambridge: Cambridge University Press).
Pitt, D. (2016), 'Conscious Belief,' in *Rivista Internazionale di Filosofia e Psicologia* **7** (1), pp. 121: 26.
Roelofs, L. (2019), *Combining Minds: How to Think about Composite Subjectivity* (New York: Oxford University Press).
Rosenberg, G. (2004), *A Place for Consciousness* (New York: Oxford University Press).
Russell, B. (1927), *The Analysis of Matter* (London: George Allen and Unwin).
Seager, W. (2016), 'Panpsychist Infusion,' in G. Brüntrup and L. Jaskolla (eds.) *Panpsychism* (New York: Oxford University Press).
Searle, J. (1992), *The Rediscovery of the Mind* (Cambridge, MA: The MIT Press).
Shani, I. (2022), 'Eden Benumbed: A Critique of Panqualityism and the Disclosure View of Consciousness,' in *Philosophia* **50** (1), pp. 233–56.
— (2015), 'Cosmopsychism: A Holistic Approach to the Metaphysics of Experience,' in *Philosophical Papers* **44** (3), pp. 389–417.
Smithies, D. (2019), *The Epistemic Role of Consciousness* (New York: Oxford University Press).
Stoljar, D. (2001), 'Two Conceptions of the Physical,' in *Philosophy and Phenomenological Research* **62** (2), pp. 253–281.
Strawson, G. (2016), 'Mind and Being: The Primacy of Panpsychism,' in G. Brüntrup and L. Jaskolla (eds.) *Panpsychism* (New York: Oxford University Press).
— (2009), *Selves* (Oxford: Clarendon Press).
— (2006), 'Realistic Materialism: Why Physicalism Entails Panpsychism,' *Journal of Consciousness Studies* **13** (10–11), pp. 3–31.

— (1994), *Mental Reality* (Cambridge, MA: The MIT Press).
Tononi G. and Koch C. (2015), 'Consciousness: Here, There and Everywhere?,' *Philosophical Transactions of the Royal Society B: Biological Sciences*, **370** (1668) (available at:
https://royalsocietypublishing.org/doi/10.1098/rstb.2014.0167).
Wakefield, J. (2018), *Freud and Philosophy of Mind* (Palgrave Macmillan).

Philip Goff
Can Experiences Sum?

It is ironic that my first publication was arguing against panpsychism, which then allowed me to pursue a career much of which I spent defending panpsychism. In fact, my views have not changed as much as this might suggest, and I still agree with the main thrust of the arguments in 'Experiences don't sum.'

The most obvious way in which my views have remained the same is that I'm still not sympathetic to *pure reductionist* forms of panpsychism, according to which my conscious mind is nothing over and above a complex arrangement of conscious particles.[1] A party is nothing over and above people partying, which is a reflection of the fact that the sentence 'There is a party' is just another way of describing people partying. In contrast, the statement 'Sara is feeling anxious' is not just another way of describing what Sara's conscious particles are up to. Statements about subjects of experience do not admit of analysis (either *a priori* or *a posteriori*) and I think this essentially rules out a pure reductionist panpsychist story.[2]

However, what I realized a few years after writing 'Experiences don't sum' is that panpsychists don't have to be reductionists. Instead, we can ascribe to particles basic capacities to combine into unified conscious wholes, where the whole is more than the sum of its parts. Of course, we would look to experimental science to tell us the conditions under which those combinational capacities kick in. If, for example, the integrated information theory turns out to be the correct view as to the correlates of macro-level experience, then we can infer that the combinatorial capacities kick in when there is more integrated information in a system than in its parts.[3]

I guess my thinking when I wrote 'Experiences don't sum' was, 'If you're not going to be a non-reductionist panpsychist, then you might

---

[1] See Roelofs (2019) for this kind of view.

[2] I have made this kind of argument in more detail in (2015a, 2016, 2017: 8.3, 2019). I also developed some of the critique of panpsychist forms of reductionism in Goff (2009).

[3] Mørch (2018) combines non-reductionist panpsychism with integrated information theory.

as well be a dualist.' But what I came to appreciate is that non-reductionist panpsychism has a clear simplicity advantage over dualism. Why believe in two kinds of fundamental property — mental and physical — when you can just believe in one? Furthermore, in contrast to the dualist, a panpsychist can embrace *partial* reductionism about the human mind. Whilst I doubt that human *subjects* can be reduced to their parts, the streams of consciousness borne by human subjects may be largely inherited from the level of fundamental physics. I say 'largely', as I agree with Thomas Nagel (Nagel 2012: Ch 4) that we cannot make coherent sense of reductionism about the forms of experience that constitute conceptual thought and understanding. My conscious grasp of a mathematical proof is not made up of lots of little partial understandings bundled together. But I see no problem with the idea that my *sensory* consciousness is made up of forms of experience at the level of basic physics. For both of these reasons, non-reductionist panpsychism is significantly more parsimonious than dualism.

Is any of this testable? It is possible in principle to empirically distinguish between the following three options:

- *Pure reductionism*: My mind and all of my experiences are reducible to facts at the level of fundamental physics. This view comes in a materialist version, according to which the nature of fundamental physical entities can be entirely captured in the language of physical science, and a pan(proto)psychist version, according to which the intrinsic nature of fundamental physical entities is constituted of forms of (proto)consciousness.
- *Pure strong emergentism*: Neither my mind nor any of my experiences are reducible to facts at the level of fundamental physics. This view comes in a dualist version, according to which the nature of fundamental physical entities can be entirely captured in the language of physical science, and a panpsychist version, according to which the intrinsic nature of fundamental physical entities is constituted of forms of consciousness.
- *Partial reductionism*: My mind and perhaps *some* of my experience (e.g. my experiences of conceptual thought and understanding) are not reducible to facts at the level of fundamental physics, but some of my experiences (e.g. my sensory experiences) are. Partial reductionism only comes in a panpsychist form.

Pure reductionist panpsychists/materialists think everything in the brain could in principle be predicted from knowledge of basic chemistry and physics, as everything in the brain is reducible to underlying chemistry and physics. Dualists think that *none* of the activity that results from my consciousness could be predicted from knowledge of basic chemistry and physics, as none of my conscious experiences is reducible to underlying chemistry and physics. Partial reductionist panpsychists occupy the middle ground of supposing that *some but not all* of the activity that results from my consciousness is thus predictable: in so far as my experiences are inherited from basic physics, their behaviour will be predictable from underlying chemistry/physics; in so far as, say, my thought experiences are not inherited from basic physics, it won't be possible to predict their behaviour from underlying chemistry/physics.

When I wrote 'Experiences don't sum', I had the impression, as many philosophers still do, that we have strong empirical grounds for accepting pure reductionism, whether of a panpsychist or a materialist form. Eighteen years later, having read a lot more neuroscience and spoken to many more neuroscientists, I now know that we're a long way away from knowing enough about the brain to test which of the above options is correct.[4] Whilst it's still an open question *empirically* which of the above options is correct, I think we have strong *philosophical* grounds for doubting pure reductionism, of either a materialist or a pan(proto)psychist form. The knowledge argument+ rules out materialist forms of reduction.[5] And the concerns I gestured at in the second paragraph of this postscript, to my mind, rule out a pure panpsychist reduction.

This leaves pure strong emergentism and partial reductionism, both of which I think are coherent possibilities. If partial reductionism does one day turn out to be true, that would confirm panpsychism over dualism. But even if pure strong emergentism turns out to be correct, i.e. if it turns out that the behaviour that results from human consciousness cannot be predicted at all from knowledge of chemistry and physics, although that would still leave open both dualist and panpsychist forms of pure strong emergentism, we would still have simplicity grounds for favouring the latter over the former (better to believe in one kind of fundamental property rather than two).

In other words, however the empirical facts turn out, pan(proto)-psychism of some form or other will always be the preferred option. If pure reductionism turns out to be true, pan(proto)psychism will be the

---

[4] See Cobb (2020) for how little we know about the workings of the brain.

[5] I use 'knowledge argument+' to refer to the knowledge argument in conjunction with a commitment to Revelation, see Goff (2015), Goff (2017, Ch. 5).

preferred option because of the philosophical problems with materialism (in this case, I would have to conclude that there is something wrong with my philosophical arguments against pure reductionism). If pure strong emergentism turns out to be true, then panpsychism has the advantage over dualism on simplicity grounds. And, given that partial reductionism only comes in a panpsychist form, if partial reductionism turned out to be true (as I suspect it will) then we will thereby have empirical confirmation of panpsychism. Given that consciousness is not a publicly observable phenomenon, it shouldn't surprise us that we might not be able to totally pin down the correct theory of consciousness experimentally.

But is partial reductionist panpsychism really coherent? What about the BIG PAIN/LITTLE PAIN argument in 'Experiences don't sum', which presses that my experience cannot be constituted of the experiences of a huge number of particles? Actually, I'm still very sympathetic to the broad thrust of these arguments too. However, in 2007 I had a chat in a bar in Canberra with Jonathan Schaffer, which led me to think that these problems go away if we adopt a 'top-down' rather than a 'bottom-up' form of panpsychism; to put it another way, if we frame panpsychism in a terms of a *field* ontology rather than a *particle* ontology. On a top-down form of panpsychism — AKA 'cosmopsychism' — the most basic forms of consciousness are the intrinsic nature of the universe-wide fields of basic physics. When a new subject, say, a brain emerges, it needn't inherit *all* of the experience from the physical fields located with it. It may rather inherit a highly structured selection of the experiences in those fields. I still think I was right in 2006 to cast doubt on the idea that what it's like to be me isn't the same as what it's like to be a huge number of particles. But I can't see any problem with the idea that what it's like to be me is the same as what it's like to be a highly structured selection of the much more complex experience of the fields in my brain.[6]

In any case, that's where I'm up to eighteen years after writing 'Experiences don't sum.' In some ways my thinking has changed a lot; in other ways, it's exactly the same.

References

Cobb, Matthew (2020), *The Idea of the Brain: A History* (Profile Books).
Goff, Philip (2009), 'Why Panpsychism Doesn't Help Us Explain Consciousness,' *Dialectica* **63** (3), pp. 289–311.

---

[6] I develop something like this argument in Goff (2017, 9.3.2), in Goff and Roelofs (forthcoming), and in more detail in Goff (MS). See also Goff (2020).

Goff, Philip (2015a), 'Against constitutive forms of Russellian monism,' in T. Alter & Y. Nagasawa (Eds.) *Consciousness and the Physical World: Essays on Russellian Monism* (Oxford University Press).

Goff, Philip (2015b), 'Real acquaintance and physicalism,' in P. Coates & S. Coleman (Eds.) *Phenomenal Qualities: Sense, Perception, and Consciousness*, (Oxford University Press).

Goff, Philip (2016), 'Fundamentality and the mind–body problem,' *Erkenntnis* 8: 4, 881–898.

Goff, Philip (2017), *Consciousness and Fundamental Reality* (New York: Oxford University Press).

Goff, Philip (2019),'Grounding, analysis and Russellian monism,' in S. Coleman (Ed.) *The Knowledge Argument Then and Now* (Cambridge University Press).

Goff, Philip (2020) 'Cosmopsychism, micropsychism and the grounding relation,' in W. Seager (Ed.), *The Routledge Handbook of Panpsychism*.

Goff, Philip & Roelofs, Luke (Forthcoming), 'In Defence of Phenomenal Sharing,' in L. Bugnon & M. Nida-Rumelin (Eds.), *The Phenomenology and Self-Awareness of Conscious Subjects* (Routledge).

Mørch, Hedda Hassel (2018), 'Is the Integrated Information Theory of Consciousness Compatible with Russellian Panpsychism?', *Erkenntnis* **84** (5), pp. 1065–85.

Nagel, Thomas (2012), *Mind and Cosmos: Why the Materialist, Neo-Darwinian Conception of Nature is Almost Certainly False* (New York: Oxford University Press).

Roelofs, Luke (2019), *Combining Minds: How to Think about Composite Subjectivity* (New York: Oxford University Press).

Georges Rey

Postscript

As a postscript to the preceding commentary on Strawson's (2006a) defence of 'panpsychism', I add here a brief reply to his (2006b) response to it, expanding on a few points.

An issue on which Strawson and I fully agree is the difficulty of explaining how what we ordinarily think of as consciousness could 'emerge' from non-mental physical phenomena in the way that we explain other macro phenomena such as liquidity. Strawson actually finds any such 'emergence' proposal for consciousness 'incoherent' (2006a, p. 12). While I wouldn't go that far, I do agree that there does seem to be a peculiarity, what Joseph Levine (2001) called an 'explanatory gap' between what is ordinarily thought of as conscious phenomena and the purely physical — in Strawson's (2006a, p. 13) terms– 'P-phenomena' of the brain.[1] I'll call the ordinary notion that concerns Strawson and Levine, 'S(trong)-consciousness', to distinguish it from a weaker, 'representational' notion, 'W(eak)-consciousness' that I'll discuss below, for which there seems to me not so great a gap.

This explanatory gap between S-consciousness and P-phenomena leads Strawson to extravagant conclusions:

> Real physicalists must accept that at least some ultimates are intrinsically experience involving. They must at least embrace micropsychism. (2006a, p. 25).

Indeed, since he

> would bet a lot against there being... radical heterogeneity at the very bottom of things ... that micropsychism is panpsychism ... All energy ... is an experience-involving phenomenon (2006a, p. 25).

What makes these claims extravagant is that there's simply not a scintilla of evidence that, for example, the kinetic energy of a falling stone involves experience, either on the part of the stone or on the part of its

---

[1] Strawson (2006a, fn6) oddly includes me among those who propose 'reductive analyses' of the experiential in nonexperiential terms. I am on record, however, for advocating *eliminativism* about at least S-consciousness in numerous places since 1983.

energy. It's not really clear what it even means to claim that energy is 'experience-involving', given that energy doesn't have a nervous system, and doesn't introspect or exhibit any other sign of experience. But, to be sure, people have historically often been quite free with the notions of consciousness and experience. Indeed, these ordinary notions seem sufficiently peculiar that I and others have argued for some years (see Frankish, 2016) for the serious possibility that nothing in fact answers to them in the actual world. That, at any rate, is a possibility that seems to me more intelligible and worth pursuing than panpsychism as an account of the explanatory gap.

One peculiarity about S-consciousness I have discussed is that, unlike most other worldly phenomena, there seems to be no non-question-begging evidence for it, that is, no evidence that doesn't simply *assume* the existence of S-consciousness as, say, a datum of introspection. In his (2006b) response to this worry, Strawson reiterated his earlier (2006a, fn6) claim that

> It is question-begging to request a non-question-begging argument for its existence, and have requested [a] non-question-begging argument for the existence of non-experiential reality. — (2006b, p. 269; see also 2006a, fn6)

Now, as I noted earlier, it's really not at all clear what question is being begged by asking for a non-question-begging argument. It's one thing to advance a *metaphysical* claim about the reality of *S-consciousness*, quite another to have *epistemological* views about the *possible evidence* for it. Evidence in general consists in effect of a phenomenon that couldn't be satisfactorily explained without it, and, per fn5 above, the demand that it be non-question-begging arises at least where there are *prima facie* reasons for accepting either of two competing hypotheses. This seems to me so in the case of choosing about S- vs. W-consciousness, as I hope what follows makes clear.

To avoid problems about characterizing 'physicalism', consider a far more modest claim, what I call:

> *Simple Bodily Naturalism* (SBN): Every human or animal bodily state and motion can be explained as the result of bodily states and events along standard bio-physio-chemical lines of current natural sciences.

Note that this makes no claim about the whole of physics, much less the entire universe, speculation about which is really beyond the pay grade of any mere philosopher of mind. It even allows that physiology and biology might not be, strictly speaking, part of *physics*: special sciences, after all, may well not be 'reducible' to physics, even if they bear reliable explanatory relations to it (cf. Fodor, 1975, chap 1). So

SBN is not, to use Strawson's (2006a, p. 4) term, 'physicSalism'; and it is certainly not any sort of article of 'faith' (2006a). SBN is a banality that is tacitly confirmed in any physiological laboratory, where one can find overwhelming non-question-begging evidence for the existence of all manner of phenomena, e.g. cell growth and repair, reflex responses, metabolism, digestion, etc. that certainly don't *seem* to require experience.[2]

Now, so stated, SBN makes no mention of any mental phenomena. But it does seriously constrain what theories we may offer of them. Any posited mental phenomena that aren't in some way included in the purview of SBN would be explanatorily superfluous, since all bodily states and motions could be explained without them. If there *are* mental phenomena, then, one way or another, they do need to 'emerge' from the bio-physio-chemical ones.

In light of SBN, one might, of course, wonder what non-question-begging reasons there are to believe in the existence of *any* mental phenomena. A thought frequently expressed in the 1950s and 60s was that such reasons couldn't be provided, since the usual *actions* one cites are not mere *physical movements,* but presuppose, e.g., *intentions.* Indeed, mental talk was thought to express more an 'attitude' towards other people than any sort of 'theory' of them (cf. Wittgenstein, 1953/68, p. 178e). Be that as it may, as I discussed in my (1997, chap 3), non-question-begging evidence for the existence of at least *some* mental states is in fact readily available in the common 'standardized' tests (such as SAT or GRE) regularly administered to university applicants: physically identical 'questions' are paired with 'answers' in the form of graphite filled circles on sheets that are then scored by machines. Some eight hundred million students have suffered through such tests since WWII, and, even putting aside declining scores, the statistical regularities are both overwhelming and counterfactually supporting (e.g. 'answers' would likely be correspondingly reversed if 'questions' were), both between 'questions' and 'answers' within tests, and between the 800 million students. It is

---

[2] The physicist Richard Feynman (1963 [2010]) made a stronger claim than SBN: '[*T]here is nothing that living things do that cannot be understood from the point of view that they are made of atoms acting according to the laws of physics'* (vol I, pp. 1-9, emphasis original), which, for all its plausibility, I see no reason to defend here. Note that neither this claim nor SBN make any mention of 'experience'. In response to Strawson's (2009b, p. 269) demand for non-question-begging arguments for such non-experiential claims, one need only note that they are both parts of the most fruitful research programmes in history. There are no idealist alternatives that are remotely plausible competitors (how is the growth of teeth to be explained by ideas?). I'll address the nature of the 'sensory' evidence for scientific theories below.

virtually impossible to imagine any serious alternative explanation of these statistics other than that the students enjoyed some sort of at least *propositional attitudes states*, such as beliefs and desires, and engaged in various forms of (sometimes fallacious) reasoning involving them.[3]

Now, propositional attitude states are the kinds of states that are studied by computational-representational theories of thought (CRTTs), whereby cognition is treated as causally efficacious computation over representations that are entokened in a creature's brain. This is a promising strategy for understanding how, within SBN, P-phenomena could in principle be capable of thought, since computational processes can be realized in P-material.[4] So there no longer appears to be an unbridgeable 'explanatory gap', at least not for those states.

But what about qualitative states, such as experiences of sensations and S-consciousness? Here there do seem to be problems, since explaining phenomena such as the SATs doesn't obviously require them: computations over, e.g., visual (analogue or digital) *representations* of stimuli and motor commands would seem to suffice. Moreover, standard psychophysical results, such as the Fechner-Steven laws relating perceived to actual amplitudes of physical stimuli, can also be understood as involving simply (meta-)representational states, of, e.g., the magnitudes of those stimuli. Introspection can be regarded as simply a process whereby certain representations are made available, perhaps in a 'global workspace' for attention and verbal reports (cf. Baars, 1997). This all suggests a promising strategy for capturing what I call 'W(eak)-consciousness', or states that play at least the functional role that consciousness plays, but patently without any in principle explanatory gap (see my 1997, chap 5 for fuller discussion). To address what might be leading Strawson (2006a, pp. 4–5) to insist that physics is committed to something more, note that all of the perceptual evidence for SBN theories can be understood in terms of just such *representations* of observed stimuli, whether or not they are accompanied by S-consciousness.

---

[3] Note that, even if every 'question' and 'answer' are physical events that individually could be explained by physical laws, that doesn't entail that *the counterfactual supporting statistical regularities* could be; cf. Fodor (1975, chap 1).

[4] The first champion of such an approach was, of course, Jerry Fodor (1975, 1987, 2008). See my (1997) for a summary of arguments for it, and Quilty-Dunn *et al* (2022) for further recent defences. I set aside Fodor's (2000, 2008) worries about CRTT capturing full human *intelligence*, as well as his (2008) and Burge's (2022) discussions of the character of its specific syntactic *form*, as orthogonal to worries about consciousness.

Many of these representations may represent *actual* phenomena: the ducks in a pond, one's extended arms, one's acidic stomach. But many representations may be *illusory*, representing, well, 'nothing'; at any rate, no *actual* thing, as in cases of standard illusions of the Kanizsa triangle, 'phantom limbs' or 'cutaneous rabbits'. These latter 'things' are (to steal a term from Franz Brentano) 'intentional inexistents', 'things' we talk about and take ourselves often to perceive, but which (we may know full well) don't actually exist. It seems to me to be at least a possibility worth examining that sensations, such as the pain in one's tooth and S-consciousness itself, are just further, illusory intentional inexistents.

Nick Humphrey (2016, p. 118) discusses illusions, but goes on to claim that such illusions cannot extend to *feelings themselves*:

> How could you . . . be experiencing a feel that 'doesn't exist'? . . . [T]he very notion of this is absurd. When the sensation represents you as feeling a certain way about the stimulation, *that is all there is to it*. The phenomenal feel arises with the representation and *thereby its existence becomes a fact*. (2016, p. 19; cf. Kati Balog, 2016, pp. 43–5)

But it's hard to detect an *argument* here. Why think *what a mental representation represents* is somehow needed along *with it*? This doesn't occur in the Kanizsa case; why here? What difference would it make? (And if it did make some (non-tendentiously described) difference, why wouldn't that be non-question-begging evidence for it?) Positing any S-conscious sensations in addition to the representations of them would seem explanatorily gratuitous, and, worse, as I have also argued elsewhere (e.g. my 1997, 2016), it would invite an unwelcome first-scepticism about whether one has what it takes beyond merely *representing* oneself as having such things.[5]

Of course, it may *seem* to many of us that the existence of the S-conscious sensations themselves is immediately obvious to introspection, but, as I mentioned in my original commentary, Nisbett and Wilson (1977) and many others (e.g. Carruthers, 2011) have provided abundant reason to think introspection is notoriously susceptible to the influence of popular belief — and what could be more popular than that the idea that there's a special 'inner world' of mental 'things' one experiences? People's temptations to reify intentional inexistents are legend, from colours and rainbows, to triangles and squares, to words

---

[5] I sometimes like to joke that I had what it takes until 1982, when I first began entertaining doubts about consciousness, after which I became a 'zombie', although most of my friends say they find this surprising. But on what basis can they second-guess me?

and phonemes.[6] Why shouldn't Humphrey's and others' similar introspections be further cases in point? Merely *insisting* that S-conscious sensations are a requisite part of *representations* of them is no reason to believe in them if there's no independent evidence of them; as I stress in my 2020 §8.7, representations can be empty and refer to nothing real whatsoever.

To be sure, however, as Balog (2016, p. 48) has rightly stressed, no one has yet provided a completely convincing explanation of how W-conscious introspected representations could give rise to such illusions (the difference between red and green experiences remaining, to my mind, a difficult case in point). But note this is no longer a problem of *(meta)physics*, explaining *some special worldly phenomenon* beyond the reach of a CRTT or SBN, but simply a problem of *psychology*, explaining the peculiar *content of the specific representations* employed in our introspections and attributions of S-consciousness; and, given SBN, there must be *some* bio-physio-chemical difference between, e.g., red and green experiences, difficult though it may be to specify. Suggestions of such explanations can be found in, e.g., Wittgenstein's (and Chomsky's 2000, pp. 32ff, 137) worries about excessively referential views of meaning, Brian Loar's (1991) 'phenomenal concepts', and Dan Dennett's (1991) 'heterophenomenology'.[7] To be sure, there's serious work still to be done here; but I don't see why it shouldn't eventually succeed.[8] *Pace* Balog (2016), I don't see why it should require 'a miracle' to ensure that people are representing 'all those wonderful qualitative and subjective properties even though nothing in the world instantiates them' (p. 49). All manner of intentional inexistents may be quite wonderful. Representations that don't refer to real phenomena need not be, as she claims, 'meaningless mental junk' (p. 49); consider, e.g., *Don Quixote*.

Another peculiarity about S-consciousness that I've explored in my (1996) is that most people find it difficult to believe that a non-living artefact merely programmed along the lines of a CRTT would have

---

[6] I defend eliminativism about these phenomena in my (2020, §§9.1–9.2) based on the usual observer variability and failures of property preservation (e.g. of word segmentation in the acoustic stream). I discuss intentional inexistents in §8.7.

[7] I pursue Wittgenstein's and Chomsky's worries in my (2016), a kind of Humean 'projectivist' view of S-consciousness in my (1996), and a suggestion about special sensory predicates in my (1997); but I'm under no illusion that any such stories are yet adequate.

[8] Colin McGinn (1991) has, of course, suggested we may simply not have the cognitive equipment to solve the problems here. But given that serious theories of mind have been pursued for only about sixty years now, and continually confounded with all manner of issues of religion, therapy, dualism — not to mention speculations of panpsychism — it is likely far too early to tell.

any actual sensations and consciousness. For recent example, Peláez and Taniguchi (2016) have proposed a model of pain which they claim

> can be initially tested in a computer environment so that a personalized strategy can be planned depending on the subject's pain condition (2016 §5).

But I doubt they or anyone thinks that the computer itself is actually in pain when the model is run on it.[9] The issue can't be evaded by claiming that a model is always different from the phenomena it models: the puzzle is what would one have to add — and why? — to an implementation of the model on a computer for it to be in genuine pain, and why they don't include it in advancing what they take to be an account of it?

Here perhaps Wittgenstein (1953/68) was right that 'only of a living human being and what resembles (behaves like) a living human being' can one say it is conscious, thinking, etc. (§§111,281, quoted with approval by Chomsky, 2000, p. 44). Although I worry that such a remark may be implicitly influenced by some kind of biological vitalism — after all, doesn't SBN teach us that living things are just floppy bio-machines? — it may well be that our concepts of consciousness and qualitative states are ineluctably biased in this way (I know mine seem to be). In which case: so much the worse for them, since, so far as I've read, there is no necessary *explanatory* connection between S-consciousness and life.

In any case, there's a lot to be investigated about how we *think* of thinking things, either of others or of ourselves, and how we understand our introspections. This seems to me a far more promising avenue of research with regard to the explanatory gap than pressing extravagant metaphysical claims of panpsychism that, as Carruthers and Schechter (this volume, pp. 38–9) point out, simply presuppose what they purport to explain.

References

Baars, B. (1997), *In the Theater of Consciousness* (New York, NY: Oxford University Press).

Balog, K. (2016), 'Illusionism's Discontent', in Frankish (2016), pp 40–51.

Burge, T. (2022), *Perception: the First Form of Mind* (Oxford: Oxford University Press).

Carruthers, P. (2011), *The Opacity of Mind: an Integrative Theory of Knowledge* (Oxford University Press).

---

[9] As Joe Levine once put it to me, no one is about to form a society for the prevention of cruelty to computers.

Chomsky, N. (2000), *New Horizons in the Study of Language* (Cambridge: Cambridge University Press).
Feynman, R. (1963/2010), *The Feynman Lectures on Physics*: *The New Millennium Edition,* vol I (New York: Basic Books).
Fodor, J. (1975), *The Language of Thought* (New York: Crowell).
— (1987), *Psychosemantics* (Cambridge MA: MIT Press).
— (2000), *The Mind Doesn't Work That Way* (Cambridge MA: MIT Press).
— (2008), *LOT 2: The Language of Thought Revisited* (Oxford: Oxford University Press).
Frankish, K. (2016)(ed.), *Illusionism,* special issue of *Journal of Consciousness Studies*, **23** (11–12).
Humphrey, N. (2016), 'Redder than Red: Illusionism or Phenomenal Surrealism?', in Frankish, K. (2016), pp116–123.
Loar, B. (1990), 'Phenomenal States', in *Philosophical Perspectives: Action Theory and Philosophy of Mind*, ed. J. Tomberlin (Ridgeview).
Quilty-Dunn, J. Porot, N. and Mandelbaum, E. (2022), 'The Best Game in Town: The Re-Emergence of the Language of Thought Hypothesis Across the Cognitive Sciences', *Behavioral and Brain Sciences*, **6** pp. 1–55, doi: 10.1017/S0140525X22002849.
Rey, G. (1996),'Towards a Projectivist Account of Conscious Experience', in Metzinger, T. (ed.), *Conscious Experience*, (Paderhorn: Ferdinand-Schöningh-Verlag), pp 123–42.
— (2016), 'Taking Consciousness Seriously — As an Illusion', in Frankish (2016), pp.197–214.
— (2020), *Representation of Language: Philosophical Issues in a Chomskyan Linguistics* (Oxford University Press).
Strawson, G. (2006a), 'Realistic Monism: Why Physicalism Entails Panpsychism', this volume, pp. 3–31.
Strawson, G. (2006b), 'Panpsychism? Reply to Commentators with a Celebration of Descartes', this volume, pp 184–281.

William Seager

From Panpsychism to Neutral Monism . . . and Back Again(?)

Since the publication of Strawson's 'Realistic Monism: Why Physicalism Entails Panpsychism' (2006) and its swarm of replies in the *Journal of Consciousness Studies*, the remarkable reinvigoration of panpsychism continues apace. See e.g. Skrbina (2009; Brüntrup and Jaskolla 2016; Seager 2020), and here, for what it is worth, is the google ngram for 'panpsychism':

This recent surge of work in panpsychism may trace back to a chapter in David Chalmers's *The Conscious Mind* (1996). This at least is what sparked my interest in the view. But panpsychism has been of perennial interest, falling out of sight for only a few decades in the mid-twentieth century (for a thorough history see Skrbina 2017).

Through these decades, various forms of reductive materialism held sway in Western analytic philosophy but there was a constant undercurrent of resistance that focused on consciousness and its strange, seemingly rather arbitrary relation to the presumably mindless entities of basic physics, which — at least in the minds of many physicists and especially their philosophical handmaidens — plumbed the true, complete, and fundamental nature of reality and which provided the ultimate constituents of absolutely everything. No place for consciousness down there and seemingly no way, and never any physics-driven need to, generate consciousness as nature evolves from the early inchoate quark-gluon plasma to the complexity we observe today for the fundamental physics driving all this is the same throughout.

What Chalmers presented as the 'hard problem' of consciousness might have been recently under-appreciated but nonetheless, as Strawson has often pointed out, it was a vivid mystery throughout the 19th century exactly when the completeness of the physical picture was taking hold. Thomas Huxley said that 'how it is that any thing so remarkable as a state of consciousness comes about as the result of irritating nervous tissue, is just as unaccountable as the appearance of the Djin when Aladdin rubbed his lamp' (1866). I think Huxley is right. We should pause to wonder at just how very weird it is that in a world fundamentally devoid of it, a state of consciousness could arise.

John Tyndall put the same point thus:

> ... the passage from the physics of the brain to the corresponding facts of consciousness is inconceivable as a result of mechanics. Granted that a definite thought, and a definite molecular action in the brain, occur simultaneously; we do not possess the intellectual organ, nor apparently any rudiment of the organ, which would enable us to pass, by a process of reasoning, from the one to the other. They appear together, but we do not know why. (1870, p. 63)

Tyndall was, of course, basically a supporter of the new scientific materialist outlook but held that at bottom both matter and consciousness, as well as their linkages, were ultimately mysterious: 'On both sides of the zone here assigned to the materialist he is equally helpless. If you ask him whence is this "Matter" of which we have been discoursing . . . he has no answer' (p. 64).

The now largely neglected — though not by Strawson — 19th century physiologist, Emile du Bois-Reymond, enunciated a version of the hard problem and went so far as to declare it unsolvable in an infamous address to the Congress of German Scientists and Physicians in 1874 (du Bois-Reymond 1874). Although couched in terms of the classical physics of the 19th century, du Bois-Reymond articulates a version of the hard problem and the explanatory gap (Levine 1983). He sketches the reasons for it whose essence still informs recent panpsychism. The core idea is that it is impossible to deduce the presence of consciousness from even a complete fundamental physical description of the world. Du Bois-Reymond works in the context of classical, pre-relativity, physics — an antediluvian world view still tacitly lurking in the thought of many philosophers working today — but enunciates quite well the principle of the causal closure of the physical. In principle, a complete fundamental physical description of the world (or any isolated part) would provide the resources to explain everything that happens in the physical world. He is not worried about the phenomena of life which was, he wrote, merely an 'exceedingly difficult mechanical problem' (1874, p. 23) but 'we cannot, by means of any imaginable movement of material particles, bridge over the chasm between the conscious and the unconscious' (p. 28). Du Bois-Reymond is happy to agree that the physical state of the brain will give rise to consciousness but this transition is completely inexplicable.

That is one ultimate mystery of two he identifies. The other is that of the 'inner nature' of matter itself. Du Bois-Reymond even enunciates the Russellian monist's deepest hope:

> ... the question arises whether the two limits of our knowledge of Nature are not perhaps identical, i.e., whether, supposing we understood the nature of matter and force, we should not also understand how the substance that underlies them could, under certain conditions, feel, desire, and think (1874, p. 32).

Du Bois-Reymond could not share this hope, and ends his address with his famous pessimistic assessment: not merely *Ignoramus* but *Ignorabimus!*[1]

The worry here can be expressed in a number of basic arguments, all of them controversial though some are very familiar and well honed. Let us briefly review the arguments.

1. Subjectivity vs. Objectivity. Thomas Nagel argued that because consciousness is an inherently subjective feature of the world it could not be explicated in purely objective terms (see Nagel 1974). The victory condition for a 'reduction' or 'naturalistic explanation' of consciousness is for the reduction to be expressed in completely objective (or 'scientific') terms. Reminiscent of du Bois-Reymond, Nagel points out the 'subjective

---

[1] This pronouncement really rubbed people the wrong way and du Bois-Reymond was widely vilified. One perhaps surprising reaction came from mathematics. David Hilbert was convinced there were no unsolvable problems and in his 1930 radio address proclaimed what became his epitaph: 'Instead of the ridiculous *Ignorabimus*, our solution is, by contrast, "We must know. We will know"' (see McCarty 2004). A year later came Gödel's incompleteness results.

character of experience ... is not captured by any of the familiar, recently devised reductive analyses of the mental, for all of them are logically compatible with its absence' (1974, p. 436). This is in stark contrast to *all* other cases of emergence we know of. For example, the non-liquidity of water (at normal temperature and pressure) is logically incompatible with the properties of hydrogen and oxygen and the laws governing them.

2. Epistemic Access. The objective–subjective gap leads to the strange feature of the phenomenal character of experience that it cannot be known except by its instantiation in the knower. Frank Jackson (1982) leveraged this into the argument that the scientific or objective description cannot be complete because one could know all of it without knowing the nature of subjective experience (e.g. what it is like to see red). This highly intuitive argument reinforces the sense that there is an unbridgeable 'gulf' or explanatory gap between consciousness and the world as studied by physics.

3. Modal Variation. The absence and unknowability of subjective character in terms of the objective view of the world seems to imply that the undeniable linkages between, say, brain states and states of consciousness are not dictated by physical law alone. Maybe, instead, there is some kind of extra psychophysical law which, as John Searle put it, underpins the fact that 'biological processes produce conscious mental phenomena, and these are irreducibly subjective' (1992, p. 98[2]). Evidently, there is nothing in the physical laws which necessitates the production or appearance of consciousness, so there could be a possible world with our physical laws but different psychophysical laws. This leads to the 'zombie' argument (see e.g. Chalmers 1996, ch. 2): there could be a world just like ours physically, but lacking the subjective character side of reality. The existence of such a possible world follows from Nagel's observation, save for a posit of bizarre, inexplicable, and ad hoc facts about absolute necessity.

4. Vagueness and Consciousness. On standard pictures of the relation between matter and consciousness, it takes complex physical states to ignite consciousness. What these might be, despite our growing knowledge of some interesting though rough corre-

---

[2] This is what Searle said; it's not clear what he meant however since he frequently claims to be a materialist and holds that the emergence of consciousness is much like the emergence of liquidity. Strange, since obviously there is nothing irreducible about liquidity.

lations between brain states and states of consciousness,[3] we have no clue. But the candidates are things like neural synchronization or large-scale recurrent organization of neural processes. Whatever the neural candidate, call it ɸ, its general complexity will mean that there will be borderline cases where it is simply unclear whether the state is a ɸ or not (e.g. is the neural synchronization extensive enough). This will be a matter of vagueness, not ignorance. Just as some people count as borderline cases of 'tall', some neural states will be borderline cases of ɸ. But there are no borderline cases of consciousness, even the slightest, most inchoate, confused and dim feeling is totally a case of consciousness. There is a basic mismatch between the ontological status of consciousness and its neural surrogates.[4]

5. Consciousness and Value. Arguably, the *only* thing which possesses intrinsic value is consciousness or rather, more precisely, states of consciousness. More arguably, 'nothing can be intrinsically good unless it contains . . . consciousness' (Moore 1912). Undeniably, at least some states of consciousness are intrinsically valuable. But extant physicalist theories of consciousness endorse a thesis of multiple realizability, by which the identity of a qualitative state of consciousness is independent of its physical realizers (that is, a computer simulation of my brain is — on many accounts — enjoying the same state of consciousness as myself). This implies that the realizers are states with instrumental value. If I had some brain disease and the doctors replaced diseased neurons with electronic surrogates, and my consciousness thereby continued unchanged, I would not care about the replacement. The surrogate neurons are merely the means to the intrinsically valuable end of consciousness. By Leibniz's law something with instrumental value only cannot be identical to something with intrinsic value.[5]

All these arguments suggest that consciousness does not fit into the otherwise smooth system of integration with the scientific-physicalist picture of the world in which every phenomenon can, in principle, be explicated in terms of or 'grounded in' the resolution of the

---

[3] For a fascinating clinical take on this see Owen (2017).

[4] For discussions of the vagueness argument see Antony (2006), Simon (2017), Tye (2021).

[5] There has been lots of work on consciousness and value lately. The argument sketched here was floated in Seager (2001). See also Cutter (2017), Lee (2019), Siewert (1998, ch. 9).

phenomenon into more or less complex systems of fundamental physical entities.

It's a fairly quick step from this core problem of consciousness to panpsychism via a straightforward argument:

P1. Consciousness is real, it exists.

P2. Consciousness cannot be physically reduced (or fully explicated in purely physical terms).

P3. Nature does not exhibit radical emergence.

C. So, consciousness is a fundamental feature of the world, presumably ubiquitous in nature, which is panpsychism.

The premises range from highly to reasonably plausible. A few philosophers — the illusionists — deny that consciousness exists, e.g. Daniel Dennett (1991)[6] and Keith Frankish (2016). Illusionism is the tribute that materialism pays to the problem of consciousness. It is the last dodge in the face of the problem, but to say the least it is highly implausible. P2 is the burden imposed by the arguments above. P3 is more controversial. The non-existence of radical emergence is certainly not *a priori* (see Wilson 2021). But nature does not seem to make any such radical leaps and the leap to consciousness must be, by argument 4 above, sudden and hugely disruptive as well as apparently absolutely unique in nature (as Strawson 2006 argued, following a long tradition upholding continuity in nature). William Clifford put it thus:

> ... we cannot suppose that so enormous a jump from one creature to another should have occurred at any point in the process of evolution as the introduction of a fact [i.e. consciousness] entirely different and absolutely separate from the physical fact (1886, p. 265).

Clifford then notes that there is no way to stop this argument working back prior to the origin of life itself:

> we are obliged in order to save continuity . . . that along with every motion of matter, whether organic or inorganic, there is some fact that corresponds to the mental fact (1886, p. 266).

Panpsychism is thus a fairly reasonable position to take in the face of the problem of consciousness. On the other hand, it faces at least two

---

[6] Where we find the remarkable claim that 'There seems to be phenomenology . . . But it does not follow from this undeniable, universally attested fact that there really is phenomenology' (p. 366). Elsewhere he adds that subjects reports of their phenomenology constitute 'a fictional world' including the 'sounds, smells, hunches' etc. which the subject 'sincerely believes to exist' (p. 98).

major hurdles. The first, of perhaps dubious significance, is the flat implausibility of assigning some kind of consciousness to the fundamental units of physical reality. Many baulk at this but as David Lewis noted, it is hard to refute an incredulous stare. It does not seem to be *impossible* for the fundamental physical entities, whatever they might be, to have some spark of inner life (this truth is part of what makes the problem of consciousness so difficult — we don't really know anything about the physical conditions that can be associated with consciousness).

The second major problem is the infamous combination problem.[7] The problem arises when the panpsychist tries to cash the promissory note that panpsychism avoids radical emergentism. To succeed at this, there needs to an intelligible account of how the elementary portions of consciousness distributed over the physical fundamentals can lawfully combine, conjoin, fuse or otherwise come together into the kinds of complex consciousnesses with which we are familiar. It is hard to find any account that does not face serious objections.[8] Although I believe there are number of promising approaches to the combination problem it is far from clear that it can be solved.[9]

But instead of further articulating the panpsychist position I'd like to sketch a somewhat different approach, one which I think has the advantages of panpsychism without the disadvantages (I concede it might have an incredulous stare issue of its own). Notice that P2 in the above argument has an obvious loophole. What would happen if we relaxed the condition that consciousness be *physically* reducible?

One way to exploit the loophole is via some version of what has come to be called Russellian monism (for a recent overview see Goff and Coleman 2020), a view which has seen a parallel explosion of interest to that of panpsychism and is often presented as a version of panpsychism. Roughly speaking, a *non-panpsychist* Russellian monist posits an intrinsic, non-relational and non-mental nature 'behind or below' the observable physical world investigated by the physical sciences and further holds that this intrinsic nature is what gives rise to consciousness. Quite independently of the problem of consciousness, this intrinsic background is needed to provide the ground for all the relational structures which science investigates, and

---

[7] The problem is most trenchantly presented by William James (see 1890/1950, especially ch. 6; see also Seager 1995).

[8] Much has been written about the combination problem. Probably the best overview is Chalmers (2016).

[9] The deepest, and rather optimistic, look at the general issue of combining minds is Roelofs (2019).

to which its investigations are restricted. As Strawson points out, Arthur Eddington championed this line of argument, but took the panpsychist path:

> ... the exploration of the external world by the methods of physical science leads not to a concrete reality but to a shadow world of symbols, beneath which those methods are unadapted for penetrating. Feeling that there must be more behind, we return to our starting point in human consciousness — the one centre where more might become known. (Eddington 1929, p. 73)

Eddington's thought is that in consciousness we already know a non-relational, intrinsic or 'categorical' feature of the world. If the relational structures given by science need a categorical basis, why not take the simple path and let something we know already fits the general bill, namely consciousness itself, serve? Russell himself did not quite go down this path but he did agree that in consciousness we have some access to the intrinsic nature of reality, as in his cryptic pronouncement that 'what the physiologist sees when he looks at a brain is part of his own brain, not part of the brain he is examining' (Russell 1927, p. 383).

However, the loophole in P2 does not require an immediate leap to panpsychism or the idea that consciousness itself is the ground of reality. Perhaps the ground is some aspect of the physical world itself, invisible to science in so far as it is restricted to revealing only the relational structure of the world (see Stoljar 2006). The obvious problem with such an approach is that it seems little more than the bald assertion that the posited physical features suffice to generate consciousness. There is no intelligible account of the nature of this new intrinsic physical feature, or how it, being devoid of consciousness, would serve to ignite consciousness in matter.

Perhaps, then, the ground is something else altogether, a neutral, neither physical nor mental, feature of the world. Proponents of such a 'neutral monism' (a term coined by Russell), 'radical empiricism' (the term favoured by William James) or 'elementary event monism' (the term of Ernst Mach) evidently still face the 'generation problem' of explaining how the neutral will enable consciousness. Typically, they try to shift the goalposts by characterizing the neutral in suspiciously mentalistic terms: Russell's 'sense data' or 'percepts', James's 'pure experience'[10] and Mach's 'sensations'.[11]

We can follow James in endorsing a truly neutral monism and perhaps not quite end up in panpsychism. Nor should we take the neutral to be a mysterious, shadowy background entity whose nature, as neither mental nor physical, is incomprehensible. We would like to explicate consciousness within this system, outline its relation to the physical world and do the latter without rebooting the problem of consciousness.[12]

In order to do all this we need but one primitive notion: presence. Take some time to appreciate what is present to you at this moment. Both the 'external' and 'internal' worlds are present in a host of different ways. Presence transcends the mental–physical split in at least two ways. One is that both the mental and physical, despite their differences, are equally present. The second is that presence is never definitively physical or mental. What is present to you now is (I expect) some more or less ordinary physical objects (in my case a desk, a room, a computer, etc.). But evidently the right dose of LSD in the right circumstances, or even just an especially vivid albeit rather pedestrian dream, could have made what is present to you now an hallucinatory experiential feature of reality, non-veridical in its presentation of the physical world, but just as much present to you. This suggests that presence is inherently neither mental or physical.

It is of course natural to take 'presence' to mean 'presence to mind', but this slide towards the subjective is not mandatory and should be resisted. Presence itself should be taken as the foundational feature of reality. Rather than taking presence to be a relation to mind, take minds to be a relation defined over presence. Similarly, let us take the physical world to be another — intersecting — system of relations defined in terms of presence. This is the core idea of neutral monism, which James expressed as:

> The one self-identical thing has so many relations to the rest of experience [i.e. James's 'pure experience'] that you can take it in disparate systems of association, and treat it as belonging with opposite contexts.

---

[10] James came to recognize this and in a 1909 notebook entry wrote that 'the constitution of reality which I am making for is of the psychic type' (1988, p. 126). James's path toward panpsychism is nicely charted in Cooper (1990).

[11] For a detailed survey of the neutral monisms of these three thinkers which assiduously avoids ascribing anything like panpsychism to any of them see Banks (2014).

[12] The view outlined below bears interesting relations to Sam Coleman's (Coleman 2014, 2017) 'panqualityism' though I think once we grasp the nature of presence and the 'presence first' programme various difficulties facing Coleman are much less pressing.

In one of these contexts it is your 'field of consciousness'; in another it is 'the room in which you sit' . . . (1912/2003, p. 7).

Russell put it this way:

> the whole duality of mind and matter . . . is a mistake; there is only one kind of *stuff* out of which the world is made, and this stuff is called mental in one arrangement, physical in the other (Russell's italics; 1913/1984, p. 15).

We just need to appreciate that this *stuff* is simply presence, something we are directly acquainted with in every experience but which, as is generally evident in experience, is not itself mental. This stuff does not 'turn into' the physical world, or the mental realm when arranged in the right way; rather the right sort of arrangement *is* the physical world or *is* the mental realm.

Just as it is natural to fall into error by assimilating presence to 'presence to mind' it is also natural to think we can *generate* presence. Just by turning my head I can change what is present. But if neutral monism is true, this too is the wrong way to think of presence. Mark Johnston, in a paper which is a gold mine of ideas helping to capture the kind of neutral monism I am proposing,[13] comes close to what I am trying to get at here when he writes:

> We are not Producers of Presence; it is not that our mental acts make things present. We are Samplers of Presence; our mental acts are samplings from a vast realm of objective manners of presentation. It is of the nature of existents to present, in all the various ways in which they can be grasped in this or that mental act of this or that individual mind (2007, p. 253).

This is not quite right because it's wrong to think of the 'modes of presentations' as dependent on objects which 'have' them. Rather, presentations are the fundamental reality and objects are relations defined over the presentations. This makes another of Johnston's remarks less cryptic. He says 'properly understood, there are no subjective phenomena' (2007, p. 248). Every experience is an entrance to the objective world of presence through, I cannot resist writing, the doors of perception.

How much presence is there? We have no idea of the limits of presence. It could be that presence is restricted to *me*, leading to a peculiar

---

[13] I hasten to point out that Johnston does not endorse even his own account, let alone the neutral monist extension of it. In fact, he begs his readers to answer 'a plea for help: Here follows a hypothesis, help me to see just why it couldn't be so!' (2007, p. 233).

kind of solipsism.[14] After all, I've never run across any that wasn't mine. Seriously, such narcissistic parochialism is no reason to restrict the range of presence; any possible form of experience will reveal more sorts of presence. But this again has it backwards: every form of presence reveals more sorts of possible experiences. Presence itself is fundamental reality and some of it is within our grasp (I wonder if all of it within *something's* grasp).

If solipsism is one extreme,[15] the other is that every possible form of presence is actual.[16] I rather favour this view since it avoids the need to impose what would seem an arbitrary limitation on what is present, based upon the extremely narrow range to which human beings have access. So imagine a hugely multidimensional space where each axis represents some feature of presence. What is present to you right now is a point in this space delimited by what we inadequately call colour, shape, position, smell, sound, distance, occlusion, etc. etc. I don't mean to suggest that presence is limited to sensory qualities; cognitive aspects are included in what is present and rich systems of interrelatedness, e.g. a 'cell phone' (yes, a cell phone can be present to me). Of course, these names are impositions on presence, which in itself is neutral: not physical, mental, technological, biological. The true *epoché* is beyond Husserl's. Do not just bracket the world, bracket mind and world to be left with reality simply as it is present.

Of note is that this space of presence does not contain an axis for time or physical space. It does not contain an axis for mind, consciousness, mass, charge or other familiar forms of existence. These are all systems of relations defined over what is present. A physical object could be, perhaps, likened to a certain 'thread' through the space of presence: one which satisfies all the conditions of physical objecthood, whatever those might be: maybe size, shape, continuity over temporal and spatial tracks, etc. plus the myriad of connections to other things constitutive of objective existence. Space and time themselves will be co-defined with these objects and with mental features. Minds form another set of 'threads' meeting other conditions

---

[14] For a fascinating examination of this kind of solipsism, though not in the context of neutral monism, see Hare (2009).

[15] It might seem that the true minimal extreme would be that there is no presence, but we know that is not so. No one can deny that something is present. Beyond, or behind, consciousness itself, it is the fact of presence that underpins the unassailable knowledge that something exists.

[16] The picture of plenitudinous presence is somewhat inspired by Julian Barbour's picture of physical reality, in which every physical possibility exists and a 'world' is set of possibilities which obey a host of constraints, some of which we know as the 'laws of nature' (see Barbour 2000).

(whatever those might be, maybe memory continuity, various sorts of coherence relations, many kinds of relations between sensory and cognitive features, etc.). Our quotidian world of objects, animals, and people is a set of these threads of presence. On this view, persons are systems of presence. Here, again, is Johnston:

> ... the modes of presentation of the items in my perceptual field are perspectival; that is, they present items to a particular viewing position, or more generally to a particular point from which someone might sense the surrounding environment. The implied position at which those modes of presentation seem to converge is the position of my head and body. To that same implied position, a bodily field, as it were a three-dimensional volume of bodily sensation, also presents. And that implied position is also one from which certain acts, presented as willed, emanate (p. 259).

But again, don't think of modes of presentation as anchored to objects. Presence is first, with all its perspectival character (other forms of presence — maybe — lack perspectival character, maybe we can even experience some of these under the right conditions of 'loss of self').

This sort of neutral monism can offload a host of traditional philosophical problems onto the quotidian world. What is causation? We might follow recent interventionist or Bayesian accounts: causation is a certain relation of events in the quotidian world which we can assess and discover as we thread our way through the space of presence. Is there a metaphysical problem of causation? Causation is a productive or sustaining relation between events which involve objects and properties. The system of presence does not include any such. The realm of presence just, so to speak, sits there all at once internally densely articulated but never changing. So no, there is no metaphysical problem of causation. What is freedom of will? Freedom is the quotidian system of personal acts which meet ordinary conditions of autonomy (whatever they may be in detail, neutral monism won't answer that question). Is there a metaphysical problem of freedom? Freedom only exists in the realm of beings who can entertain options and make choices. The system of presence does not include any such, so no, there is no metaphysical problem of freedom.

More exotic philosophical issues can also be offloaded. Scientific realism? As James celebrated in the name of his theory, neutral monism is a kind of empiricism. It accords primacy to what is 'observable' or 'experienceable'. But that does not entail in any straightforward way scientific anti-realism, though it is compatible with it. An anti-realism such as Bas van Fraassen's (1980; 2002) takes the quotidian world as ontologically fundamental and observability within the quotidian world thereby takes primacy. Neutral monism's brand of empiricism is more revisionary. So, an object in the *quotidian* world is real if it meets the standards of evidence needed to establish its existence (whatever those standards might be). So, do electrons exist? I'd say we have pretty good evidence they do. Are they part of the fundamental nature of reality? Not at all — they don't even come close to being a candidate.

Given that some sense has been made of the picture of reality aimed at here, two questions naturally arise. The first is: what about the original problem of consciousness? The theory of neutral monism is designed to solve (or perhaps dissolve) this problem. The world is made of 'what it is like' stuff. If consciousness is apprehension of what is present by a mind then there is no problem of consciousness. Some of the threads of presence constitute minds which, essentially by definition, are apprehending what is present and so automatically there is something it is like for them. There is no traditional problem of consciousness, no problem of explaining how the physical world generates, realizes or constitutes consciousness, because that problem stems from a basic mistake about the ontology of reality.

Perhaps there is a question about why we find ourselves in a world where there are stable psychophysical relations; in fact there is a question about why we find ourselves in a stable world at all. I suspect a Kantian answer is available to answer these sorts of questions. The threads of presence which constitute minds will meet constraints akin to Kant's conditions of experience which will include a stable world and stable relations between mental features and the world. Of course, almost all randomly selected threads of presence will not abide by these constraints and won't constitute anything like a mind in a world, or even just a mind (if there is a genuine difference between minds and minds-in-worlds).

There will be sceptical possibilities of course, but these live in the quotidian world. One might try to think of a problem rather akin to the philosophical version of the Boltzmann Brain issue. Which is more likely: that my experience is part of a large thread of presence or simply an appropriately structured sub-thread? It might seem, in some *a priori* sense, the latter. But it is a mistake to think of a conscious being as 'moving' along its thread of presence. To mangle Santayana's beautiful expression of the temporal now, my consciousness is *not* like a spark running along a fuse of presence. What I am is not a point on a thread but the whole thread with its relational ramifications extending through the whole world. Could 'I' be nothing but *this* quasi-instantaneous bout of presence? That is *not* the same as the quotidian worry that the entirety of existence is the recent 300 milliseconds of my experience. I don't see how to definitively disprove that kind of quotidian solipsism of the present moment but obviously that offers no reason to believe in it. Things look different, however, from the perspective of the complete system of presence. In the arena of presence, this quasi-instantaneous bout of presence is part of a world-involving thread and I am that thread. It is a weird kind of error to try to pin 'me' down to one point in the thread as if there is some worry that there are no other parts to it. Those parts are real and timelessly constitute the thread of presence which I am, as part of the giant set of threads which make up *our* world.

The second natural question: is this form of neutral monism a kind of panpsychism after all? It is obviously closely related to panpsychism, in so far as presence is more or less equated to the 'what it is like' aspect of conscious experience. And while it does seem to me — though it is controversial — that William James ended up concluding that his radical empiricism was a kind of panpsychism, that his 'pure experience' *was* a kind of experience, this does not seem to be an inevitable conclusion. Presence can be without being presence to mind. Perhaps it could be argued that 'mind' should be taken in some ultra minimal sense and that therefore presence resolves into infinitesimal sparks of consciousness. Perhaps this is only a verbal dispute, but such sparks are not what one would call conscious minds. Consciousness is a familiar part of the quotidian world and is typically taken to be something like the apprehension, usually under some categorizing guise, of what is present. Such apprehending parts of the world are quite special and a little bit rare. The hope behind panpsychism is to solve the problem of consciousness by adding consciousness to the physical world at a fundamental level. Why not instead take the world to be the hugely various relational structures organizing what is present? This makes the world open to us, leaves it scientifically investigable and eliminates any problem of consciousness.

References

Antony, Michael V. (2006), 'Vagueness and the Metaphysics of Consciousness', *Philosophical Studies: An International Journal for Philosophy in the Analytic Tradition* **128** (3), pp. 515–38.
Banks, Erik (2014), *The Realistic Empiricism of Mach, James, and Russell: Neutral Monism Reconceived* (Cambridge: Cambridge University Press).
Barbour, Julian (2000), *The End of Time: The Next Revolution in Physics* (Oxford: Oxford University Press).
Bois-Reymond, Emil du (1874), 'The limits of our knowledge of nature', *Popular Science Monthly* 5: 17–32.

Brüntrup, G., and L. Jaskolla, eds. (2016), *Panpsychism* (Oxford: Oxford University Press).
Chalmers, David (1996), *The Conscious Mind: In Search of a Fundamental Theory* (Oxford: Oxford University Press).
— (2016), 'The Combination Problem for Panpsychism.' In *Panpsychism*, edited by G. Brüntrup and L. Jaskolla, 229–48 (Oxford: Oxford University Press).
Clifford, William (1886), 'Body and Mind', In *Lectures and Essays*, edited by L. Stephen and F. Pollock, 2nd ed., 244–73 (London: Macmillan).
Coleman, Sam (2014), 'The Real Combination Problem: Panpsychism, Micro-Subjects, and Emergence'. *Erkenntnis* **79** (1), pp. 19–44.
— (2017), 'Panpsychism and neutral monism: How to make up one's mind'. In *Panpsychism: Contemporary perspectives*, edited by G. Brüntrup and L. Jaskolla, 249–82 (Oxford: Oxford University Press).
Cooper, Wesley (1990), 'William James's Theory of Mind', *Journal of the History of Philosophy* **28** (4), pp. 571–93.
Cutter, Brian (2017), 'The metaphysical implications of the moral significance of consciousness', *Philosophical Perspectives* **31** (1), pp. 103–30.
Dennett, Daniel (1991), *Consciousness Explained* (Boston: Little, Brown & Co.).
Eddington, Arthur Stanley (1929), *Science and the Unseen World* (New York: Macmillan).
Fraassen, Bas van (2002), *The Empirical Stance* (New Haven: Yale University Press).
Frankish, Keith (2016), 'Illusionism as a theory of consciousness', *Journal of Consciousness Studies* **23** (11-12), pp. 11–39.
Goff, Philip and Coleman, Sam (2020), 'Russellian Monism', in *The Oxford Handbook of the Philosophy of Consciousness*, edited by U. Kriegel, 301–27 (Oxford: Oxford University Press).
Hare, Caspar (2009), *On Myself, and Other, Less Important Subjects* (Princeton, NJ: Princeton University Press).
Huxley, Thomas (1866), *Lessons in Elementary Physiology* (London: Macmillan).
Jackson, Frank (1982), 'Epiphenomenal Qualia.' *Philosophical Quarterly* **32**, pp. 127–36.
James, William, (1912/2003), *Essays in Radical Empiricism* (Mineola, NY: Dover).
— (1890/1950), *The Principles of Psychology*. Vol. 1 (New York: Henry Holt & Co.).
—(1988), *Manuscript Essays and Notes* (Cambridge, MA: Harvard University Press).
Johnston, Mark (2007), 'Objective mind and the objectivity of our minds.' *Philosophy and Phenomenological Research* **75** (2), pp. 233–68.
Lee, Andrew Y. (2019), 'Is consciousness intrinsically valuable?' *Philosophical Studies* **176** (3), pp. 655–71.
Levine, Joseph (1983), 'Materialism and Qualia: The Explanatory Gap.' *Pacific Philosophical Quarterly* **64**, pp. 354–61.
McCarty, David Charles (2004). 'David Hilbert and Paul du Bois-Reymond: Limits and Ideals', in *One Hundred Years of Russell's Paradox*, edited by G. Link, 517–32 (Amsterdam: De Gruyter).
Moore, George Edward (1912), *Ethics* (Oxford: Oxford University Press).
Nagel, Thomas (1974), 'What Is It Like to be a Bat?' *Philosophical Review* **83** (4), pp. 435–50.
Owen, Adrian M. (2017), *Into the Gray Zone: A Neuroscientist Explores the Border Between Life and Death* (New York: Scribner).

Roelofs, Luke (2019), *Combining Minds: How to Think about Composite Subjectivity* (Oxford: Oxford University Press).

Russell, Bertrand (1913/1984), 'Theory of Knowledge: The 1913 Manuscript', in *The Collected Papers of Bertrand Russell, Volume 7*, edited by E. Eames (London: Routledge).

— (1927), *The Analysis of Matter* (London: K. Paul, Trench, Trubner).

Seager, William (1995), 'Consciousness, Information and Panpsychism.' *Journal of Consciousness Studies* **2** (3), pp. 272–88.

— (2001), 'Consciousness, value and functionalism.' *Psyche* **7** (20).

— ed. (2020), *The Routledge Handbook on Panpsychism* (London: Routledge).

Searle, John (1992), *The Rediscovery of the Mind* (Cambridge, MA: MIT Press).

Siewert, Charles (1998), *The Significance of Consciousness* (Princeton, NJ: Princeton University Press).

Simon, Jonathan A. (2017), 'Vagueness and Zombies: Why 'Phenomenally Conscious' Has No Borderline Cases', *Philosophical Studies* **174** (8), pp. 2105–23.

Skrbina, David, ed. (2009), *Mind That Abides: Panpsychism in the New Millennium* (Amsterdam: John Benjamins).

— (2017), *Panpsychism in the West*, revised edition (Cambridge, MA: MIT Press).

Stoljar, Daniel (2006), *Ignorance and Imagination* (Oxford: Oxford University Press).

Strawson, Galen (2006), 'Realistic Monism: Why Physicalism Entails Panpsychism', *Journal of Consciousness Studies* **13** (10-11), pp. 3–31.

Tye, Michael (2021), *Vagueness and the Evolution of Consciousness*, (Oxford: Oxford University Press).

Tyndall, John (1870), *Essays on the Use and Limit of the Imagination in Science* (London: Longmans, Green).

van Fraassen, Bas (1980), *The Scientific Image* (Oxford: Clarendon Press).

Wilson, Jessica M. (2021), *Metaphysical Emergence* (Oxford: Oxford University Press).

David Skrbina

On the Present and Future of Panpsychism

To say that panpsychism has undergone a resurgence in recent years is an understatement. By most any measure, the topic has truly blossomed, in both academic discourse and even among the broader public — a remarkable feat for any philosophical concept.

Let's take a brief look at some specific numbers. If we scan the database WorldCat for the keyword 'panpsychism', we find entries rising from an annual average of 5.9 in the 1990s to 28 in the 2000s, to 70.4 in the 2010s, to 96 in the 2020s. Checking the results of Google Scholar for the same time intervals, and including citations, we see entries per year increasing as follows: 92, 189, 577, 1072. A search on Amazon books yields about one title per year in the 2000s, rising to 9 in the 2010s, to a remarkable 41 in the year 2023 alone. Finally, a look at Google's NGram chart shows a general prevalence of the term 'panpsychism' rising from a normalized figure of 26 in 2000, to 44 in 2010, to 123 in 2019. This latter metric is especially significant because it indicates something of the broader usage of the term across disciplines, not strictly philosophy. Panpsychism has indeed become something of a 'hot' concept these days, and scholars everywhere are taking notice.

What accounts for this dramatic ascent? As the most prominent philosophical advocate, Galen Strawson's open endorsement of panpsychism in his now-classic essay 'Realistic monism' in 2006, and subsequently in several other published pieces, surely is a large factor. But if we look back in time several decades, we find a number of milestones along the way. All the way back in 1967, historian Lynn White published his seminal and widely-cited essay 'The historical roots of the ecologic crisis'. White's focus was the Christian underpinnings for the environmental crisis, but often overlooked was his advocacy of St. Francis's 'unique sort of panpsychism' (1967: 1207) as a remedy to that crisis — a point which holds to this day. Gregory Bateson's *Steps to an Ecology of Mind* (1972) pushed a kind of panpsychism to the fore, as did another classic in environmental thinking, Christopher Stone's article 'Should trees have standing?' (1972). A scientific and technical variation on panpsychism appeared in David Bohm's *Wholeness and the Implicate Order* (1980). The following year witnessed the publication of Morris Berman's populist work *The Reenchantment of the World* (1981); his discussion of animism introduced the idea of universal ensoulment to a wide audience.

By the late 1980s, philosophical interest was accelerating as well. David Ray Griffin's anthology *The Reenchantment of Science* (1988) addressed panpsychism seriously, as did David Chalmers in his book *The Conscious Mind* (1996). In that same year, David Abram's *The Spell of the Sensuous* again brought a form of animism to wider public notice. In 1998, Griffin published his important work *Unsnarling the World-Knot*; this stands as the first serious, book-length philosophical treatment of panpsychism, albeit from a process perspective. Into the 2000s, we find the appearance of Christian De Quincey's *Radical Nature* (2002), and my own work, *Panpsychism in the West* (2005), which was the first comprehensive historical study.

All this set the stage for Strawson's milestone essay. It was, and still is, one of the most succinct, compelling, and analytically rigorous presentations of a non-emergence argument for panpsychism. Today, some eighteen years later, it withstands the test of time. Any real physicalism — and indeed, any real monism — does seem to entail some form of panpsychism, if we are to avoid the disaster of a brute and magical emergence of mind or experientiality from an utterly nonexperiential substrate. As Strawson says, experientiality is the *primum factum* of philosophy; it is our first and most certain metaphysical truth. Experientiality is an intrinsic aspect of the real, and we ourselves, and our immediate experiences, are proof of this. Furthermore, we know *only* of experiential reality, and we simply speculate on the existence of other, nonexperiential forms of reality. Unfortunately, we have neither data nor evidence nor logical argument to support this speculation. We have no good reason to believe in nonexperiential reality. The logical inference, and most plausible conclusion, is that *all* of reality is experiential reality. And thus we arrive at panpsychism.

Strawson's milestone essay appeared, of course, as a special issue of the *Journal of Consciousness Studies* (v. 13, no. 10–11, 2006). It was the target paper of that issue, and was accompanied by 17 philosophical commentaries, including one by myself ('Realistic panpsychism'). Among the commenters were many prominent individuals, including Frank Jackson, William Lycan, Colin McGinn, and J.J.C. Smart. The overall picture was striking: Of the 17 commentaries, fully 15 were either critical or highly critical. Only two were supportive: my piece, and that of Sam Coleman. Commentaries by Goff and Seager were sceptical at best, but all other papers were flat-out negative. Lycan mocks the 'sensory experiences' of a photon: 'What would be the contents of its beliefs or desires? Perhaps it wishes it were a quark' (70). McGinn sarcastically calls panpsychism 'surely one of the loveliest and most tempting views of reality ever devised'; sadly, for him, 'it's a complete myth, a comforting piece of utter balderdash' (93). Georges Rey was all but apoplectic; 'we haven't the slightest reason, independent of this argument, for any such fantastic conclusion [as panpsychism]. No one has produced the slightest evidence that anything but certain animals (and maybe certain machines) have experiences' (110). Indeed, he says, 'there is obviously not a shred of reason to take [panpsychist claims] seriously'. Obviously not — and we need only refer to such historical dupes as Empedocles, Heraclitus, Plato, Campanella, Spinoza, Leibniz, Herder, Schopenhauer, James, Nietzsche, Bergson, and Whitehead — panpsychists all. Perhaps McGinn, Lycan, and Rey should spend a bit of time reviewing their history; my *Panpsychism in the West* would be a good place to start.

Critics notwithstanding, in the 17-plus years since that time, panpsychism has flourished, both within philosophy and without. Obviously, many people have found at least 'a shred of reason' to take the idea seriously. As one measure, consider the philosophical books published since 2006 on the subject. We have my anthology *Mind That Abides* (2009; Benjamins), which was the first-ever collection of new essays on panpsychism. There is *Consciousness in the Physical World* (2015; Oxford UP), edited by Alter and Nagasawa; *Panpsychism: Contemporary Perspectives* (2016; Oxford UP); Phil Goff's *Consciousness and Fundamental Reality* (2017; Oxford UP); Luke Roelofs' *Combining Minds* (2019; Oxford UP); Bill Seager's anthology *The Routledge Handbook of Panpsychism* (2020); and *Panentheism and Panpsychism* (2020; Brill). In 2017, MIT Press published an expanded and revised edition of my *Panpsychism in the West*. Such output would not happen without a clear and marked trend toward panpsychist thinking, or at least an openness and sympathy toward it.

History and recent scholarship, then, argue for taking panpsychism seriously. Still, many resist it, some viscerally. Most often, one finds a series of silly remarks, jokes, or *ad hominem* attacks against panpsychism and its defenders; the more polite opponents will simply claim that the theory is 'counterintuitive' or 'implausible', and leave it at that. But then again, it was not so long ago that black holes, quarks, and dark energy were 'counterintuitive' and 'implausible' — yet today they are accepted as objective reality. History is filled with such examples. Obviously this cannot count against panpsychism.

What is required, then, are actual philosophical arguments against panpsychism. And when we search for these, we find them lacking. In my review of the literature, I find seven substantive objections; it is worthwhile summarizing these here.[1] The most significant objection is the 'combination problem': if, for example, the atoms in our brains are in some way enminded, how can these micro-minds fuse or combine into our single, higher-order mind? This is an important question, one that has vexed philosophers at least since the mid-17th century. William James wrestled with this issue, ultimately concluding that 'the self-compounding of mind in its smaller and more accessible portions seems a certain fact' (1909: 292) — in other words, that combination does in fact occur. But others are less convinced. McGinn argues that 'we cannot envisage a small number of experiential primitives yielding a rich variety of phenomenologies' (2006: 96). Lycan similarly cannot imagine 'in what way a mental aggregate [could] consist of a host of smaller mentations' (2011: 362). But in fact, we do have plausible models for such fusion or combination. Field theory, for example, is very amenable to combination, and there is good reason to expect that mind, like all real phenomena in the universe, is at some level a field phenomenon. Notions in quantum physics, especially superposition, provide another possible means for combination of mental states. There are options here. At worst, the combination problem is a call for details; it is not a decisive argument against panpsychism.

A second objection we might call 'the tenability of brute emergence'; that is, perhaps mind does in fact, in some unknown way, simply emerge, in a miraculous fashion, from a material substrate that is utterly devoid of mind. This is a stance that all non-panpsychists — i.e. virtually all conventional philosophers — must assume. And yet, as Strawson has so ably shown, it is an unacceptable alternative. It demands the acceptance of magic, of miracles. Until the objectors can present a viable theory of brute emergence, the objection is moot.

Third, some object to panpsychism as an 'inconclusive analogy' in the sense that any comparisons between, say, enminded animals and inanimate objects are invalid. Any putative intrinsic natures — as in, for example, Russellian monism — are not necessarily mental, say the

---

[1] For a more detailed discussion, see Skrbina (2020, pp. 108–113).

objectors. Lycan (2011) suggests that perhaps things have no such intrinsic nature at all; and even if they do, he asks, why assume that this is an experiential or conscious nature? But analogies do hold, as everyone admits, between humans and 'higher animals' at least, all of which are experiential. What, then, causes the analogy to fail at some arbitrary point on the phylogenic scale? The objectors have no answer here.

A fourth objection can be called the 'untestable' or 'no signs' objection: inanimate things give no apparent indication of enmindedness, nor can we devise a test to confirm or invalidate it. In a Popperian sense, then, the thesis of panpsychism is unverifiable, irrefutable, and hence 'non-scientific'. McGinn (1999: 97) says 'regular matter gives no sign of having mental states ... If electrons have mental properties, these properties make no difference to the laws that govern electrons'. Paul Churchland (1997: 213) argues that the 'explanatory successes' of modern physics are vast, whereas that of panpsychism 'is approximately zero'. Lycan (2011: 361) says simply, 'panpsychism's most obvious liability is the absence of scientific evidence'. Sometimes the point is pressed to absurdity; Churchland (1997: 212) insists that panpsychism must 'construct...theoretical proposals and testable hypotheses' and must achieve 'systematic successes in experimental predictions and technological control'. This is obviously a ridiculous demand. Were this to be necessary, no one would ever be able to construct a plausible theory of consciousness or experientiality.

Fifth is the question of supervenience: do the experiential or mental properties of, say, atoms supervene on their non-mental properties, or do they not? If they do, then it implies a kind of emergence after all. If they do not, then it implies some form of epiphenomenalism, or at least a disconnect between mental and non-mental properties. But there are other possibilities here that the objector overlooks. Mental and non-mental properties could exist in a condition of 'equal status' (to use Strawson's words), with neither supervening on the other. Alternatively, the non-mental could supervene on the mental — a position that reverses the claim of the objector. This would be close to an idealist metaphysic, and not far from Strawson's recent thinking on the topic.

A sixth objection would be epiphenomenalism itself. Lycan (2011: 362) calls this 'a more worrying difficulty for the panpsychist'. The causal closure of the physical realm would seem to leave no role for any panpsychist mentality or consciousness. But of course, this is a universal issue for all theories of mind; it holds no special weight against panpsychism.

Finally, a seventh objection: any nominal panpsychist 'minds' out there in nature are utterly irrelevant for us, one might say. Objectors claim that atomic minds, for example, would have to be as rich and diverse as our own, at least in principle. Atoms would have to have not only experiences but qualitative states, beliefs, desires, emotions, and the like. 'This is a game', say McGinn (2006: 95), 'without rules and without consequences'. But of course, this is a false presumption. Elementary particles need not have anything like our complex mental states or properties. There maybe certain core and universal aspects of mind — intentionality, broadly conceived, or experientiality in an elemental sense — that exist at vastly lower levels of complexity than in us. Nagel famously argued that we cannot know what it is like to be bat; clearly, we have no burden, then, to be able to imagine what it is like to be a quark. There may be epistemological limitations to panpsychism, but that does not invalidate the theory.

These are the technical objections to panpsychism, such as they are. On the other side, there seem to be increasing reasons for philosophers and others to view it credibly. Mechanistic materialism, or conventional physicalism, which posits intrinsically nonexperiential, nonmental matter, seems to be falling ever more out of favour. Idealism, dual-aspectism, and even variations on dualism seem to be coming back into fashion. Physicalism is increasingly viewed as having failed to explain mind, mental causation, and qualia. When conventional physicalism goes, the way is clear for a fresh look at panpsychist variations. (I emphasize here that panpsychism is best viewed as a meta-theory of mind, as a theory about theories. It simply extends a given theory of mind to all beings. Even dualism can exist in a panpsychist form.)

People in many disciplines seem increasingly open to animal minds, even the so-called 'lower' forms of animal life. Insects are viewed by some now as intelligent and conscious (e.g. Barron and Klein, 2016). Researchers are moving beyond animal life, finding strong evidence of intelligence in plants (e.g. Trewavas, 2017; Calvo et al., 2020). Viruses are now seen as exhibiting a 'sociology' (Diaz-Munoz et al., 2017). Normally staid social scientists are finding evidence of collective mind and intelligence in human social groups (e.g. Wendt, 2015). Mathematicians have resurrected the old idea that atoms may possess something like free will (e.g. Conway and Kochen, 2009). Giulio Tononi's quasi-panpsychist IIT theory continues to draw considerable attention. Mind, it seems, is popping up everywhere.

And there are increasingly pragmatic reasons to endorse something like panpsychism. With the planetary environmental crisis looming in many people's minds, eco-philosophers and environmentalists are seeking evidence of intrinsic value in nature and in the planet Earth. If the basis for human intrinsic value is our awareness, sentience, and ability to suffer, and then if animals, plants, and perhaps all of nature possess similar characteristics, then we have a new ground for seeing all things in nature as inherently valuable — and thus deserving of moral consideration. If human personhood is the basis for our moral, civil, and legal rights, then perhaps the personhood of things in nature is likewise a basis for their rights. Panpsychism offers an entirely new form of defence for the environment. And these days, we need all the defences that we can muster.

Such is the standing of panpsychism in the present. For literally thousands of years, it was a respected and dignified concept, openly held by some of the brightest minds of our past. For a relatively short period of time — say, from about 1900 to 1990 — a newly-emerging school of analytic philosophy and mechanistic materialism dismissed the view. But in the last two or three decades, the tide has turned once again. Strawson's outstanding work, and frankly his intellectual courage and integrity, have contributed in no small part to the growing prestige of this briefly-maligned view. I have no doubt that history will look very favourably upon his work of the past eighteen years, at least, as it will upon panpsychism generally.

Despite all this, panpsychism's detractors are unrelenting. I can't recall a single instance of a prominent committed physicalist 'converting' to a panpsychist view. If anything, they are becoming more intransigent. But I suppose this is unsurprising. The establishment philosophers, the non-panpsychists, tenaciously hold their ground, even as their graduate students and younger colleagues remain open-minded on the subject. We have here the makings of a truly Kuhnian paradigm shift, one in which the Old Guard, like old soldiers, simply fade away; soon enough, a new cohort of philosophers will emerge, one that is much more open to a very old idea.

References

Abram, D. (1996), *The Spell of the Sensuous* (New York: Vintage).
Alter, T. and Nagasawa, Y., eds. (2015), *Consciousness in the Physical World* (Oxford: Oxford University Press).
Barron, A. and Klein, C. (2016), 'What insects can tell us about the origins of consciousness', PNAS **113** (18), pp. 4900–4908.
Bateson, G. (1972), *Steps to an Ecology of Mind* (New York: Ballantine).

Berman, M. (1981), *The Reenchantment of the World* (Ithaca, NY: Cornell University Press).
Bohm, D. (1980), *Wholeness and the Implicate Order* (London: Routledge).
Bruntrup, G. and Jaskolla, L., eds. (2016), *Panpsychism: Contemporary Perspectives* (Oxford: Oxford University Press).
Bruntrup, G. et al, eds. (2020), *Panentheism and Panpsychism* (Leiden: Brill).
Calvo, P. et al (2020), 'Plants are intelligent, here's how', *Annals of Botany* **125** (1) pp. 11–28.
Chalmers, D. (1996), *The Conscious Mind* (Oxford, UK: Oxford University Press).
Churchland, P. (1997), 'Panpsychism: A brief critique', in *History as the Story of Freedom* (Amsterdam: Rodopi): 211–216.
Conway, J. and Kochen, S. (2009), 'The strong free will theorem', *Notices of the American Mathematical Society* **56** (2), pp. 226–232.
De Quincey, C. (2002), *Radical Nature* (Chicago, IL: Invisible Cities Press).
Diaz-Munoz, S. et al. (2017), 'Sociovirology: Conflict, cooperation, and communication among viruses', *Cell Host Microbe*, **22** (4), pp. 437–441.
Goff, P. (2017), *Consciousness and Fundamental Reality* (Oxford: Oxford University Press).
Griffin, D., ed. (1988), *The Reenchantment of Science* (New York: SUNY Press).
Griffin, D. (1998), *Unsnarling the World-Knot* (Berkeley, CA: University of California Press).
James, W. (1909/1996), *A Pluralistic Universe* (Lincoln, NE: University of Nebraska Press).
Lycan, W. (2006), 'Resisting ?-ism', *Journal of Consciousness Studies* **13** (10–11), pp. 65–71.
Lycan, W. (2011), 'Recent naturalistic dualisms', in *Light Against Darkness* (Gottingen: Vandenhoeck & Ruprecht): 348–363.
McGinn, C. (1999), *The Mysterious Flame* (New York: Basic).
McGinn, C. (2006), 'Hard questions', *Journal of Consciousness Studies* **13** (10–11), pp. 90–99.
Rey, G. (2006), 'Better to study human than world psychology', *Journal of Consciousness Studies* **13** (10–11), pp. 110–116.
Roelofs, L. (2019), *Combining Minds* (Oxford: Oxford University Press).
Seager, W., ed. (2020), *The Routledge Handbook of Panpsychism* (London: Routledge).
Skrbina, D. (2005), *Panpsychism in the West* (Cambridge, MA: MIT Press).
Skrbina, D. (2006), 'Realistic panpsychism', *Journal of Consciousness Studies* **13** (10–11), pp. 151–157.
Skrbina, D., ed. (2009), *Mind That Abides* (Amsterdam: John Benjamins).
Skrbina, D. (2017), *Panpsychism in the West*, revised edition (Cambridge, MA: MIT Press).
Skrbina, D. (2020), 'Panpsychism reconsidered', in Seager (2020), pp. 103–115.
Stone, C. (1972), 'Should trees have standing?' *Southern California Law Review* **45**, pp. 450–501.
Trewavas, A. (2017), 'The foundations of plant intelligence', *Interface Focus* (21 April).
Wendt, A. (2015), *Quantum Mind and Social Science* (Cambridge, UK: Cambridge University Press).
White, L. (1967), 'The historical roots of our ecologic crisis', *Science* **155** (3767), pp. 1203–1207.

Daniel Stoljar

Underestimating the World

In 2006 I contributed a commentary on Galen Strawson's impressive paper 'Realistic Monism: Why Physicalism Entails Panpsychism'; see (Strawson 2006, Stoljar 2006a). As I said in the commentary, there is much that seems right and important in the paper. I agree with Strawson that people tend to mistakenly assume that they know in outline what the physical world is like, and I agree also that this assumption, which seems so innocent and natural, is largely the cause of our troubles over the metaphysics of consciousness. If we take a very different view of the physical, if we do not 'underestimate' it, to adopt a wonderful phrase Strawson uses in a later paper (Strawson 2019), the problems go away or at any rate are transformed into something quite different; see, e.g., Stoljar (2006b, 2020a), Kind and Stoljar (forthcoming).

But, as I also indicated, my agreement with Strawson on this central point is tempered in several ways. One of Strawson's main aims in his paper (you can see this from the subtitle) is to draw out a consequence from the kind of physicalism he and I both find plausible — the kind that drops the idea that the physical world is exhaustively accounted for using physical theory of the kind available to us now. His thesis is that this form of physicalism entails a kind of panpsychism.

I don't and still don't see that this entailment thesis is true. Strawson's main consideration in support of it is that if consciousness in the form that we humans have it is derivative, then the things on which it is derivative must have a nature such as to yield consciousness. I agree with that, though I don't think that it begins to show that panpsychism is true. At best it shows what David Chalmers (2015) calls 'panprotopsychism' is true. But as I have argued in other places (Stoljar, 2018, 2020b), this is just a misleading name for any view on which the world contains some fundamental elements that somehow or other have the capacity to combine together to yield consciousness, just as it contains some fundamental elements that somehow or other have the capacity to combine together to yield every other derivative, existing thing, such as chicken salt or the Promenade des Anglais.

Perhaps this modest kind of world-view is deniable in principle but it is a long way from panpsychism.

Lying behind this disagreement about whether physicalism entails panpsychism is another disagreement, or to put it more accurately, a suspicion I had that Strawson has not given a proper account of the relation between two different ideas in the paper. One idea is that consciousness in some form or other is fundamental (i.e. not derivative on anything else); if correct, this entails that the only world-views that could possibly be right are classical dualism, panpsychism or idealism — positions that entail it is fundamental. The other idea is to reject the widespread assumption that we know exactly what the physical is, if not in detail then in outline.

Strawson wants, if I understand him correctly, to combine these two ideas together, suggesting that if we don't know something important about the physical, we can't rule out it includes fundamental consciousness. The problem with this, as I tried to bring out in my commentary, is that the two ideas are in significant tension. It is not that there is any logical problem in conjoining them. It is rather that, if it is really true that we are ignorant of some relevant features of the physical world, we lose whatever reason we had to endorse the fundamentality of consciousness in the first place. To put it another way, Strawson underestimates underestimation: he misses the radical consequences of the view that we underestimate the physical. In principle you can combine that view with the claim that consciousness is fundamental, but there is no philosophical point in doing so.

In this follow-up commentary I thought I would step back from these particular criticisms, and ask Strawson a question that I am sure he has heard before but which seems to me to become prominent when you bring out the tensions in his 2006 paper in the way I tried to do. The question concerns, not Strawson's attitude to physicalism and what it entails, but rather his contrasting attitude to two issues: consciousness, which is of course the main topic of 'Realistic Monism', and free will, something on which he has written extensively; see, e.g., Strawson (1986, 2002, 2018).

The contrast I have in mind is this. Regarding consciousness, Strawson is extremely hardboiled about the view that consciousness does not exist. He says it is the 'greatest woo-woo of the human mind' (Strawson, 2006), and elsewhere 'the silliest claim' (Strawson 2018). I'm not sure I would initiate this language myself, but I do agree with the sentiment.

But let's consider the parallel position on free will. Just as there are people who deny consciousness, there are people who deny free will.

For my part, I regard this as just as implausible as the denial of consciousness. When people say that free will doesn't exist, I don't believe them. In fact, I don't think even they believe it, though they may believe they believe it. If we compare the likelihood of what they are saying with the likelihood that they are confused or mistaken or imagining things, the answer I think is quite clear.

However, while the denial of free will seems to me just as implausible as the denial of consciousness, this is not how it seems to Strawson. He thinks free will doesn't exist; it is 'provably impossible' (2018, p. 97). The options that confront us when we think about free will, he thinks, are well understood, and when we consider them dispassionately, the only acceptable one is to deny free will. Of course, Strawson agrees that it seems to us that we are free — that is a psychological fact. It is just that in reality we are not.

So the question is this: what if anything justifies this asymmetry of attitude? How can Strawson (or anyone) say it is a woo-woo to deny consciousness but not a woo-woo to deny free will?

Before trying to answer this question directly, it is worth reminding ourselves of some of the grand figures. One is Emil du Bois-Reymond, recently described as the 'the most important forgotten intellectual of the 19th century' (Finkelstein 2013; see also du Bois-Reymond 1886, 1886a). Du Bois-Reymond's overall position is in outline similar to Strawson's. He says, as regards consciousness or sensation, that while there is nothing in 19th century science that can explain it, it doesn't follow that there is nothing at all that can explain it; all that follows is that we are ignorant of the explanation and will remain so to the extent that we remain confined to the epistemological framework of 19th century science — 'ignoramus et ignorabimus', in du Bois-Reymond's memorable phrase. As regards free will, however, which he lists along with consciousness as one of the seven biggest problems confronting science and philosophy, du Bois-Reymond takes a different view. In this case, as in the case of consciousness, there is nothing in 19th century science that could explain it. But here du Bois-Reymond concludes that free will doesn't exist. Not only is there nothing known that could explain it, there is nothing that could explain it all.

An even grander figure is Noam Chomsky. Chomsky doesn't discuss consciousness in quite the way that Strawson or du Bois-Reymond do, but it is natural to read into what he says on related topics a position that in general terms is similar; see, e.g. Chomsky (2016). What about free will? Here Chomsky takes a position very different from Strawson and du Bois-Reymond. For him free will exists

(Chomsky, 1988). Indeed, its existence, he thinks, is completely undeniable; giving it up is not a psychological option, but it is not an epistemological option either.

Is Chomsky therefore a compatibilist in the sense that, Dennett (e.g. 2021) is a compatibilist? No; Chomsky thinks the existence of free will is incompatible with determinism, which we may assume to be the thesis that every actual event is determined, i.e. necessitated by the past and the laws. But he also thinks it is incompatible with the thesis that most people move to when they deny determinism, namely, that every actual event is either determined or random. What Chomsky thinks instead is that these theses don't exhaust the options. It is true, in fact it is a logical truth, that every event is either determined or not determined. But it is not true, and it is certainly not a logical truth, that every event is either determined or random. Some events are neither determined nor random; among these are the free actions.

Chomsky doesn't think we can easily understand what these events are. On the contrary, not only does he say we can't at present understand what they are, it is a real possibility we will never do so; if so, the situation here too is one of ignoramus et ignorabimus. But none of this, he thinks, undermines the reality of free will.

Whatever else we may say about it, Chomsky's attitude has the attractions of uniformity, while that of du Bois-Reymond and Strawson does not. What then justifies their asymmetrical position, or, if we set aside our colleague from the 19th century, what justifies Strawson's?

One answer to this question is the noble one: this is where the argument leads. In other words, Strawson may insist that there is an argument that leads decisively against free will but there is no similar argument in the case of consciousness.

Now Strawson does indeed have a distinctive argument against free will. It is contained in several of his works but comes out most clearly for me in 'Luck Swallows Everything', reprinted in Strawson (2018). The leading idea is that you are free, and so responsible for what you do, only if you are responsible for what you are; but since you are not responsible for that, you are not free. In premise and conclusion form, the reasoning is something like this:

P1. You do what you do because of the way you are. [Presumed fact]

C1. You are responsible for what you do only if you are responsible for the way you are. [From P1]

P2. You are not responsible for the way you are. [Presumed fact]

C2. You are not responsible for what you do (and so are not free). [From C1 and P2]

I think there are several questionable features of this argument. For one thing, P1 as stated is subject to counterexample. Suppose mild Mildred swears at her neighbour. Clearly her behaviour is out of character. 'That's not the way I am', Mildred might remorsefully say afterwards. In this case, there is something Mildred did, namely swear at her neighbour; in normal circumstances we would describe this as something she did freely, and something for which she can be held responsible. Nevertheless, she did not do it because of the way she is; on the contrary she did it in spite of the way she is.

Strawson might reply that the phrase 'the way you are' is not to be understood as limited to somebody's character, something that may or may not be reflected in their action. Such phrases are certainly slippery; in context, they could attribute almost any property at all. Let us therefore read P1 in a more general way, as follows: you do what you because you have some property, any property, which explains what you do. Now the premise is unlikely to be subject to counterexample. But the argument remains unpersuasive since, if we understand P1 in that way, both the inference from P1 to C1 and C1 itself look very implausible.

For C1 now entails that you are responsible for doing what you did only if you are responsible for having the property, whatever it is, that explains what you did. That is an eminently deniable claim. For example, the property that explains what Mildred did might be incredibly complicated, including all manner of features of her constitution and history, things neither she nor anyone else has any inkling of. Is she responsible for having that property? Surely not. Yet she remains, as she herself may well agree, responsible for swearing at her neighbour. If so, we should reject C1, and moreover reject that it follows from the first premise of the argument.

There is also a different way to bring out the problem for C1. Suppose for the sake of argument I am a free agent. If so, I am responsible for what I do, at any rate for what I do freely; that is what it means to be free. But more than this, being free is surely part of 'what I am' in any reasonable sense; being free if you are free isn't just some arbitrary property, it is built into your nature. Am I therefore responsible for my being a free agent? Certainly not. I was born this way, to echo both Rousseau and Lady Gaga. If anybody (or anything) is responsible my being a free agent, it is certainly not me. Hence we again have a good reason to reject C1: I am responsible for what I do (since I am free) but am not responsible for what I am.

I don't mean these brief remarks about Strawson's subtle argument to constitute any quick refutation. What I do think, though, is that they motivate an attitude to this and similar arguments that I would argue is the correct one, namely, that since the conclusion is so implausible — since it leads, as we might say, to one of the greatest woo-woos of the human mind — there must be a flaw in the reasoning somewhere, even if it is a challenge to say exactly where, and even if in this particular case I am mistaken to be so suspicious of the phrase 'the way you are'.

I asked above why Strawson defends his asymmetrical position. One answer we have just been considering is that this is where the argument leads. But we may take the question in a slightly different way, in which what is at issue is not so much what philosophical or scientific arguments lead Strawson to hold his view, but why he finds it a rational position in the first place. I suspect the reason is again that he underestimates underestimation, i.e. he has not quite appreciated how claims about ignorance cast a shadow over all these issues.

As regards his 2006 paper, as I said, this lack of appreciation comes out in the argument that physicalism entails panpsychism, and in the attachment to the view that consciousness is fundamental. In the case of free will, I think it comes out in the view that, if you reject determinism, you must be committed to the view that every event is either determined or random. It is certainly plausible that *if* that is the only way to deny determinism, free will would be provably impossible; it would then be incompatible both with the truth of determinism and with its falsity. On the other hand, to assume that this is our only option if we deny determinism is to underestimate the world — not so much its physical nature, as in the case of consciousness, but its dynamics: it is to underestimate the forms that indeterminacy can take.

So it seems to me that, in both his work on consciousness and on free will, Strawson underestimates the world, or at least underestimates the significance of doing so. If you underestimate the world, and if you correlatively overestimate your level of insight into the world, you will almost inevitably think there is no place in it either for consciousness or free will. In turn, you will almost inevitably adopt either fundamentalism or eliminativism with respect to both. Having arrived at this crucial choice point, it may be that there are reasons to follow Strawson in going one way in one case, and the other way in the other case. Perhaps, for example, appeals to fundamentality don't work in the case of free will; or perhaps appeals to intrinsic features of matter have more force in the case of consciousness; or perhaps the argument above about 'what you are' can avoid the objections I raised. Perhaps; but the more important thing is that you should never have arrived at this choice point in first place. The way to avoid doing so is to not underestimate the world.

References

Bois-Reymond, Emil Heinrich du (1886), 'The Nature of Scientific Knowledge', in *Reden von Emil du Bois-Reymond*, edited by Emil Heinrich du Bois-Reymond (Leipzig: Erste Folge, Verlag von Veit & Comp.).

Bois-Reymond, Emil Heinrich du (1886a), 'The Seven Riddles of the World', in *Reden von Emil du Bois-Reymond*, edited by Emil Heinrich du Bois-Reymond (Leipzig: Erste Folge, Verlag von Veit & Comp.).

Caruso, Gregg, and Dennett, D.C. (2021), *Just Deserts: Debating Free Will* (Polity).

Chalmers, David J. (2015), 'Panpsychism and Panprotopsychism', in *Consciousness in the Physical World: Perspectives on Russellian Monism*, edited by Torin Alter and Yujin Nagasawa, pp. 246–276 (New York: Oxford University Press).

Chomsky, Noam (1988), *Language and Problems of Knowledge: The Managua Lectures* (Cambridge, MA: MIT Press).

Chomsky, Noam. (2016), *What Kind of Creatures are We?* (New York: Columbia University Press).

Finkelstein, Gabriel (2013), *Emil du Bois-Reymond: Neuroscience, Self and Society in Nineteenth Century Germany* (Cambridge, MA: MIT Press).

Kind, Amy, and Daniel Stoljar (forthcoming), *What is Consciousness?* (New York: Routledge).

Stoljar, Daniel (2006a), 'Galen Strawson's Realistic Monism', *Journal of Consciousness Studies* **13** (10–11), pp. 170–176.

Stoljar, Daniel (2006b), *Ignorance and Imagination: The Epistemic Origin of the Problem of Consciousness* (New York: Oxford University Press).

Stoljar, Daniel (2018), Review of Philip Goff's 'Consciousness and Fundamental Reality', *Notre Dame Philosophical Reviews*, https://ndpr.nd.edu/news/consciousness-and-fundamental-reality/.

Stoljar, Daniel (2020a), 'The Epistemic Approach to the Problem of Consciousness', in *The Oxford Handbook of the Philosophy of Consciousness*, edited by Uriah Kriegel, pp. 482–498 (Oxford: Oxford University Press).

Stoljar, Daniel (2020b), 'Panpsychism and Non-Standard Materialism: Some Comparative Remarks', in *The Routledge Handbook of Panpsychism*, edited by William Seager (Routledge).

Strawson, Galen (1986), *Freedom and Belief* (Oxford: Oxford University Press).

Strawson, Galen (2002), 'The Bounds of Freedom' in *The Oxford Handbook of Free Will*, edited by Robert. H. Kane, pp. 441–460 (Oxford: Oxford University Press).

Strawson, Galen (2006), 'Realistic Monism: Why Physicalism Entails Panpsychism', in *Consciousness and its Place in Nature: Does Physicalism Entail Panpsychism?*, edited by Anthony Freeman (Exeter, UK: Imprint Academic).

Strawson, Galen (2018), *Things That Bother Me: Death, Freedom, the Self etc.* (New York: New York Review Books).

Strawson, Galen (2019), 'Underestimating the Physical', *Journal of Consciousness Studies* **26** (9-10), pp. 228–240.

Galen Strawson

Blockers and Laughter:
Panpsychism, Archepsychism,
Pantachepsychism

1. Introduction

There's been hot and heavy talk of panpsychism since the first edition of the collective volume *Consciousness and its Place in Nature* was published in 2006.[1] Much of it has been confusing. There's been the usual terminological chaos, but some progress has — I think — been made.

I endorsed the core idea of panpsychism in 1994, in *Mental Reality* (ch. 3): the *archepsychist* idea (see p. 354 below) that consciousness or experience must be among the fundamental features of concrete reality. I haven't changed my basic view since then, but I have changed the way I put it. I've also read more of what was written on the subject in the past, in the Western tradition; especially between 1870 and 1940. I've found that almost everything I've wanted to say has been said before, often somewhat differently, but often very well.[2] William Kingdon Clifford's defence of panpsychism is relatively well known (Clifford 1874, 1875). So also, now, is Arthur Eddington's (see e.g. Eddington 1928, chs. 12 and 13). Few, however, know of the work by Morton Prince, a full-on panpsychist who was one of the earlier fully explicit exponents of the so-called mind–brain identity theory: 'states of mind and neural activities are identical', he wrote in 1885; 'consciousness and the brain process are identical'.[3] Gerard

---

[1] When I cite a work I give the date of first publication or occasionally the date of composition, while the page reference is to the published version listed in the bibliography.

[2] I've recently been reading a posthumous collection of papers by Sprigge (Sprigge 2011), in which he also says, excellently, many of the things I say. See also Sprigge (1983).

[3] 1885, p. 44; 1904, p. 447. As Sprigge says, 'Anything going for the identity theory is evidence for the truth of panpsychism, as was realized long ago by philosophers such as Josiah Royce' (1983: 102). Some think that the so-called mind–brain identity theory,

Heymans (1905, third edition 1921), who helped to win William James over to panpsychism, is also little known. Again, few know of Durant Drake or Charles Augustus Strong; or indeed of James's late endorsement of panpsychism.[4] I've been particularly struck by Drake and Strong, two remarkable philosophers in the first half of the twentieth century who have been forgotten.[5]

The first thing I want to do is to recommend these authors. The admirable Roy Wood Sellars (father of Wilfrid Sellars), who agreed with Strong about almost everything except his panpsychism, wrote of 'Strong's magnificent attempt to carry panpsychism through'.[6] Sellars saw that 'panpsychism must be considered a species of naturalism' (1927: 218) even as he resisted it. He continued to be troubled by the strength of its claims throughout his long career.[7]

For the rest, I want to set down a couple of notes, correct a couple of misapprehensions, quote again some passages I've quoted in other work, provide a well-stocked bibliography, and speculate a little. Before that, a few somewhat plodding remarks about terminology, which may cast some light on some of the current confusion.

2. A Remark on Method

I've given up trying to persuade anyone of anything. This means I can say straight out what I think is true. This tends to be perfectly counterproductive when one is trying to persuade philosophers who disagree with one. I don't, however, think I could make much difference to the debate even if I tried my hardest. There are many things I take for granted, quite independently of the question of panpsychism, that other philosophers reject (I'm putting aside the fact that when people tell me about the kind of panpsychism they think I favour, they're

---

according to which conscious experience (C) is brain activity (B), was a daring product of the second half of the twentieth century. Chomsky makes fun of this effectively (Chomsky 1995). Some in the last sixty years have called themselves 'identity theorists' even as they deny the existence of one of the seemingly two things — but really only one thing — that they claim to be the same thing. In effect, they argue as follows: [1] C = B; [2] B certainly exists; [3] but (so!!) C doesn't really exist ....

[4] James (1909). In 1890, and in spite of the difficulties he foresees, James already holds that panpsychism must be correct in some version (1890, p. 152). Prince in his early work has grotesque views about race, but they don't affect his views about panpsychism.

[5] See e.g. Drake (1925), Strong (1930). I sketch Strong's panpsychist metaphysics in Strawson (2023).

[6] Sellars (1932), p. 296. Sellars and Strong demolished the 'The Myth Of The Given' — they showed in detail that the idea of the merely sensory given was a myth — long before W. S. Sellars gave it that name; and they weren't the first.

[7] See for example Sellars (1960). Skrbina (2005) is a good source of quotations from past panpsychists (I don't always agree with his glosses).

almost always wrong). What I hope to do is to furnish some more materials — a quotation or a thought of my own — for those who are already sympathetic. Deep blockers stand in the way of taking panpsychism seriously, even if one is generally favourably disposed. I hope that eventually someone else will come along and put the case in a way that catches on. Until then I follow the advice of the political consultant Frank Luntz:

> It's not what you say; it's what people hear that matters . . . There's a simple rule. You say it again, and you say it again and you say it again, and you say it again, and you say it again, and then again and again and again and again, and about the time that you're absolutely sick of saying it is about the time that your target audience has heard it for the first time.[8]

3. Terminology

3.1 Experience, experientiality, conscious experience, consciousness

I use these terms interchangeably, as I always have, to denote any sort of conscious experience, any 'experiential what-it's-likeness', however complex, however primitive: anything that others have in mind when they speak of 'phenomenal experience', or 'qualia', or 'phenomenology' (as this word is used in present-day analytic philosophy of mind). Experiences of colour, taste, smell, pain, fear, and so on are offered as paradigm cases of experience, but conscious thought is also a case of experience. Experience is something whose nature we know by what some call 'immediate acquaintance'. We all have experiences of many different kinds, and there's a fundamental sense of 'know' in which, in the case of experience, the having is the knowing.[9] And although we only have certain specific types of experience, we know what experience is in an entirely general way. This is why we can already, as children, grasp the idea that there may be creatures who have experience that is — unimaginably — different from our own. We know what general kind of thing is in question. I'll call it 'ψ', as well as 'consciousness' or 'experientiality', 'ψ' for 'psychical' (ψ has nothing to do with the Φ of IIT, integrated information theory).

---

[8] Luntz (2003). Descartes writes that 'confident assertion and frequent repetition are two ploys that are often more effective than the most weighty arguments' (1641, p. 358).

[9] There's a minor blocker here: some think that the use of the word 'have' in the phrase 'the having is the knowing' requires one to think that the *haver* is essentially ontically distinct from the *had*. This isn't so.

It's important to see the sense in which ψ is something concrete. It is, as I like to say — detaching the word 'stuff' completely from any connotations of thickness or solidity or clumpiness and repurposing it as an utterly general word for concrete being that abstracts from all traditional categorial distinctions (substance, quality, event, process, and so on) — a kind of *stuff*.[10] It's concretely existing, spatiotemporally located stuff; something of which there can be more or less. Consider human, canine, and feline ψ: there's a lot more of it in India, in any given twenty-four hour period, than there is in Antarctica.[11]

Some think ψ can't be thought of as *stuff*. It can only be a *property* or *quality* of something, a property that needs some *object* or *substance* or *stuff* that it is a property *of*, if it — not itself stuff — is to exist.

This is one of the blockers. I can't address it in this brief postscript, except to say that I believe that the object–property or substance–quality distinction, which is second nature to us in everyday thought and language, and has many completely unexceptionable uses, leads eventually, in fundamental metaphysics, to catastrophe. Many philosophers have seen that this is so, including Spinoza, Descartes, Kant, Ramsey, and Whitehead, but the mistake recurs again and again, and always will, as new generations arise. D. C. Williams, who taught David Lewis, is clear on the point that

> any serious philosophical theory of mind... challenges [i.e. demands] a serious revision of fundamental ontology (1959, p. 239).

The point applies equally and more generally to any serious metaphysics.[12]

---

[10] I take it that energy, understood in Heisenberg's way as concrete being, is stuff. (The standard scientific use of the word 'energy' is different.)

[11] A few side comments. (i) Philosophers have objected to the use of the term 'what-it-is-likeness' in philosophy of mind. It's certainly not ideal, but its established meaning is clear. (ii) The word 'quale' (plural 'qualia') has recently been taken to refer only to qualities of sensory experience, but its original use is quite general; it covers absolutely any kind of experiential what-it's-likeness. (iii) There's a narrower use of the word 'consciousness', popular in the mid-twentieth century (Russell and Strong favour it), given which mere or bare sensation — feeling, experientiality — is *not* a matter of consciousness. According to this use, *consciousness* is essentially cognitive (in addition to being essentially experiential), essentially intentional with respect to something other than itself (other than its own purely qualitative content and the type of qualitative content of which it is a token instance). It's very important to be aware of this when reading older texts, just as it's important to be aware that 'thought' and 'thinking' are often still being used in their wide Cartesian sense to mean absolutely any sort of conscious episode.

[12] See further Strawson (2024).

3.2 Materialism

This (unsurprisingly) is the view that everything that concretely exists is material or physical. I use the word in its original sense to denote a view that is *wholly realist about* ψ. Historically speaking, materialism's claim that ψ — *consciousness, of all things!* — is wholly material or physical (since everything is) has been its most striking claim. It's this claim that has made many find it a thrilling view — even while it has made many others find it threatening.

It's unfortunate, therefore, that a good number of people now use 'materialism' to mean a view that denies the existence of ψ — directly contrary to materialism's original stand-out claim![13] This second and more novel use can be traced back to the second half of the nineteenth century, when some began to think of materialism as the obviously false view that everything can be fully characterized (and even explained) in wholly mechanistic terms. Since then the second use has coexisted with the original use, and there's been an extraordinary amount of misunderstanding. Already in 1885, Morton Prince notes that 'the term materialism has come to be clothed with a meaning which does not belong to it, and has been used simply as a term of vituperation and abuse' (1885, p. 151). Haldane in 1940 is one of many who tries to get the word back on track: 'when I say that I am a materialist ... I do not mean that consciousness does not exist, or has a lesser reality (whatever that means) than matter' (1940, p. 27).

It's only terminology, you say. True. But words work deep into one's thinking (see 3.8), and can render crucial points invisible.

3.3 Physical

I use this term as a purely referential term that applies to the concrete reality that is the subject matter of *physics*, whatever its intrinsic stuff nature may be. I'll say more about it after saying something about 'physics' and 'physicalism'.

3.4 Physics

This term, evidently, denotes the science of physics, about which it's important, here, to say one fundamental thing. It's something that was a commonplace 100 years ago: given its descriptive resources, physics says (and can say) nothing about the non-structural, non-logico-mathematically characterizable, nature of concrete goings-on. It

---

[13] For this widespread use of 'materialism', see e.g. Goff, this volume p. 300f. One might call the view it names 'Looney Tunes materialism'.

follows immediately that panpsychism is wholly compatible with everything true in physics.

This is an elementary point. Drake makes it in 1911: 'Science studies ... the reality that makes up the universe ... but it can tell us only its order, not its substance' (p. 43). Russell stresses it over and over again; hundreds of thinkers have made it. Physics, in Stephen Hawking's words, 'is just a set of rules and equations'.[14] What is the fundamental 'stuff being' or 'stuff nature' — the *categorical being* — of the concrete reality, the concrete *goings-on*, that the equations hold true of? It's a hard question, but we know at least one thing for certain, given that materialism is true. We know that these goings-on include conscious goings-on — ψ.

This, in fact, is as far as certainty goes: physics adds (and can add) nothing to our stock of knowledge about the nature of concrete goings-on, beyond saying what its equations say.

Some find this hard to take. It requires a little reflection, after a standard upbringing in Western philosophy. But it already shows the falsity of the idea that physics gives us some reason to doubt or deny the existence of ψ. How many in philosophy have been misled, and for so long, in the last sixty years! What a sad circus it has been. It helps a little, perhaps, to know that physicists like Planck, Einstein, Lorentz, Schrödinger, de Broglie, Dirac were never misled in this way. This doesn't, however, help the reputation of philosophy.[15]

As materialists, then, who hold that the universe is entirely physical, we are bound to acknowledge that the only categorical feature of the physical world whose nature we know is — ψ. R. W. Sellars puts it sweetly: 'in consciousness ... alone are we on the inside of nature' (1927, p. 225). Is it not theoretically rash, in this epistemic position, to assume the existence of irreducibly non-ψ stuff? Panpsychism is certainly the more economical theory.

To say this — that it's theoretically rash to suppose that there is any non-ψ stuff — will still seem wild to many. There's another deep blocker here, and it's interlocked with yet another, which I'll mention in §6.

---

[14] 1988, p. 174. 'If you want a concrete definition of matter, it is no use looking to physics' (Eddington 1928, p. 95); physics can't reveal at 'its inner unget-atable nature' (*ibid.*, p. 257).

[15] D. C. Williams is funny: 'it is a famous anomaly of recent science that while an influential number of physicists, once supposed to be students of physical nature, are suggesting that only conscious experience exists, an equally influential number of psychologists, once supposed to be students of consciousness, have suggested that only physical nature exists' (1934: 23).

3.5 Physicalism

The view that physics gives us a reason to doubt or deny the existence of ψ is today often called 'physicalism', and many embrace 'physicalism' so understood. The view is simply a mistake (an elementary mistake) and it contains a fine irony. For no one would be quicker to agree that it's a mistake than the Vienna-Circle philosophers who invented the term 'physicalism'! Physicalism, for them, has *nothing to do* with doubt or denial of the existence of ψ. No one is a more robust realist about ψ than, say, Moritz Schlick. What a circus.

'Physicalism', then, is another term that has been brutalized in the terminological Barad-dûr of recent analytic philosophy of mind,[16] along with 'materialism', with which it is standardly taken to be synonymous.[17] Neither of these two positions, properly (or best) understood, has anything to do with denial of the existence of ψ. This is fortunate, given that the existence of ψ is certain.

3.6 Physical

Back now to 'physical', understood as a purely referential term that refers to everything that is the subject matter of physics.[18] I take this subject matter to be everything in the universe. Plainly ψ goings-on are physical, if materialism is true. Equally plainly, to say that ψ goings-on are physical is not to say that they aren't really real, or not really ψ after all. It's simply to say that the reality of ψ goings-on is physical reality.

To say this is not to say that *physics* can say anything about (or explain the existence of) the actual categorical phenomenal character of ψ goings-on considered specifically as such — as experiential. We know that physics can't do this — unsurprisingly, since it can't say anything non-logico-mathematical about the categorical character of *anything* (3.4). It's just to say — again — that ψ goings-on are physical goings-on. If your current conception of the physical is incompatible with this idea, or in *any* tension with it, you need to change your conception of the physical; if, that is, you want to be a materialist — a real materialist. As Prince says:

> To show that matter is not what it is supposed to be by the vulgar and ignorant, that it is something far removed from the ordinary conception of it, is not to remove it in any way from the field of materialism.

---

[16] For rich detail, see e.g. Uebel (1995, 2021). For a brief account, see Strawson (2023, pp. 437–8).

[17] See Lewis (1994, p. 293), for a justification of the synonymy claim.

[18] On the different uses of 'physical', see, famously, Stoljar (2001).

> As long as anything is the resultant of the forces of nature it belongs to materialism ... To show, then, that matter is something else than what we have supposed it to be, is not to remove it to the realms of spiritualism, for it is still something which is conditioned by natural laws. And consequently because we have reason to believe that mind is identical with this real matter ..., and is not identical with the vulgar conception of matter, we do not in any way escape from the bonds of materialism.[19]

3.7 Mental vs. physical

It follows from the above gloss of 'physical' that 'from the standpoint of philosophy the distinction between physical and mental is superficial and unreal' (Russell 1927, p. 402). I think that the failure to see this has been the principal blocker preventing progress in the so-called mind–body problem. If materialism is true, then, once again, and of course, the mental is physical:

> if one is a materialist, to say that there is a fundamental distinction between mental or experiential [ψ] phenomena and physical phenomena is like saying that there is a fundamental distinction between cows and animals — that on the one hand there are cows and on the other hand there are animals (Strawson 1994, p. 57).

It hardly follows that ψ is anything other than or less than what we know it to be. Materialism doesn't somehow diminish ψ — or our understanding of it. On the contrary: 'materialism elevates our conception of matter and our appreciation of the powers of nature' (Prince 1885, p. 153).

Some argue that the mental–physical distinction is a time-honoured distinction, enshrined in the traditional debate, and necessary to it, and should for that reason not be abandoned. Aïe aïe aïe! This is the reverse of the truth. Preservation of the standard mental–physical distinction (as opposed to the conscious–non-conscious or ψ–non-ψ distinction, which is perfectly respectable) is perhaps the principal reason why the debate is stuck.

Here there's a practical difficulty — another blocker. It consists in the fact that 'physical' is standardly opposed to 'mental' in everyday thought and talk. It is to that extent taken to have a certain descriptive (as opposed to merely referential) force. It's taken, in particular, and as already remarked, to mean — or at least imply — 'non-mental'.

This is all well and good — in everyday life. Problems arise only when this use is carried over into philosophy, and in particular into philosophy of mind. This is as disastrous as it is common. It's another

---

[19] Prince (1885, pp. 151–2); he uses the word 'bonds' to represent his opponents' view rather than his own.

deep blocker, as Russell, for one, is well aware. (Do I think I can stop it? No.)

It's hardly surprising that it's disastrous, given that many in present-day philosophy of mind do indeed hold that the mental is in fact physical. For if 'physical' is taken to mean or imply 'non-mental', then these people are arguing for the truth of a contradiction.

What can I say? Only that I myself think that the mental — in particular the experiential — is physical, and that I'm certainly not arguing for the truth of a contradiction. Like Russell and a host of others, I know that there's a great deal more to the physical than ordinary thought supposes, and that physics — with its extraordinarily thin, abstract descriptive resources — describes or can describe. I also know that this isn't any sort of failure of or restriction on physics. To think this, even for a moment, is to show that you don't know what physics is. A theory isn't defective when it doesn't do something it isn't meant to do.

3.8 Real materialism

In the attempt to avoid some of the confusion I sometimes call materialism-as-originally-understood '*real* materialism'. The name is justified as follows: the existence of ψ is certain; it follows that any remotely *realistic* or *serious* version of materialism must acknowledge its existence. 'Real' compresses 'realistic' and 'serious' into a single word. No one who is an 'eliminativist' or 'illusionist' about ψ is a real materialist.

3.9 Real naturalism

Just as real materialism is realistic materialism, so real naturalism is realistic naturalism. It is in other words naturalism that fully acknowledges the existence of the (wholly natural) phenomenon of ψ; which is, after all, the only thing we know absolutely for certain to exist. As the physicist Lee Smolin says:

> qualia ... must ... be understood as aspects of nature. *That is our commitment to naturalism* — the philosophy that asserts that all that exists is part of the natural world science studies (Smolin 2013).

It's unfortunate that many use the word 'naturalism' to mean, precisely, a view that denies the existence of ψ — the only natural thing that we know with certainty to exist.

Objection. 'Look, it's only a matter of terminology.'

Reply. No; 'only' is a great mistake. The frolicsome Francis Bacon makes the point:

> clumsy and inept applications of words bedevil our intellect in extraordinary ways, and the various definitions and explanations with which scholars have habitually fortified and protected their views haven't improved things at all. Plainly words do violence to the intellect. They muddy everything and seduce people into innumerable and inane controversies and theoretical constructions.[20]

When it comes to the controversies of the last sixty years or so, we should follow Prince. We should

> let the dead bury the dead, and ... dip at once into the present-day problem of consciousness and treat the question in the light, if there be light, of present-day knowledge. Let us make a *tabula rasa* of our minds, if we can, and wipe out, as with a sponge, all inherited traditional concepts which, ingrained as accepted formulae and beliefs, necessarily, in accordance with psychological laws, conflict with, repress and make us blind to new points of view and new interpretations of old facts in the light of new ones. A poor figure of speech, I fear, that of the sponge, for modern psychology has taught us that when concepts are deeply scratched and grooved into the mind, it takes more than a sponge, rather a grindstone, to polish it off into a *tabula rasa*. And this is particularly true when the accepted concepts are tinged with feeling tones, and, it may be, vibrate with deep and often overpowering emotion (1928, p. 3).

Do I think that well-intentioned analytic philosophers, reading this, will take it on board? I'm not optimistic. Even if they accept it in the moment, their acceptance is likely to be occluded — overridden — by the way words like 'physical' and 'physicalism' are grooved in their minds and chiselled deep in the great mountain of writing on this topic to which they are exposed.[21]

3.10 Panpsychism

I take panpsychism to be a form of materialism, following the *Oxford English Dictionary*, according to which panpsychism is 'the theory or belief that there is an element of consciousness in all matter'. Margaret Cavendish seems an exemplary materialist panpsychist in this sense, as does Anne Conway.[22] So are the all-out nineteenth- and twentieth-century panpsychists Prince, Royce, late James, Eddington, Drake, Strong, and Whitehead. Strong states explicitly that his panpsychism 'is only a revised materialism' (1930, p. 270). Drake makes

---

[20] 1620: §1.43 (my translation).

[21] Maxwell (1978, p. 365) made a similar unheeded terminological plea (unhelpfully, in my view, he used 'materialism' in the consciousness-denying sense and 'physicalism' in the consciousness-realist sense). I had no better luck in Strawson (1994: chs. 3 and 4).

[22] See Cavendish (1664), Conway (1690). I'm no longer sure about this, after reading Peterman (2023).

essentially the same move (1925, p. 243). It's helpful, given his authority with analytic philosophers, that David Lewis makes a related point: 'a thesis that says [that] panpsychistic materialism... is impossible... is more than just materialism'.[23]

So the root way of understanding panpsychism is as a form of materialism (perhaps I should change 'is' to 'was', given the current terminological winds). One way to make this explicit is to speak of *hylopsychism*.[24] It isn't, however, obligatory. One can say instead that one is, most fundamentally, a *monist*, i.e. a *stuff* monist: in addition to being an out-and-out realist about ψ, one takes it that there is only one kind of fundamental stuff.

A *stuff monist*: it seems clear that one can't be a *neutral* monist, given several of the ordinary understandings of this hideously shifting term, inasmuch as one takes it that ψ is part of the intrinsic — and fundamental — nature of reality.

— 'How then can *Russell* call himself a neutral monist, given that he accepts that ψ is at least part of the intrinsic — and presumably also fundamental — nature of reality?'

This is another terminological teratoma, but it's plain enough why Russell can call himself a neutral monist even if he takes it that ψ is part of fundamental reality.[25] First, his neutrality is quite specific: it's neutrality with respect to (what he sees as) the traditional mental–physical distinction. Second, he takes it (this may surprise) that ψ — e.g. bare sensation — *isn't intrinsically mental*. Mentality, for Russell, is essentially complex in a certain way. It essentially involves cognition, intentionality, and what he calls 'mnemic' causation, as bare sensation does not. It's worth adding that late Russell sees no reason to stop calling himself a neutral monist even when he's prepared to call himself a materialist — or a physicalist... .[26] To think that he can't do this is to fail to have understood him. (If one wants to cite Russell, one should prepare well before one starts. I speak as one who has erred.)

---

[23] 1983, p. 36. I expect I can add Ward and Stout to this list, but I don't know enough about them. Note that although Cavendish and Conway are out-and-out materialists with respect to the natural universe, they're not all-out materialists inasmuch as they believe in a transcendent (non-material) God; and this may be true of some others on the list. Whitehead is a real materialist, in my terms — someone who thinks that the mental is physical — although he would not call himself a materialist, given the associations of the word in his time.

[24] '*Hyle*' means matter in Greek philosophy. See e.g. Montague (1912, p. 281), Sellars (1916, p. 204), Peterman (2023, p. 421). Matthews (2003, p. 1880) considers the term and puts it aside.

[25] No doubt some will dispute this.

[26] See e.g. Russell (1944, 1945), Eames (1967), Strawson (forthcoming).

4. I'm Not a 'Smallist'

I've laid out some terms. Let me now try to cancel a misunderstanding stemming from the first edition of this book. In the two papers I contributed (pp. 3–31, 184–280), I assumed the truth of what is sometimes called *smallism* — the view that concrete reality is made up of tiny little bits or 'ultimates'.[27] I didn't, however, *endorse* smallism (contrary to what most readers and commentators supposed). I introduced it as an explicit *assumption*, made specifically and only for the sake of argument (p. 9), and immediately made three qualifying remarks.

The first was that I didn't think I needed to assume it in order to make the case that realistic materialism leads to some form of panpsychism (p. 9). The second was that 'if anything, [the assumption of smallism] makes things more difficult for me' (*ibid.*). The third flagged the metaphysical view I actually favoured at that time (and still do): the 'powerful rival view that there is at bottom just one thing or substance' (*ibid.* n.); or at least, if you baulk at this, the *field-theoretic* view of things that treats fields, not particles, as the ultimate existents. Straight-up 'smallism' gives rise to the classic form of the 'combination problem' for panpsychism, and it's arguable that it should be rejected for that reason alone, independently of its scientific inadequacy. William James, whose statement of the combination problem has been cited hundreds of times in the last twenty years (1890, p. 162), saw this clearly, even as he continued to suppose that panpsychism is the best option for genuinely naturalistic thinkers, and must surely be true in some form (1890, p. 152, 1909, ch. 5).

5. Archepsychism, Not Micropsychism

A connected point. In 'Realistic monism: why physicalism entails panpsychism', I introduced 'micropsychism' as a name for the view I now call '*archepsychism*' — the view that, when it comes to concrete reality, ψ must in some manner be at the bottom of things. It must be among the fundamental qualities of concrete reality even if, contrary to all-out or *pure* panpsychism, there are *also* non-ψ fundamental qualities.[28]

This is the view I first endorsed in *Mental Reality*: 'real materialists must hold at least some experiential properties to be fundamental

---

[27] See Coleman (2006).

[28] The argument rests on two claims: (i) there is no *radical emergence* in nature (a fundamental principle of methodological naturalism); (ii) the emergence of experientiality from wholly and utterly nonexperiential being would be a case of radical emergence.

physical properties, like electric charge' (1994, p. 60). The choice of the term 'micropsychism' in 'Realistic monism' was unhelpful, because it was tailored specifically to fit the assumption of smallism, which I'd made for the sake of argument, although I didn't myself favour it.[29]

With *archepsychism* in place, one can taxonomize further. *Pure* or all-out panpsychists, *panexperientialists* who thinks that the fundamental 'stuff nature' of everything is wholly ψ, are obviously archepsychists; but archepsychists needn't be pure panpsychists. Archepsychists may (for example) favour Keith Campbell's view that all that exist are a few 'tropes'; that the fundamental tropes are the fundamental physical fields; and that one of them is a ψ field.

Some, like Margaret Cavendish (but see note 22), think that archepsychists can be *pantachepsychists* or 'everywhere' panpsychists, holding that ψ is strictly ubiquitous in the universe, absolutely everywhere, without being *pure* panpsychists. One might perhaps hold this view if one thought that there was a fundamental Keith-Campbellian ψ field (or fields) that 'lived in' or 'inhabited' the 'space' of concrete reality along with various other fundamental non-ψ fields (see e.g. Campbell 1990, pp. 150–1). I'm not sure that it is in the end a coherent position, taken straight, but a field-theoretic approach may well make it seem so, inasmuch as it seems to allow that different physical fields 'living on the same space' can be supposed to have non-zero excitation values in exactly the same place. My own preference is for the thing-monist ('supersubstantivalist') view according to which the universe consists of just one fabulously complex field (its complexity displayed by the different fundamental equations of physics) which does not 'live on' a 'space' that is in any sense ontologically distinct from it.[30]

6. Panpsychism and Space

Strong writes that

> the difficulty of making people believe that there is in suns and atoms anything of the nature of feeling is so mountainous that I sometimes wish I had devoted my energies to something else, such as writing poetry or helping to bring about the millennium (1936, p. v)

---

[29] My first attempt to replace 'micropsychism' with a better name was a failure: I tried 'psychism', defining this as 'panpsychism without the *pan*', and noting that the name was far from ideal (Strawson 2017, pp. 384, 388–9).

[30] *Thing monism* is the view that there's a level of description at which it's non-trivially correct to say that there's only one thing (non-trivial: there's a use of 'the universe' according to which it's trivial that there's only one).

It's not hard to sympathize with him. There are many deep blockers. We tend, for one thing, to think [1] that space — space *as it is in itself* (or: whatever it is about reality that gives rise to the spatial aspect of our experience) — is at least *something* like what we ordinarily experience it to be. And we think [2] that occupying space *obviously* requires some sort of non-ψ being — non-ψ 'stuff being'. Our natural commitment to [2] lacks, in fact, any respectable warrant, but it doesn't weaken even when we master the complexities of the theory of relativity's account of space-time. The central point in reply is that granted that space-occupation is itself a nonexperiential property, it doesn't require the possession of *stuff-constituting* nonexperiential properties. The energy that constitutes stuff may be experiential in nature.[31]

It's pretty hard (for many) to get past this blocker, but I think one can get past it even if one accepts [1]. One doesn't need to invoke the contrary view, currently widespread among theoretical physicists, that [3] what we experience as space is, in itself, not at all like what we experience it to be. Nor — putting physics aside — need one think that the general case for panpsychism supports [3]; although one certainly can think this (see this volume, pp. 244–5).

7. Panpsychism and Power

Here is another blocker. I take it [1] that the *power* being of concrete reality is wholly grounded in — it is in fact nothing over and above — the *categorical* being of concrete reality.[32] And it seems at first natural to suppose that [2] there is far more difficulty in supposing that all the power being that we find in concrete reality is grounded in the categorical being of ψ stuff than in the categorical being of non-ψ stuff. We tend to think we know what ψ (experientiality) is in such a way that we know that it couldn't possibly ground all the power being that we see in the world (up to volcanoes and nuclear explosions).

So it may seem. And yet we have no good reason to suppose that we know enough about ψ stuff to know that it is less well fitted than non-ψ stuff to ground, or rather be, the power being of the world. To know the general nature of ψ in the special having-is-the-knowing way we do is not *ipso facto* to know the whole power being of any

---

[31] I address this issue briefly in Strawson (2023, p. 451) and at more length in Strawson (in preparation). See also this volume, pp. 244–5.

[32] This is I think extremely close to the common-sense view of power, and it's forcefully endorsed by Descartes, Locke, and many others. Philosophers have of course questioned it, and it has been recently rebranded as the 'powerful qualities' view.

particular portion of it. To know the specific nature of a particular portion of our own ψ in the having-is-the-knowing way we do is not *ipso facto* to know all there is to know about its power to affect other things. It's not to know the effects it is disposed to have on all other portions of reality in all circumstances.

The pre-critical Kant put this point well:

> every substance, including even a simple element of matter, must ... have some kind of inner activity as the ground of its producing an external effect, and that in spite of the fact that I cannot specify in what that inner activity consists. Leibniz said that this inner ground of all its external relations and their changes was a power of representation

where (*nb*) this power is to be understood as an 'inner activity', i.e. actual experiential goings-on, not just a possibly unexercised disposition.[33] 'This thought', Kant continued,

> was greeted with laughter by later philosophers. They would, however, have been better advised to have first considered the question whether a substance, such as a simple part of matter, would be possible in the complete absence of any inner state. And, if they had, perhaps, been unwilling to rule out such an inner state, then it would have been incumbent on them to invent some other possible inner state as an alternative to that of representations and the activities dependent on representations (1766: 315, Ak. 2.328)

The basic point is simple, but can seem difficult (I don't know how many will see its force). We have in fact no good reason to think we know anything about concrete reality that favours non-ψ stuff over ψ stuff as (a ground for) the power being of the world. Certainly physics doesn't favour any such view (it's just mental habit). But the laughter continues.

8. Power: Objection and Reply

Objection. 'Whatever the prospects for some view of this sort, your own position is incoherent. For you hold

> [1] we know the categorical being of an experience just in having it (the having is the knowing)

> [2] the power being of an experience is (identical with, nothing ontically over and above) its categorical being

---

[33] It's wildly anachronistic to think that when Leibniz talked of 'unconscious' perceptions, he supposed that there could be perceptions without any experientiality at all, rather than perceptions (necessarily intrinsically experiential) of which the human subject (the 'top' or in his terms 'dominant' subject) wasn't, in his sense of 'conscious', consciously or 'apperceptively' aware.

[3] we do not know the power being of an experience just in having it.

But [2] and [3] jointly entail

[4] we do not know the categorical being of an experience just in having it

which is the negation of [1].'

Reply. One response to this objection is to be found on pp. 250–62 of this book; I'll make a few other comments here. It's worth noting, first, that some may be reminded of the *masked man* fallacy ('I don't know the identity of this masked man; I do know the identity of my father; therefore this masked man is not my father'). But I'll put this aside.

Secondly, and more importantly, we can challenge [3] head on. Hedda Mørch argues forcefully that we do directly experience something of the power being of pain (for example) simply in having it.[34] This requires changing [3] to (at least)

[3*] we do not know *all there is to know* about the power being of an experience just in having it.

This, in fact, is all that I claimed in the last section. But [3*] may still be thought to couple with [2] to contradict [1].

Let me try to clarify. Although I define 'experience' in §3.1 in such a way that we may say of a particular experience that the having is the knowing, I don't for a moment think that I know the *total being* of a particular experience-occurrence E (a two-second experience of middle C played on an oboe) in having the experience I have when E occurs. I take it, for one thing, that E has a complex true description in physics terms, and I have no sense of this simply in having it.[35]

A further thought: we standardly think of power as essentially dispositional. It is, however, quite unclear what it would be to experience something dispositional; in fact the idea makes no sense. The power being of a thing is simply misconceived if thought of as directly experiencable when unexercised. Even if we experience the categorical being C of E simply in having E, and even if C is identical with the total power being P of E (even if P is nothing ontically over and above C), we do not *ipso facto* experience the total power being of C. More

---

[34] See e.g. Mørch (2017, 2020).

[35] Note that panexperientialists (but not archepsychists) who accept this must take it that E has complex *experiential* aspects that I, the human-being subject, do not experience in any way. Fortunately, Nature has given us a streamlined 'user interface' or GUI.

precisely (given note 35): even if I experience the whole categorical being c of that *part* or *aspect* of E — call it e — that is me-the-human-subject having the oboe experience, and even if c is identical with the whole power being — call it p — of e (even if p is nothing ontically over and above c), I do not in having e *ipso facto* experience the whole power being of e.

There's another very different response. Premiss [3] is false: I *do* experience the whole power being of E, or more precisely e, in having the experience that I have; for the categorical being is the whole power being. I experience the whole power being *in a certain way* — the way I experience it simply in having e. It's just that I don't experience the whole power being *as power* — even if I do experience *some* of it in that way (see, again, Mørch *op. cit.*). One simple point is this: I certainly don't have to know, in virtue of my experience, that the power being of e — e is a tiny amount of physical stuff — is such that feeding it into a nuclear fusion reactor could produce a significant quantity of electric current.

It's remarkably hard, I find, to think clearly about power (I don't think it's just me).

9. Orgel's Second Rule

There are of course further difficulties — difficulties that incline me to favour the theory that locates animal ψ in the brain's electromagnetic field rather than in the basic neuronal processes.[36] For now, in summary and conclusion, let us suppose that pure panpsychism or panexperientialism is correct. The fundamental stuff of concrete reality — the stuff that has, however mysteriously, all the powers that physics, in its own relational-structural way, says so much about — is ψ (noun, not adjective).

ψ is, to repeat, stuff. Its existence is not to be thought of as dependent on the existence of subjects of experience that are in some manner ontically independent of it.[37] On the present (Jamesian) view, there are no such things as subjects, thought of as entities whose existence transcends the existence of ψ.

Consider this planet five billion years ago. It's made of physical stuff (ψ), like everything else. ψ is the subject matter of physics, about which physics says many true things. Differently put: the categorical

---

[36] See e.g. Pockett (2012), Jones (2013), Jones and Hunt (2023).

[37] It has nothing to do with Berkeleyan ψ, which essentially depends for its existence on the existence of minds that do not consist of ψ. (Another blocker: the mistaken idea that panpsychism must have something to do with Berkeley-style idealism.)

being of energy, thought of in Heisenberg's way as the fundamental stuff — the actual concrete being — of reality, is ψ. There is, however, or so we suppose, five billion years ago, no *interesting* ψ. How are we to explain the coming into existence of interesting ψ, complex ψ, animal ψ in all its variety — right up to human ψ?

The *general* answer is easy: evolution. Given the intrinsic nature of the material it works with (ψ), and a great deal of time, it's no harder for evolution to arrange matter in brains in such a way that some of its processes are conscious experiences like ours than it is for us to arrange matter in a pocket calculator in such a way that it is good at sums. Evolution, evidently, needs something to work with. It works the spatial nature of the physical stuff (= ψ) that it finds into remarkable shapes, adaptive spatial forms — eyes, limbs, opposable thumbs.[38] So too it works the experiential nature of the physical stuff (= ψ) that it finds into remarkable adaptive forms — vision, smell, hearing, conscious thought. In so doing it evolves a simple 'user interface' for individual organisms: it pushes almost all the extraordinarily complex information-bearing workings of physical stuff in the brain (= ψ) below the level of creature consciousness, so that they're no part of the content of the experience of the 'top subject', by which I mean the subject of experience that we ordinarily experience ourselves to be.[39]

How does it do this? I like to cite Orgel's Second Rule: *evolution is cleverer than you are*. It's not, however, clever enough — so say I, as a sober methodological naturalist — to spin ψ out of wholly and utterly non-ψ material, by a process of 'radical emergence' (see this volume, pp. 12–24). Postulation of radical emergence of this sort is appealed to nowhere else in science, and is ruled out by any respectable variety of methodological naturalism.

Objection. 'How do you know that evolution isn't clever enough to do this? It may be so clever that it can do just this.'

Reply. I can't prove that it's impossible. Perhaps it's enough to say that it's an entirely unnecessary supposition, a huge theoretical extravagance that blatantly contravenes methodological naturalism.

---

[38] We can say this whatever the ultimate nature of that which we think of as space. Recall the difficult claim (see §6 and reference there) that physical stuff doesn't have to have non-ψ 'stuff being' in order to have spatial shape.

[39] I think it plausible — quite independently of any panpsychist hypothesis, and even putting aside evidence from cerebral commissurotomy — that there are many subsidiary and some relatively high-level subjects in a human brain and body (note that this view has nothing to do with the idea that panpsychism involves belief in a multitude of fundamental-particle-sized 'microsubjects').

Even if it happened, it could never be theoretically — scientifically — reasonable to suppose that it had.

Panpsychism is put forward simply as the least implausible version of materialism, or if you like naturalism. It offers nothing to science in the way of a research programme.[40] It remains a difficult view, if only because one consequence of the rejection of radical emergence is that fundamental ψ must already involve — in seed form, as it were, and as Margaret Cavendish surmises — something that can be worked up not only into simple sensation, but also into positive and negative affect, and indeed conscious thought and rationality (this last thing is not, I think, a hard problem; logic is instilled in us by the demands of adaptive agency). The base claim remains: given the empirical evidence, panpsychism in some form is the most parsimonious, conservative, down-to-earth, hard-nosed, plausible, realistic, even plodding view there is about the fundamental nature of reality. This has been clear since Darwin published his theory of evolution, as James and many others soon saw. Do I repeat myself, and others? Very well then I repeat myself, and others.

References

Bacon, F. (1620/2000), *The New Organon*, trans. L. Jardine and M. Silverthorne (Cambridge: Cambridge University Press).

Chomsky, N. (1995), 'Language and Nature', *Mind* **104**, pp. 1–61.

Clifford, W. K. (1874/1901), 'Body and Mind', in *Lectures and Essays*, vol. 2, ed. L. Stephen and F. Pollock (London: Macmillan), pp. 1–51.

Clifford, W. K. (1875/1879), 'The Unseen Universe; or, Physical Speculations on a Future State', in *Lectures and Essays*, vol. 1, ed. L. Stephen and F. Pollock (London: Macmillan), pp. 228–53.

Conway, A. (1690/1996), *The Principles of the Most Ancient and Modern Philosophy*, ed. and trans. A. Coudert and T. Corse (Cambridge: Cambridge University Press).

Dennett, D. (1991), 'Mid-Term Examination: Compare and Contrast', *The Intentional Stance* (Cambridge, MA: MIT Press), pp. 339–51.

Descartes, R. (1641/1985), *Meditations* and *Objections and Replies, The Philosophical Writings of Descartes,* vol. 2, trans. J. Cottingham et al. (Cambridge: Cambridge University Press).

Drake, D. (1911), *The Problem of Things In Themselves* (Boston, MA: Ellis).

Drake, D. (1925), *Mind and Its Place in Nature* (New York: Macmillan).

Eames, Elizabeth (1967), 'The Consistency of Russell's Realism', *Philosophy and Phenomenological Research* **27** (4), pp. 502–11.

Eddington, A. (1928), *The Nature of the Physical World* (New York: Macmillan).

Haldane, J. B. S. (1940/1968), 'Why I am a Materialist', in J. B. S. Haldane, *Science and Life* (London: Pemberton Publishing).

Hawking, S. (1988), *A Brief History of Time* (New York: Bantam Books).

---

[40] See Strawson 2006, this vol, pp. 27-8. That said, panpsychism may perhaps favour electromagnetic field theories of consciousness; see e.g. Pockett (2012), Jones (2013).

Heymans, G. (1905/1921), *Einführung in die Metaphysik auf Grundlage der Erfahrung* 3nd edition (Leipzig: Barth).Ï

Jones M. and Hunt T. (2023), 'Electromagnetic-field theories of qualia: can they improve upon standard neuroscience?' *Front Psychol* **14**:1015967 doi: 10.3389/fpsyg.2023.1015967

Jones, M. (2013), 'Electromagnetic-Field Theories of Mind', *Journal of Consciousness Studies* **20** (11-12), pp. 124–49.

Lewis, D. (1983/1999), 'New Work for a Theory of Universals', in D. Lewis, *Papers in Metaphysics and Epistemology* (Cambridge: Cambridge University Press), pp. 8–55.

Luntz, F. (2003), Interview, *PBS*, 15 December 2003.

Mathews, F. (2003), *For Love of Matter: A Contemporary Panpsychism* 3rd edition (Albany, NY: SUNY Press).

Maxwell, G. (1978), 'Rigid Designators and Mind-Brain Identity', in *Perception and Cognition: Issues in the Foundations of Psychology*, ed. C. Wade Savage (Minneapolis: University of Minnesota Press), pp. 365–403.

Montague, W. P. (1912), A 'Realistic Theory of Truth and Error', in *The New Realism*, ed. E. B. Holt et al. (New York: Macmillan), pp. 251–300.

Mørch, H. (2017), 'The Evolutionary Argument for Phenomenal Powers', *Philosophical Perspectives* **31** (1), pp. 293–316.

Mørch, H. (2020), 'The Argument for Panpsychism from Experience of Causation', *The Routledge Handbook of Panpsychism* ed. W. Seager (London: Routledge).

Peterman, A. (2023), '"Actions of a body sentient": Cavendish on the Mind, and against Panpsychism', *Oxford Studies in the Philosophy of Mind* 3, pp. 399–431.

Pockett S. (2012), 'The Electromagnetic Field Theory of Consciousness', *Journal of Consciousness Studies* **19** (11-12), pp. 191–223.

Prince, M. (1885), *The Nature of Mind and Human Automatism* (Phildelphia: Lippincott).

Prince, M. (1891), 'Hughlings-Jackson on the Connection Between the Mind and the Brain', *Brain* **14** (2–3), pp. 250–269 https://doi.org/10.1093/brain/

Prince, M. (1904), 'The Identification of Mind and Matter', *The Philosophical Review*, pp. 444–451.

Prince, M. (1928), 'Why the Body Has a Mind and the Survival of Consciousness After Death', *Mind* **37**, pp. 1–20.

Russell, B. (1927/1992), *The Analysis of Matter* (London: Routledge).

Russell, B. (1944/1946), 'Reply to Criticisms', in Schilpp. P. (ed.) *The Philosophy of Bertrand Russell* (Chicago: Northwestern University Press), pp. 681–741.

Russell, B. (1945/1997), 'Mind and Matter in Modern Science', in Slater, J. (ed.) *Collected Papers of Bertrand Russell, Volume 11* (London: Routledge).

Sellars, R. W. (1916) *Critical Realism*, Chicago: Rand, McNally & Company.

Sellars, R. W. (1927), 'Why Naturalism and Not Materialism?', *Philosophical Review* **36**, pp. 216–225.

Sellars, R. W. (1960), 'Panpsychism or Evolutionary Materialism', *Philosophy of Science* **27** (4), pp. 329–350.

Skrbina, D. (2005), *Panpsychism in the West*, 2nd edn. (Cambridge, MA: MIT Press).

Smolin, L. (2013), 'Free will, determinism, quantum theory and statistical fluctuations: a physicist's take', *Edge*, 8 July, 2013 https://www.edge.org/conversation/carlo_rovelli-free-will-determinism-quantum-theory-and-statistical-fluctuations-a)

Sprigge, T. L. S. (1983), *The Vindication of Absolute Idealism* (Edinburgh: Edinburgh University Press).

Sprigge, T. L. S. (2011), *The Importance of Subjectivity* (Oxford: Oxford University Press).

Strawson, G. (1994/2010), *Mental Reality* 2nd edition (Cambridge, MA: MIT Press).

Strawson, G. (2006a), 'Realistic Monism: Why Physicalism Entails Panpsychism', in *Consciousness and Its Place in Nature*, ed. A. Freeman (Thorverton: Imprint Academic), pp. 3–31.

Strawson, G. (2006b), 'Panpsychism? Reply to Commentators, With a Celebration of Descartes', in *Consciousness and its place in nature* ed. A. Freeman (Thorverton: Imprint Academic), pp. 184–280.

Strawson, G. (2016), 'Mind and Being: the Primacy of Panpsychism', in *Panpsychism: Contemporary Perspectives*, ed. G. Brüntrup and L. Jaskolla (New York: Oxford University Press).

Strawson, G. (2017), 'Physicalist Panpsychism', in *The Blackwell Companion to Consciousness* 2nd edn, ed. S. Schneider and M. Velmans (New York: Wiley-Blackwell).

Strawson, G. (2023), 'Charles Augustus Strong: Real Materialism, Evolutionary Naturalism, Panpsychism', in *Oxford Studies in the Philosophy of Mind* **3**, pp. 432–461.

Strawson, G. (2024), *Stuff, Quality, Structure: The Whole Go* (Oxford: Oxford University Press).

Strawson, G. (forthcoming), '"The Problem of the Relation of Mind and Matter Can Be Completely Solved" (Russell)'.

Strawson, G. (in preparation), *Being and Consciousness*.

Strong, C. A. (1930), *Essays on the Natural Origin of Mind* (London: Macmillan).

Strong, C. A. (1936), *A Creed for Sceptics* (London: Macmillan).

Uebel, T. (2020), 'Intentionality In the Vienna Circle', in *Franz Brentano and Austrian Philosophy*, ed. D. Fisette et al. (Switzerland: Springer Nature), pp. 135–168.

Uebel, T. (2021), 'Carnap, Knowledge of Other Minds, and Physicalism', *Philosophers' Imprint* **21** (34), pp. 1–27.

Index

Aaron, R. I., 78
Abstract phenomena, 3n, 10, 21, 29, 53, 73, 91, 115, 140, 142, 194, 244
Access consciousness, 35
Acquaintance, 55, 102, 104-5, 223, 240n, 250-4, 256, 262-5, 290, 345
Aggregation, 26n, 62-64, 69, 181
Analytic truth, 66, 190n, 269
Aristotle, 152, 180, 244, 264
Armstrong (David M), 123, 159-61, 196
Armstrong (Louis), 5n, 111, 120n, 264
Arnauld, A., 4n, 5n, 26, 200, 201n, 211, 225n
Australian zombies, 22n, 170, 271-2. See also Zombies.
Ayers, M. R., 78n

Bateson, G., 328
Bateson, W., 152, 157
Bergson, H., 122, 152, 157, 330
Berkeley, G., 229n, 289, 359n
Bermúdez, J., 78n, 82
Blake, R., 119n
Block, N., 5n, 35, 38, 56n, 120n
Bohm, D., 152, 157, 329
Bradley, F., 131n, 159, 196n
Braithwaite, R., 143
Breitmeyer, B., 119n
Brentano, F., 149, 308
Broad, C. D., 11n, 13n, 25n, 161, 185, 205
Bruno, G., 152
Buddhism, 26n, 191, 192n

Calef, S., 82
Campanella, 152, 330
Cardano, G., 157
Carnap, R., 8, 268n
Carruthers, P., **32-9**, 240n, 248, 264, 308, 310
Cartesianism, 5n, 19, 55-6, 92, 114, 151, 181, 192, 201, 204-5, 207-8, 214, 216, 222, 226, 229-30, 234, 239, 247, 258, 263, 271, 346n
See also Dualism, Cartesian.
Cartesian intuition, 222, 224, 226, 230, 234, 240, 241
Caston, V., 263n
Categorical vs. dispositional, 141, 195, 262n, 319, 348-9, 356-9
Causal completeness of the physical, 44, 52
Causation, 44, 141, 143n, 245, 257-8, 260, 262, 323, 333, 353
  problem of, 51-2, 275-6
  structural analysis of, 141
Chalmers, D., 1, 22, 33, 37, 70, 102, 143n, 157, 185-6, 208, 225, 256n, 286n, 287n, 288n, 312-3, 315, 318n, 329, 336
Chomsky, N., 11n, 170-2, 176, 210n, 309-10, 338-9, 344n
Clark, A., 126
Clarke, D., 5n, 199n, 202, 205n, 207, 212n, 215, 216, 239
Cognitive closure, 134
Coleman, S., **40-52**, 185, 189n, 193, 226n, 228-9, 232-3, 243, 250n, 256, 260n, 263-4, 272n, **285-98**, 318, 320n, 330, 254n
Combination problem, 49, 50-1, 96, 155-6, 248-250, 253, 255-6, 261, 286-7, 293, 318, 331, 354
Composition problem. See Combination problem.
Consciousness, 3, 10-12, 20, 26n, 29, 32-5, 38, 44, 50-1, 54, 63-4, 66, 68, 73, 93, 98, 101, 104, 105-6, 108-114, 120, 133-7, 139, 143-4, 148-9, 151-5, 159, 165-7, 178, 179, 181, 182, 183, 190, 192, 202, 206, 209, 214, 246, 248, 249, 250, 252, 260, 270, 271, 285, 286, 287-96, 300-2, 304-10, 312-25, 332, 336-9, 341, 343, 345-8, 352, 360
  distribution of, 179
Craig, E., 19n, 215n
Crane, T., 73-5, 187n

366 INDEX

Deiss, S., 157
Demopoulos, W., 143
Dennett, D., 5-6, 38, 65, 112-13, 234, 249n, 265, 309, 317, 339
De Quincey, C., 157, 329
Derivation problem. See Combination problem.
Descartes, 2, 5, 11n, 51, 55, 77n, 117, 136, 151, 168, 170-1, 175, 178, 181, 184-5, 190-1, 192n, 193-4, 196n, 199-216, 218, 224-5, 226, 230, 233, 236-7, 238-40, 243-4, 257, 264, 271, 345n, 346, 356n (see also Cartesianism)
Determination, 40, 45, 63, 69, 75, 131-2, 134-5, 137, 139, 140-1, 166-7, 233
Diderot, D., 152
Dipert, R., 138
Direct realism, 201-2, 215, 291
Downward causation, 44. See also Causation.
Dretske, F., 5n, 65, 113
Dualism, 5, 7-8, 24-26, 28, 42-3, 51, 72, 76, 87-8, 104, 110, 136, 148, 154, 160, 176, 270, 300-2, 309n, 333, 337
   Cartesian, 207, 216
   macro property, 43n
   micro property, 43
   property, 28, 43, 51, 67, 72, 79, 80-4, 87-8, 140, 193, 209, 216, 221n, 242n
   substance, 5, 26, 28, 79-81, 84-5, 87, 140, 193, 221n, 237
   stuff, 76, 221n, 237

Eddington, A., 4, 9-12, 20, 29, 56n, 67-8, 93, 97, 112-13, 122, 129, 134, 143-4, 152-5, 228n, 242-5, 258, 263, 319, 343, 348n, 352
Eliminativism, 7, 24, 65, 70, 75, 114-5, 123, 172-3, 219, 231, 242, 246, 286-7, 293, 304n, 309n, 341, 351
Elizabeth of Bohemia, 200, 209, 211-13
Emergence, 12-14, 16-24, 27, 33, 41-3, 45-8, 51, 53-4, 59-60, 62-3, 68, 73, 82, 85-89, 92-95, 98-9, 107-109, 117-118, 121-2, 133-137, 146-153, 161, 172-4, 177-8, 180-182, 232, 246, 266, 272-4, 287, 300-2, 304, 315, 329, 332
   brute, 18-20, 43, 53-4, 60-1, 63, 90, 98, 133, 141, 146, 153, 250, 266, 270, 331
   Modest Kind, 14n
   natural, 148-150
   radical, 22-3, 151, 156, 223, 231-2, 241n, 246, 266, 270, 273, 317-8, 354n, 360-1
   Radical Kind Emergence, 14n
Emergentism. See Emergence.
Emergent properties, 12, 15-7, 21, 62-3, 133, 148. See also Emergence.
Entailment, 18, 25-6, 33-4, 39, 46, 48, 65, 90, 92-3, 130, 136, 164-5, 178, 182, 234, 270, 329, 336-7, 341
Epiphenomenalism, 1, 69, 94, 101, 166, 276, 292, 332
Essence, 4, 25, 52, 56n, 90, 92, 98, 100-102, 106, 114, 118, 163-5, 206, 251n, 264n, 270, 294, 313
Experience, 3-10, 12, 14-16, 18-20, 22-9, 32-8, 41-51, 53-61, 64-5, 68-70, 73, 82, 85-6, 88, 90-8, 100-1, 104, 106-8, 110-5, 117-9, 121-7, 129, 146-150, 156, 158-161, 163, 165-9, 172-3, 177-182, 187-192, 194n, 202-5, 207-11, 218-9, 222-4, 227, 229n, 231, 240n, 242-5, 247-258, 260-2, 264-7, 269-272, 274, 276, 288-91, 295, 299-302, 304-7, 309, 315, 320-2, 324-5, 330, 333, 343, 344n, 345, 348n, 356-60
   argument, 174
   composition of universe by, 243-4
   content of, 201, 242-3, 255, 258, 290
   denial of, 5, 148-9, 168, 172, 226, 270
   distinction between experience and experiencer, 254, 256, 258, 267
   emergence of, 14, 16, 20, 24, 32, 121, 188, 231-2, 241n, 250, 263, 266, 272-3
   essence of, 4, 106, 118, 164, 223, 240n, 251-4, 264-5, 270
   as essential property of mind, 204
   existence of, 4-5, 65-6, 146-7, 149, 168, 188, 192, 244, 266, 269
   future, 168-9
   human, 27, 59, 166-7, 178, 187n, 249, 271
   inexplicability of, 20, 110, 179
   irreducibility of, 20, 98, 230
   knowledge of, 55, 98, 102, 112, 119-20, 122, 127, 159, 188n, 243, 250-4, 256, 262-5, 268, (see also Revelation)
   laws of, 245, 253, 261
   nature of, 66, 92, 159, 171, 253, 302
   phenomenal, 140, 345 (see also Phenomenal consciousness)
   problem of, 20, 244, 273

realism about, 188, 202, 215
role of in physics, 240
sensory, 14, 111, 251n, 300, 330
subject of. See Subjects
subject without, 192, 222, 224, 256
summing of, 58, 60
understanding of, 160 visual, 115
without an experiencer, 50, 189-191
Euler, L., 138
Experiential-and-non-experiential ?-ism, 7, 187
Experientiality, 4, 22-3, 26-28, 32, 36, 38, 46-9, 52, 59, 88, 107, 140, 153, 186, 220, 243-5, 260, 293, 329, 332-3, 345-6, 354n, 356
Experiential phenomena, 4, 6-8, 12-17, 21-24, 33, 47, 53, 55, 57, 59-60, 68, 72-3, 83, 115, 118, 127, 153, 188, 216, 254n, 268, 272n
Explanation, 32-9, 42, 57, 107-8, 111-3, 134, 216, 276, 314
Explanatory gap, 33-35, 37, 39, 68, 87-89, 133-4, 139-40, 144, 149, 304-5, 307, 310, 313, 315
Extension, 15-17, 28, 46, 178, 180, 202-205, 207, 209-211, 214, 216, 222, 224-5, 234, 237, 240-242, 244, 274
Externalism, 102, 143n, 193, 201-2, 215
Extrinsic natures, 52
Extrinsic properties, 129-32, 143

Falkenstein, l., 141n
Feigl, H., 11n, 267-8, 287n
Feyerabend, P., 66n, 114
Fichte, J., 192n
Frege, G., 26n, 190
Friedman, M., 143
Functional complexity, 97
Functional concepts, 34
Functional explanation, 34
Functional properties, 33, 56-7, 98, 195n, 273
Functional role, 35, 307
Functional states, 160
Functionalism, 38, 93, 98, 139, 160, 272
Functionalist theory. See Functionalism.

Galileo, 103, 165, 212n, 215n, 285
Gardner, S., 7n
Goff, P., **53-61**, 208-9, 226n, 248-55, 262-3, 266, 286n, 287n, 290n, **299-303**, 318, 330, 347n
Goodman, N., 126
Greene, B., 9n, 244n

Griffin, D., 156, 189n, 286n, 329

Haldane, J., 152, 157, 347
Hartshorne, C., 152, 156-7
Hilbert, D., 161, 314n
Hill, C., 56n
Humberstone, L., 129
Hume, D., 4-5n, 11n, 17-19, 25n, 111, 149, 182, 191-2n, 201-2, 215n, 245, 250n, 255, 258n, 309n
Huxley, T., 152, 313

Idealism, 5n, 83, 92, 113-5, 154-5, 167-8, 229n, 243, 267, 275n, 287-91, 306n, 332
Intelligibility, 15, 33, 45, 47, 59
Intensional semantics, 114
Intentionality, 50-1, 70, 193, 201, 260, 292, 295, 333, 353
Interaction problem, 52
Intrinsic nature, 10, 52, 67, 91, 93, 98, 123, 129, 131-2, 134, 136-41, 143-4, 154, 188n, 224n, 243, 245, 258, 262, 268, 300, 302, 318-9, 331-2, 360
  argument, 139, 140, 143
Intrinsic properties, 67, 129-44, 187n, 196, 239
Introspection, 36, 54, 56-7, 97, 99, 105, 111-12, 251-2, 263, 293, 305, 307-10

Jackson, F., **62-4**, 102, 232, 234, 240n, 26970, 272n, 315, 329
James, W., 21n, 26, 28-9, 36n, 49, 54, 58, 136, 149, 152, 155-7, 192n, 197n, 248-50, 253-4, 256, 259n, 263, 287n, 318n, 319-20, 323, 325, 330-1, 344, 352, 354, 359, 361
Jeans, J., 152

Kant, I., 4n, 11n, 79, 140-3, 178, 183, 185, 190-1, 196, 198, 211n, 266-7, 324, 346, 357
Kim, J., 130
Kim, C.-Y., 119n
Koestler, A., 157
Kripke, S., 252n, 294

Ladyman, J., 138
Langton, R., 130, 141n, 143n, 194n
Laws of nature, 18, 22n, 42, 64, 322n
Leibniz, G., 11, 26-7, 131-2, 138, 140, 142, 146, 150, 152, 178-80, 182, 192, 194, 244, 257n, 262n, 273-5, 289, 316, 330, 356-7
Levels, 41, 43, 49, 74, 98, 135, 333

INDEX

Levine, J., 56n, 68, 103, 112n, 304, 310n, 313
Lewis, D., 130, 194n, 318, 346, 349n, 353
Lichtenberg, G., 191n
Life, 20, 63, 122, 153, 181, 310, 314, 317
Liquidity, 13-16, 19, 22-24, 27-8, 33, 37-8, 41, 43, 46-7, 59, 63, 68, 88, 106-108, 117, 121-2, 133, 147-8, 153, 161, 173-4, 304, 315
Loar, B., 34, 53, 56, 104n, 309
Locke, J., 4, 8, 11, 20, 72, 76-87, 126, 178, 185, 200-1, 204, 213, 263, 271-2, 356n
Lockwood, M., 10n, 243n, 244n
Looking-glassing, 5n, 113, 202n
Loux, M., 196n
Lucretius, 177, 271
Lycan, W. G., 5n, 56n, **65-71**, 113, 141, 159, 232n, 234, 260n, 269, 329-32

Mach, E., 159, 257n, 287n, 319-20
Mackie, J. L., 18n
Macpherson, F., **72-89**, 187n, 193, 228, 242n, 263, 271n
Macro conscious experience. See Macro-experientiality.
Macro-experience. See Macro-experientiality.
Macro-experientiality, 32-3, 36, 48-51, 54, 59-60, 69, 88, 107, 246, 263, 266-7, 288, 299
Manzotti, R., 157
Marcel, A., 119n
Materialism, 3n, 5, 56, 160, 170, 186-7, 210-11, 224, 231, 236, 271, 302, 312, 317, 333-4, 347-54, 361 (see also Physicalism)
a posteriori. See Physicalism, a posteriori
real, 29n, 211, 236
Maxwell, G., 56n, 143n, 352n
McGinn, C., **90-99**, 112n, 134, 187n, 228-9, 232n, 234, 252, 255, 260n, 266, 270-1, 309n, 329-33
McLaughlin, B. 13, 25, 56,
Mellor, D. H., 73-5, 187n
Melnyk, A., 104n
Mental causation, 27, 44, 69, 275
Micro conscious experience. See Micro-experientiality.
Micro-experience. See Micro-experientiality.

Micro-experientiality, 26-7, 32, 36, 49-51, 53-4, 57, 59-60, 69, 88, 181, 246, 263, 266
Micropsychism, 24-5, 48, 85-6, 107, 146, 158, 178, 182, 186, 222, 227, 244n, 246, 304, 354-5
Mill, J. S., 159
Mind–body problem, 33-35, 37, 184, 193, 199-200, 208, 212n, 217-18, 229-230, 238, 243, 246, 265, 270, 275-6, 286, 350
Modality, 46, 132n, 141
Monism, 7, 11n, 76-7, 80, 84, 153, 187-8, 213, 231, 235, 237, 239, 246, 275, 287, 318-21, 323-4, 329, 331, 355n
 dual-aspect, 153, 257n
 equal-status, 187-8, 241
 ESFD, 223, 240-2, 246, 256, 273-4
 experiential-and-non-experiential, 7, 187
 fundamental-duality, 223, 234, 236-238
 micro-property, 52
 neutral, 23, 188, 238, 240, 242-3, 245
 panpsychist, 188, 256, 273
 realistic, 9, 215, 221-2, 224, 230, 234-6
 stuff, 221-2, 229, 234-239
 thing, 221-2, 239
 substance, 72, 76, 79, 81, 83-4, 87-8, 92, 221n, 239, (see also Monism, stuff)
 type F, 186
Mozart, W., 148, 150

Nagarjuna, 28n, 196, 244n
Nagel, T., 3n, 49, 67n, 133, 140, 152, 187n, 189, 221n, 246n, 300, 314-5, 333
Naturalism, 4, 246, 249n, 257, 261-2, 305, 314, 344, 351, 354, 360-1
Natural-kind term, 8, 42, 76, 91-2, 224n, 244, 271
Natural laws. See Laws of nature.
Nietzsche, F., 28n, 115n, 152, 196, 198, 330
Nisbett, R., 112, 308
Neurophysiology, 7, 118, 169
Newman, M., 142-3
Normand, C. 13n

Occam, 11, 19, 29, 67, 213, 266-7
Ontology, 40, 43-4, 48-9, 51, 75, 91, 151, 155, 167, 190, 194, 207, 233, 247,

INDEX   369

259, 262n, 302, 316, 324, 346
 micro, 48-50, 52
Operationalism, 93, 98, 113n, 122

Panexperientialism, 25, 40, 48, 50-52, 189, 256, 289, 355, 358n, 359
Panpsychism, 8-9, 25-6, 32, 35-40, 48-51, 53-4, 57, 61, 63, 65-7, 69-70, 80, 83, 85, 90, 92-8, 100, 106-9, 118, 133-6, 140, 148, 151, 153, 155-8, 161, 164-5, 168, 181-2, 186, 189, 193, 219-220, 222, 227, 229, 236, 238, 244-5, 250, 253, 255-9, 261-2, 267, 269, 270-1, 273, 275, 286-96, 299-302, 304-5, 309n, 310, 312, 314, 317-20, 325, 328-34, 336-7, 341, 343-5, 348, 352-6, 359-61
 Anaxagorean, 273
 arguments against, 69-70
 argument for, 131, 133, 138, 140, 146
 attractiveness of, 132
 case for, 66, 129
 and causation, 262
 defence of, 117
 entailment of by physicalism, 25-6, 163
 evidence for, 152
 falsity of, 150
 hostility to, 151
 and intrinsic properties, 129
 and James, 155
 Leibniz's, 131, 180
 naturalistic, 258
 pressure for, 109
 pure, 227-9, 235, 237, 239, 241, 245-7
 and Real Materialism, 105
 smallist, 249, 252-6
Pascal, B., 185, 200
Peirce, C., 152-3, 157
Perry, J., 56n
Perspicuity, 44-48, 50-52
Phenomenal concepts, 34, 39, 45-47, 104, 250n, 264, 309
Phenomenal concepts strategy, 47, 264
Phenomenal consciousness, 35, 37-9, 134, 144n. See also Consciousness
Phenomenology, 3, 6n, 50, 95-6, 98, 143, 147, 159-161
Physicalism, 3-5, 8-9, 12, 22-26, 34, 38, 40, 42, 52-3, 56, 63, 70, 72-3, 75, 83-4, 87-8, 90, 92-3, 100, 105-6, 186-8, 211-12, 226, 231, 236, 245, 286-7, 291, 305, 316, 333-4, 336-7, 341, 347, 349, 352
 ad hominem, 186, 233, (see also Physicalism, conditional)
 a posteriori, 45-47, 56, 215, 238n, 240, 250n, 263-6
 a priori, 50, 232, 240n
 conditional, 186
 conventional, 24
 experiential, 222, 226
 and Identity Theory, 65
 macro non-reductive, 42-44, 47
 mainstream, 72-76, 86-88
 non-reductive, 74n, 91
 object conception of, 42, 51
 old-style, 90, 92
 panpsychist, 227, 232
 physics-based definition of, 42
 real, 4, 7-9, 12, 22, 25-6, 73, 101, 104-106, 212-13, 219, 222, 236, 245, 270, 304, 329
 reductive, 5n, 74n
 straightforward, 100-1, 105-6, 108-9, 258
 Strawsonian, 72-3, 76, 83-4
 zombie argument against, 46, 208, (see also Australian zombies Zombies)
Papineau, D., 45n, 56n, 68n, 74-5, **100-09**, 240n, 252n, 257, 263-5, 270
Physics, 4, 7, 9, 10-11, 13-16, 19-20, 24, 33, 41-42, 51-52, 56, 67, 73-75, 93-94, 97-101, 106, 113-115, 117-18, 120, 122-124, 134-135, 147-148, 150, 153-154, 158-159, 161, 163-5, 167, 169, 199, 202n, 216, 228n, 232, 244, 270, 271, 275-6, 294, 300-2, 305-7, 313-5, 332, 347-9, 351, 355, 357-9
 causal closure of, 44, 52, 69
 causal completeness of. See Physics, causal closure of classical, 165-6, 168
 laws of, 9n, 13, 19, 24, 91, 151, 225n, 245, 247-249, 261
 microphysics, 67, 69
 particle, 276
 quantum, 16n, 66n, 98-9, 142n, 148, 157, 165-9, 221n, 227n, 228n, 331
PhysicSalism, 4-5, 56-7, 90, 100-1, 147, 153, 163-165, 270, 306 (see also Physicalism, straightforward)
Plato, 152, 330
Pluralism, 8, 231n
P phenomena, 13, 14, 15, 20, 22-24, 47, 83n, 153, 304, 307
Priest, G., 229
Priestley, J., 6, 11, 20, 152, 217

Qualia, 35-38, 50, 68, 95-6, 110-11, 113-14, 156, 158, 160-1, 291, 294, 333, 345, 346n, 351
Quantum theory. See Physics, quantum.
Quine, W.V., 10, 114

Ramsey, F., 28n, 138, 198, 346
Recognitional concepts, 34
Reduction, 13, 24, 32-3, 63, 74n, 87, 113, 131, 159, 166, 269, 299-302, 314
Reductive explanation, 36-39, 42, 314
Reductive analysis, 5n, 65, 113, 304n, 315
Reference, 45, 51n, 76, 101-6, 264
Regius, 205, 214-15
Relationalism, 138-43, 263. See also Relational properties.
Relational properties, 67, 130-2, 134-5, 137-41, 143, 296. See also Extrinsic properties.
Revelation, 223-4, 250-6, 259, 262, 301n
Rey, G., 5n, 65, **110-16**, 234, 269, 304-11, 330
Ro, T., 119n
Robinson, H., 77n
Royce, J. 153, 267, 343n, 352
Rosenberg, G., 27n, 65n, 290n
Rosenthal, D., 81, 112, **117-28**, 159, 234, 267-8, 290n
Russell, B., 11, 52, 56, 67-8, 93, 122-3, 129, 134-5, 142-3, 156, 258, 171, 176, 228, 243, 245, 251n, 263, 286-7, 289, 291, 293-4, 314, 318-21, 331, 346n, 348, 350-1, 353

Schechter, E., **32-9**, 240n, 248, 263
Schopenhauer, A., 152-4, 286n, 330
Seager, W., **129-45**, 157, 228, 229n, 263, 287n, **312-27**, 330
Searle, J., 156, 259, 292-3, 315
Sellars, W., 67n, 344, 348, 353n
Sesmet, 247-8, 251, 257-262
Sherrington, C., 152
Shoemaker, S., 119n, 126, 189n
Silliness, Eddington's notion of, 11, 20, 29, 67
Simons, P., **146-50**, 260n, 270
Singhal, N., 119n
Skrbina, D., **151-7**, 189n, 229, 242n, 243n, 248, 257n, 263, 312, **328-35**, 344n
Smallism, 40-45, 47-8, 50n, 52, 223, 233, 246, 248-9, 252-7, 263, 286-7, 290, 354-5

Smart, J., 11, 123, **158-62**, 234, 266-8, 330
Socrates, 272
Solomon, G., 143
Spacetime, 3n-n, 9n-n, 221n, 224n, 244, 260
Spinoza, B. de, 11, 152-3, 178-80, 188, 205, 213, 217, 224, 229, 231, 238-42, 245-7, 249n, 257n, 273, 330, 346
Sprigge, T., 267, 268n, 343n
Stalnaker, R., 56n
Stapp, H., **163-9**, 189, 227n, 244, 246, 270
Stoljar, D., 42, 56n, **170-6**, 224n, 226n, 232n, 234, 251n, 271-3, 287n, 319, **336-42**, 349n
Strawson, G., **3-29**, **184-280**, **343-63** and *passim*
Strawson, P. F., 184, 196, 214, 245
Structure, 10, 27, 52, 59, 69, 132, 136, 139, 141-143, 268
Subjects, 26, 35-6, 48-50, 53-4, 57-60, 69-70, 85-6, 88, 103, 130-2, 143n, 154, 179, 189-93, 204, 206, 214, 219-20, 222, 224, 229n, 243-4, 247, 249, 252n, 254-6, 259, 266-7, 270, 287, 290, 296, 299-300, 302, 310, 317n, 343, 357n, 358n, 359-60
  macro, 219
  micro, 219-20, 249, 255, 260
  thin, 191-3, 224n, 242n, 247, 267
Subject-combination problem, 49, 255. See also Combination problem.
Subjectivity, 68, 112n, 118, 151, 156, 191, 295, 309, 314-5
Substance, 8-9, 72, 76-85, 87-8, 92, 141, 153, 178-180, 182, 190, 195, 202-207, 209n, 210-11, 214, 216, 221n, 237-9, 243-4, 247n, 257-8, 271, 314, 346, 348, 354, 357
Supervenience, 18, 41-2, 44, 62-3, 74, 90-1, 94, 131, 216-18, 220, 233, 332

Teller, P., 142n
Terrasson, Abbot, 185
Topic-neutral construal, 123-4, 268-9
Transparency of the mental, 55, 106, 251n, 263. See also Revelation Transparency of the experiential.
Transparency of the physical, 55, 106
Transparency of the experiential, 54-5, 57-60, 102, 105. See also Transparency of the mental Revelation.
Tye, M., 5n, 34-5, 37, 56n, 65, 113, 316n

Ultimates, 9, 12-13, 15-16, 22, 24-27, 32, 35-39, 41-44, 46-51, 53-4, 57-59, 73, 82, 84-87, 93, 95, 117-18, 121-125, 193, 220, 222-3, 227-8, 232-3, 247-8, 260-1, 274, 290, 295-6, 304, 354
Upward determination, 41-2, 44-5, 49. See also Determination.

van Gulick, R., 13n, 14n, 82
Velarde, M. 13n
Vitalism, 20, 122, 310. See also Life.

Waddington, C., 152
White, N., 26n
Whitehead, A.N., 11, 146, 152, 156-7, 161, 228n, 245, 330, 346, 352, 353n
Whyte, L., 115
Wilber, K., 157
Willemarck, F., 43n
Williams, B., 203n
Wilson, C., **177-83**, 185n, 260n, 271, 273-4
Wilson, T., 112
Wittgenstein, L., 110, 111, 114, 306, 309-10
Woolhouse, R., 78
Worrall, J., 138

Yablo, S., 199n, 212n, 225n

Zombies, 22n, 38, 46, 170-1, 272, 308n, 315. See also Australian zombies Physicalism, zombie argument against

ia Journals in the Philosophy of Mind

imprint-academic.com

Journal of Consciousness Studies

- How does the mind relate to the brain?
- Can computers ever be conscious?
- What do we mean by subjectivity and the self?

The *Journal of Consciousness Studies* (JCS) is a peer-reviewed journal which examines these issues in plain English.

As late as the 1980s, the very mention of the word 'consciousness' sent scholarly eyeballs rolling to the ceiling. But in 1995 *Nature* described the journal's 1994 launch as a 'defining moment'.

These questions are now debated in fields as diverse as cognitive science, neurophysiology and philosophy, and JCS is the leading forum. The journal was launched at the same time as the conference series *Towards a Science of Consiousness* at the University of Arizona, and shares a common broad perspective — nothing is off-limits and the journal regularly includes contributions from diverse fields including anthropology, physics and theology, as well as the core disciplines of philosophy, psychology and neurobiology.

The journal helped establish David Chalmers' 'hard problem' of phenomenal experience as the principal focus of the field, and special issues have been published on topics ranging from emotion experience, panpsychism and altered states to the singularity.

Google Scholar ranking: 19th in philosophy journals worldwide.

Editor
Valerie Gray Hardcastle, Northern Kentucky University

Articles should be submitted to managing editor Graham Horswell at graham@imprint.co.uk. All articles are peer reviewed and open-access options are available.

> "There is no other journal quite like it, and one day we shall, I think, look back to its appearance as a defining moment"
> **Jeffrey Gray**, Nature, 377, p. 265, 21 September 1995.

> "With JCS, consciousness studies has arrived."
> **Susan Greenfield**, *Times Higher Education*

Contact ellie@imprint.co.uk for a free sample.

imprint-academic.com/jcs

Journal of Consciousness Studies

Volume 30, No.9-10 (2023)

Introspective Systems

Cybernetics and Human Knowing

A Journal of Second Order Cybernetics,
Autopoiesis & Cybersemiotics

Cybernetics and Human Knowing is a quarterly international multi- and transdisciplinary journal focusing on second-order cybernetics and cybersemiotic approaches.

The journal is devoted to the new understandings of the self-organizing processes of information and signification in living and artificial systems as well as human knowing that have arisen through second order cybernetics and autopoiesis and their relation to and relevance for other interdisciplinary approaches such as C.S. Peirce's semiotics and biosemiotics.

This new development within the area of knowledge-directed processes is a non- or trans-disciplinary approach. Through the concept of self-reference it explores: cognition, communication and languaging in all of its manifestations; our understanding of organization and information in human, artificial and natural systems; and our understanding of understanding within the natural and social sciences, humanities, computer, information and library science, and in social practices like design, education, organization, teaching, medicine, therapy, art, management and politics.

Because of the interdisciplinary character, articles are written in such a way that people from other domains can understand them. Articles from practitioners will be accepted in a special section. All articles are peer-reviewed and open-access options are available.

Editor
Carlos Vidales, University of Guadalajara, Mexico

Founding Editor: Søren Brier

Submissions to carlos.vidales@academicos.udg.mx

Contact ellie@imprint.co.uk for a free sample.

imprint-academic.com/chk

CYBERNETICS & HUMAN KNOWING

a journal of second-order cybernetics
autopoiesis and cybersemiotics

Volume 30, No. 3-4, 2023

30th 1992-2022 ANNIVERSARY CYBERNETICS & HUMAN KNOWING JOURNAL

30 YEARS OF CYBERSEMIOTICS

Mind and Matter

Mind and Matter is aimed at an educated interdisciplinary readership interested in all aspects of mind–matter research from the perspectives of the sciences and humanities. It is devoted to the publication of empirical, theoretical, and conceptual research and the discussion of its results. The main subject areas of the journal are:

- neuroscience, cognitive science, behavioral science
- physical approaches, mathematical modeling, data analysis
- philosophy of science, philosophy of mind, applied metaphysics
- cultural and social studies, history of ideas

Topics combining approaches from all kinds of different disciplines are particularly encouraged. Appropriate review articles, commentaries, interviews, book reviews, and conference reports will be published occasionally. Each submission is professionally refereed and open-access options are available.

Editors

Harald Atmanspacher, ETH Zurich, Switzerland

Robert Prentner, Center for Mathematical Philosophy, Munich University, Germany

email: editor@mindmatter.de

Mind and Matter is the publication organ of the Society for Mind–Matter Research (ISSN 1611-8812 print, ISSN 2051-3003 online). It is published biannually.

Subscription including online access to all back issues at Ingenta Connect is free for members of the Society. Online access codes will be provided upon request.

The journal adheres to the ethical publication guidelines of COPE (Committee On Publication Ethics).

Contact ellie@imprint.co.uk for a free sample.

mindmatter.de/journal

MIND and matter

AN INTERNATIONAL INTERDISCIPLINARY
JOURNAL OF MIND-MATTER RESEARCH

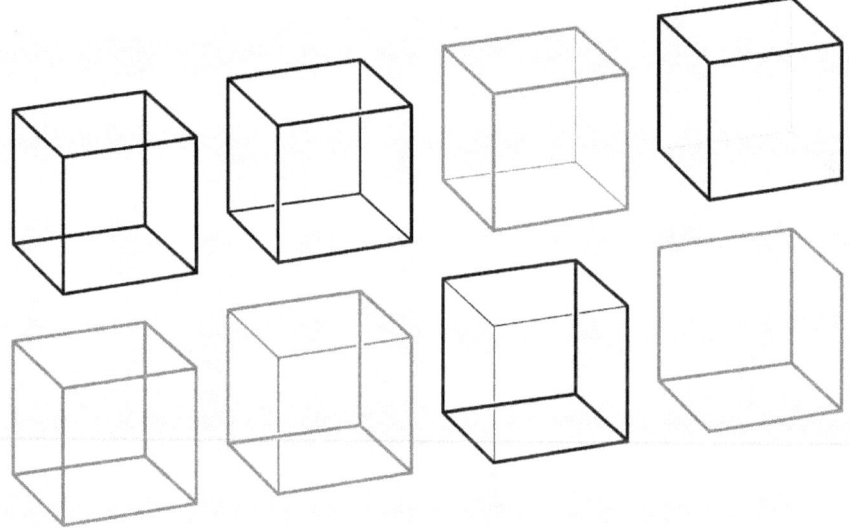

ARTIFICIAL INTELLIGENCE
AND BEYOND

ISSN 1611-8812

Volume 21
Issue 2
2023

www.ingramcontent.com/pod-product-compliance
Lightning Source LLC
Chambersburg PA
CBHW071225230426
43668CB00011B/1306